Principles of
Statistical Radiophysics 1

S. M. Rytov Yu. A. Kravtsov
V. I. Tatarskii

Principles of Statistical Radiophysics 1

Elements of Random Process Theory

With 28 Figures

Springer-Verlag Berlin Heidelberg New York
London Paris Tokyo

Professor Sergei M. Rytov
Corresp. Member of the USSR Academy of Sciences,
Department of General Physics and Astronomy, USSR Academy of Sciences,
SU-117901 Moscow, USSR

Professor Yurii A. Kravtsov
General Physics Institute, USSR Academy of Sciences,
SU-117942 Moscow, USSR

Professor Valeryan I. Tatarskii
Corresp. Member of the USSR Academy of Sciences, Inst. of Atmospheric Physics,
USSR Academy of Sciences, Pyzhevsky Per., SU-109017 Moscow, USSR

Translator

Alexander P. Repyev
Samarkandsky Boulevard 13-3-63, SU-109507 Moscow, USSR

Title of the original Russian edition: *Vvedenie v statisticheskuyu radiofiziku I,* sluchainuie pnotsessui
2. revised and enlarged edition

ISBN-13:978-3-642-69203-1 e-ISBN-13:978-3-642-69201-7
DOI: 10.1007/978-3-642-69201-7

Library of Congress Cataloging-in-Publication Data. Rytov, S.M., 1908-. Principles of statistical radiophysics. 1. Translation of: Vvedenie v statisticheskuiu radiofiziku. I. Bibliography: p. Includes index. 1. Radio waves–Mathematics. 2. Stochastic processes. I. Kravtsov, IUurii Aleksandrovich. II. Tatarskii, V. I. (Valerian Ilich) III. Title. IV. Title: Principles of statistical radiophysics. 1.
QC661.R9213 1987 537.5'34 87-4644

Softcover reprint of the hardcover 1st edition 1987

2153/3150-543210

Foreword

Principles of Statistical Radiophysics is concerned with the theory of random functions (processes and fields) treated in close association with a number of applications in physics. Primarily, the book deals with radiophysics in its broadest sense, i.e., viewed as a general theory of oscillations and waves of any physical nature[1]. This translation is based on the second (two-volume) Russian edition. It appears in four volumes:

1. Elements of Random Process Theory
2. Correlation Theory of Random Processes
3. Elements of Random Fields
4. Wave Propagation Through Random Media.

The four volumes are, naturally, to a large extent conceptually interconnected (being linked, for instance, by cross-references); yet for the advanced reader each of them might be of interest on its own. This motivated the division of the *Principles* into four separate volumes.

The text is designed for graduate and postgraduate students majoring in radiophysics, radio engineering, or other branches of physics and technology dealing with oscillations and waves (e.g., acoustics and optics). As a rule, early in their career these students face problems involving the use of random functions. The book provides a sound basis from which to understand and solve problems at this level. In addition, it paves the way for a more profound study of the mathematical theory, should it be necessary[2]. The reader is assumed to be familiar with probability theory.

In progressing from one volume to the next, the reader will see that the physical problems under consideration become more complex and the mathematical machinery more involved. This results quite naturally from the course-oriented origin of the book, based as it is on summarized lecture notes. Their extensive teaching experience has convinced the authors that such an approach is preferable to a more uniform presentation. Each chapter is followed by a num-

[1] It should be noted that certain questions in statistical radiophysics are not covered in this book. For example, it is not concerned with quantum radiophysics and quantum electronics, nor with statistical phenomena related to the propagation of waves in nonlinear media, see [1].

[2] Treatments of the mathematical theory are given, for instance, in [2–9], and more specialized applications of the theory to radiophysics and engineering are to be found in [10–19].

ber of problems worked out in detail. The problems not only serve as exercises, but in many cases contain additional theoretical material and further literature references.

Moscow, March 1986

S.M. Rytov
Yu.A. Kravtsov
V.I. Tatarskii

Preface

This volume concentrates almost exclusively on univariate random functions, i.e., random processes. The main physical questions considered are random pulse Poisson processes, Brownian motion, fluctuations in Thomson self-oscillatory systems and the action of random forces on dissipative dynamic systems, both linear and non-linear. These applications help to introduce the reader to the central concept of random processes, and to several classes of these processes: Gaussian processes, continuous and discontinuous Markov processes, stationary processes, processes with stationary increments, and also to the elements of stochastic convergence and ergodicity, and stochastic differential equations.

Volume 1 was written by S.M. Rytov. The author is indebted to Professors A.M. Yaglom, M.L. Levin, M.A. Isakovich, V.N. Tutubalin, Ya.I. Khurgin, and V.P. Yakovlev, whose valuable suggestions and comments contributed immensely to the quality and arrangement of material in some chapters.

Moscow, March 1986 *S.M. Rytov*

Table of Contents

1. General Introduction

The higher sensitivity of measuring and receiving devices, and hence the improved accuracy of measurement, have resulted in *fluctuations* playing an ever-increasing role in various branches of physics and technology. Fluctuations are known to be random deviations of macroscopic quantities from their mean (for example, thermodynamic equilibrium) values. Along with the theory of equilibrium states and the kinetics of physical processes, the theory of fluctuations is one of the key subjects in statistical physics. The mathematical tools of statistical physics are not only the classical theory of probability concerned with random events and variables, but also the much more general *theory of random functions*. It includes both random *processes* occuring in time (these are considered in Volumes I and II of this work, i.e., this volume and [1.1]), and random *fields* depending on time and space (Volumes III and IV [1.2,3]). The aim of *Principles of Statistical Radiophysics* is to introduce the reader to the theory of random functions at a physical level of rigor, drawing on specific physical problems taken mostly from the field of radiophysics.

A word of clarification on what is meant by "physical level of rigor". We are familiar with the difficulties encountered in seeking a compromise between the requirements of a rigorous presentation of a mathematical theory and the clarity desirable for a first encounter with the subject. The faultless reasoning of a mathematician is on occasion seen by a physicist or engineer as nothing more – just less convincing – than simple and tangible arguments, while these may lack the rigor demanded by the mathematician. The present book furnishes a sufficiently sound mathematical basis for independent work, but is in no way intended as a substitute for a more profound study of the mathematical theory.

As mentioned above, radiophysical problems are used to illustrate applications of the theory, and sometimes to substantiate the need for some of its techniques. In this context, radiophysics as such is understood in a wider sense implying the following.

As early as the 1930s L.I. Mandelshtam substantially extended the heuristically very powerful principle referred to as the "isomorphism of laws". Crudely speaking, it implies that laws discovered in various branches of physics are (under the appropriate conditions) also applicable to other fields. This principle, known in linear optics and acoustics since the time of Rayleigh, has led to the creation and development of the general theory of oscillations and waves (including nonlinear ones) that has encompassed not only the radio-frequency range and optics and electromagnetic processes in general, but also oscillatory and wave phenomena of *any* physical nature, including macroscopic mechanics (in particular, acoustics), quantum mechanics, chemistry, and biology. This general theory of oscillations and waves is naturally concerned with both deterministic processes and fields, and random (stochastic) oscillations and waves, irrespective of their fundamental physical nature, or rather with a wide variety of their physical origin. This is what is meant by "radiophysics understood in the wider sense".

In fact, the very use of the term "radiophysics" is to a certain extent a tribute to the generalization of the principle of isomorphism due to Mandelshtam.

Of course, fluctuations do not exhaust the enormous variety of stochastic phenomena. Yet random variations that occur in completely *deterministic* systems have long been known. By way of example we may cite turbulence in a liquid or gas, developing under the appropriate conditions, but still within a purely deterministic description of the medium. Similar examples, along with the kinetic theory of gases and liquids, strengthened for a long time the conviction that such an "intrinsic" stochasticity (one of the major problems of the so-called ergodic theory) presupposes the presence of an immense number of degrees of freedom in the system considered. About fifteen years ago it was found, however, that this is not required at all: autostochastic behavior appeared to be possible in dynamic systems as well, if the phase space had a dimensionality of three or more. We mention this in order to stress that this book is not concerned with the issues of autostochasticity in dynamic systems. It only considers those random phenomena (including fluctuations) in dynamic systems that occur as a result of external random actions. Mathematically, this implies that the phenomena to be considered are from the very beginning described by random functions explicitly introduced into the equations of motion for the system. This implies the transition from dynamic equations to *stochastic* ones.

Stochastic equations are introduced in this volume, but a more profound development of the theory and applications of these equations are to be found in Volumes III and IV [1.2,3].

Let us take a closer look at physical objects and phenomena covered by stochastic radiophysics. Two types of problem can be distinguished. Probabilistic properties of random actions on a system are either derived from a micromodel, i.e., from stochastic assumptions at a molecular level, or they are from the outset selected at the macroscopic level, i.e, within a phenomenological approach tailored to account properly for the experimentally observed behavior.

Problems of the first kind are, for instance, concerned with the *thermal motion* of microparticles, such as microcharges – electrons, ions, and so on. At the macrolevel this results in *thermal fluctuations* of a wide variety of quantities, e.g., density, pressure, temperature, current, voltage, the strength of electric and magnetic macroscopic fields, etc., etc. Thermal fluctuations give rise to

Brownian motion, molecular scattering of light (both within and on the surface of a medium), so-called thermal noise in radio-engineering, optics and acoustics, thermal radiation of bodies, and other phenomena.

Another example is the random variation of the number of particles in the electron fluxes in thermal- and photo-emission – the so-called *shot effect* ensuing from the discrete nature of microcharge carriers.

Worthy of special mention is a fairly common type of fluctuation – *the flicker effect* – which occurs in vacuum tubes (where it is associated with local variations in the emission of electrons from the cathode), contacts, semiconductors, electrolytes, and nonmetal resistors. Evidently, the prevalence of the flicker effect cannot be associated with a general cause such as, for example, the finite dimensions of a macrosystem (the importance of its dimensions being the greater the slower the fluctuations).

One more example is *magnetic noise* (including the Barkhausen effect) due to chaotic remagnetization of domains in ferromagnets subjected to time-dependent magnetic fields.

Problems of the second kind, where the stochastic behavior of random actions is simply specified at the macrolevel, also involve either random processes or random fields. In the case of *discrete systems*, random functions may be introduced to describe fluctuations of the parameters of the system and/or of the external forces. We deal here with stochastic *ordinary* differential equations. Under this heading is included the important special case of self-oscillatory systems (the present book and Volume II [1.1]). If the oscillations are periodic but not harmonic then the presence of fluctuations results in non-monochromaticity of the self-oscillations. This nonmonochromaticity is to a large extent associated with the stability of the frequency and the accuracy with which it can be measured (and hence the accuracy of time measurements).

Such a "macrostochastic" approach can also be applied to *continuous* systems described by stochastic *partial* differential equations which include a large variety of problems concerned with wave propagation in media with random inhomogeneities which cause wave scattering, random pulsation of refraction, fluctuations of the intensity and phase of a wave at the point of observation, and so forth. In these cases, the inhomogeneities of the medium are described by random fields with given stochastic characteristics.

Similar problems occur not only in wave propagation in free space, but also along guided lines (wire feeders, wave-guides, optical fibers), whose inhomogeneities can be thought of as random. However, though Volumes III and IV [1.3,4] are mainly devoted to wave propagation in random media, they do not refer to these particular questions. The same holds for the important problems arising from random inhomogeneities in emitters (in particular, in composite antenna systems; see [1.4]).

It should be emphasized that autostochasticity does not exclude such a "macrostochastic" approach. For instance, media may be random owing to turbulence. But the emergence and development of turbulence as such are the subject of the ergodic theory. However, when turbulence is considered from the point of view of its influence on the propagation of mechanical or electromagnetic waves, then the inhomogeneities arising from it may be described by introducing aptly chosen random fields into wave equations.

For transverse waves, questions related to the interference of oscillations and waves, their temporal and spatial coherence, and also polarization, constitute a rich and important field of interest. In this context the so-called *correlation theory* of random functions is widely used. Though it represents a fairly restricted part of the general theory of random functions, it is so important that the whole of Volume II [1.1] is devoted to it. The second volume deals with the correlation theory of random processes, but some of its generalizations to random fields are contained in Volume III [1.2].

When receiving radio signals, the quality of the signal is influenced by fluctuations (noise) in the receiver and the measuring device, the transformation of external and internal noise (in a wide variety of signal transducers), the noise immunity of the receiving systems, and so on. Some of these problems, though of primary practical importance, are of no physical interest. Nevertheless, a number of them are discussed to illustrate the techniques of the theory of random functions (this volume and [1.1]). A further reason for not going into these questions at length is that they are amply dealt with in a multitude of existing monographs and textbooks. The reader is referred to this literature, see also the references given in the Preface.

The preceding comments allow us to put into the proper perspective the motives underlying the selection of the subject matter and structure of the book [1.1–3].

On the one hand the above, rather incomplete list of pertinent problems demonstrates that the field of statistical radiophysics covers a tremendous variety of phenomena – whether the term is understood in its wider or narrower sense. It is almost impossible to reflect all the diversity of the statistical problems the subject covers. Moreover, such an attempt could hardly be justified since an extensive overview would necessarily be superficial – this alone rendering it practically useless. On the other hand, the mathematical techniques of statistical radiophysics are more involved and powerful than those of the classical theory of probability. A thorough treatment of the theory of random functions would constitute a special mathematical course.

In view of all this, we believe it appropriate to single out a few radiophysical issues interesting and important in their own right, which also introduce the reader to the elements of the theory of random functions, its techniques and applications. The treatment presupposes an acquaintance with the classical theory of probability, although in the first two chapters of Volume I we recall some of its fundamental concepts and theorems for the benefit of the reader.

2. The Bernoulli Problem

This and the following chapter provides an introduction to the theory of random processes. Therefore, in these two chapters we confine ourselves to the classical theory of probability and deal only with random events and quantities. The purpose of this chapter is two-fold:

First, to clarify the connection between the mathematical concept of probability and the ways in which this concept is applied. One frequently encounters a lack of clear understanding of the fact that the measurement of the probability of an event through the relative frequency of its occurrence is based on an independent postulate not included among the axioms of the mathematical theory. At the same time, this postulate is necessary for any applications of the mathematical theory, to physics in particular (Sects. 2.1,2).

Second, referring to the binomial law (Sect. 2.3), a simple example of a discrete probability distribution, it is demonstrated that a common approach can be used for some seemingly different problems such as: the fluctuations of gas density, the fluctuations of the intensity of the sum of harmonic oscillations with random amplitudes, the one-dimensional Brownian motion, and the shot effect (Sects. 2.4,5). Some of these problems involve processes evolving in time.

The binomial law has the additional feature of having extremely important asymptotics: the discrete Laplace and Poisson distributions (Sects. 2.5,6) and the continuous Gauss distribution (Sect. 2.7).

2.1 The Physical Concept of Probability

Experience has shown that a purely mathematical presentation of probability theory does not always lead to a clear understanding of the principal concepts important for the physicist. We shall, therefore, dwell briefly on what probability is and how the physicist uses this concept.

It is well-known that probabilities had been calculated long before the true meaning of probability was perceived. The theory originated from attempts to calculate the chance of success, of fair stakes, etc., in so-called games of chance such as the throwing of dice, the tossing of coins, and some card games, i.e., those games in which the outcome depends on chance alone and in no way on the player's ability and dexterity. In this sense these games might be called frivolous. But, as mathematicians like to remark, the intellectual game of chess has made practically no contribution to science, whereas the frivolous game of dice has contributed so much. Why?

The truth is that chess obeys its own special rules, and all the situations encountered here lie within the domain of these rules, i.e., they cannot be generalized. In contrast, the throwing of dice is elementary and exhibits a pure form of an extremely general statistical law – the stability of relative frequencies with increasing number of tests. If n throws of a die yield n_i times i points, then the relative frequency n_i/n shows a surprising stability as n increases. This empirical fact is independent of whether the die is good or bad. An a priori false die, e.g., one containing a piece of lead in one of its faces, exhibits stability of the quotient n_i/n, though different faces have different relative frequencies, not equal to $1/6$.

Thus, for n_i/n to be stable does not require a "good" die, and specific "asymptotic" values of n_i/n are not predetermined by the very fact of stability. This follows either from a *statistical experiment* especially performed or from data accumulated previously. By saying that for a good (i.e., suitable for the game) die the values of n_i/n are about $1/6$, we give in essence a definition of the die that is to be regarded as good.

It is frequently believed that the value $n_i/n \approx 1/6$ follows from the classical definition of probability due to Laplace

$$P = \frac{\text{favorable outcomes}}{\text{equally likely outcomes}} \quad .$$

This definition is too restrictive since it does not cover those cases where the possible outcomes constitute an infinite set, countable or continuous. But even if we do not wish to expand the scope of this definition, we easily see that this is circular reasoning, because equally likely here means nothing but equally probable, or *equiprobable*. Hence, the probability P is here "defined" through probability. This is simply a rule for calculating the probabilities of favorable outcomes using a preassigned *uniform* distribution of probabilities of all possible outcomes. Although we think that the Laplace definition yields the probability $P = 1/6$ of the appearance of each face of the die, in fact, we get from this definition only what we have incorporated into it, namely, the equally likely appearance of each face. In exactly the same way we assume that (in a single draw from a deck of cards) one is equally likely to obtain any one of its 36 or 52 cards.

What is behind these self-evident assumptions? They are sometimes supported by the use of the so-called *principle of insufficient grounds*, which states that if the die is geometrically accurate and made of uniform material, etc., then there are no grounds to expect $P \neq 1/6$.

Intuitively, this may be clear. A person who has not the slightest idea of what a statistical experiment involves can deduce the equiprobability of the faces of the fair dice or of either of the coin's sides appearing by simply exercising common sense. Here the equiprobability is natural, if not self-evident. This is suggested by the finite number of possible outcomes and the *symmetry*.

But without the symmetry or, say, with continuous possible outcomes, common sense jibs or simply fails. With the false die, only a statistical experiment with the *given* die enables us to estimate the probability of various faces. Difficulties encountered in the selection of equally likely outcomes, constituting a continous set, were demonstrated marvellously by *J.L.F. Bertrand* in a number of geometric problems in his book *Calcule des Probabilitiés* (1888). These so-called *Bertrand's* paradoxes with geometric probabilities are another illustration of the inadequacies of common sense or intuition in the more complex situation under consideration [2.1]. Moreover, when we say that in simple cases an intuitive prediction is *correct*, this very "correctness" only implies that a test by statistical experiment will bear out the prediction. Thus, *in the final analysis* even the self-evident assumption of equiprobability implies an enormous number of tests to be actually performed with certain conditions observed, i.e., accumulated statistical experience.

A detailed history of the evolution of probability theory is beyond the scope of the book. This long history has witnessed the emergence of many varied and conflicting approaches to the concept of probability. The development of probability theory into an axiomatic branch of mathematics took about three centuries.

A mathematical theory originating from some (usually fairly simple) ideas taken from everyday life tends to free itself from its empirical roots, thus striving to attain the status of an axiomatic theory. In the case of the theory of probability the process only terminated in the 1930s, when *A.N. Kolmogorov* formulated his axioms that made probability theory a branch of the measure theory. It should be noted that as early as 1917 S.N. Bernstein stressed the need for an axiomatization of probability theory and suggested a concrete set of axioms, which eventually turned out be appreciably less convenient than those proposed by *A.N. Kolgomorov*.

Today, the mathematician treats probability as a non-negative, unit-normalized, completely additive set function, defined on an algebra of subsets. An event A is represented by a set A of points of the space of possible elementary outcomes of the experiment or test under consideration. The probability $P\{A\}$ of event A is a set function, such that if it is defined on sets $A_1, \ldots, A_n$, it may also be defined on a set consisting of all the points constituting at least one of the sets $A_1, \ldots, A_n$, and also on the set of points belonging simultaneously to all the sets $A_1, \ldots, A_n$ [2.2,3]. This function obeys the following three axioms:

Axiom I. The probability $P\{A\}$ of an event A obeys the inequality $P\{A\} \geq 0$.

Axiom II. For a certain event U the equality $P\{U\} = 1$ is valid.

Axiom III. For mutually exclusive events A_k ($k = 1, 2, \ldots, n$; n being arbitrarily large) the relation

$$P\left\{\sum_{k=1}^{n} A_k\right\} = \sum_{k=1}^{n} P\{A_k\}$$

holds. The summation signs on either side of the equality have different meanings. On the left, the sign relates to the probability of at least one of the events A_k, so that $\sum_{k=1}^{n} A_k$ is an event (A_1 or $A_2, \ldots,$ or A_n), whereas on the right it is a conventional sum of non-negative numbers $P\{A_k\}$.

The certain event mentioned in axiom II corresponds to the set U consisting of *all* imaginable elementary outcomes, and hence all the other events A are subsets of U.

Axiom III (the additivity axiom) covers also the case of $n = \infty$. That is why P is referred to as a *completely additive* function.

Axioms I–III, supplemented by the definitions of a number of related concepts (e.g., those of random variable or of mathematical expectation), lie at the foundation of the whole theory of probability.

Of course, examination of these axioms reveals the connection between the abstract quantity P and the empirical relative frequency, but this is a *genetic* connection that relates only to the origin of the axioms, and not to their content, from which all traces of the underlying empirical facts have already been removed. Therefore, after being exposed to this abstract definition of probability, a physicist, an engineer, an economist, etc., i.e, one dealing with real problems and events, is bound to ask what he should do with such a completely additive function, and how he should relate the formulas of this mathematical theory to the real world. The situation here is the same as in any physical theory.

Equations and formulas for certain mathematical quantities (symbols) do not (in themselves) constitute a physical theory. The latter requires a knowledge of how to determine from real things and events the *numbers* to be inserted into mathematical relations. That is, the values of the respective mathematical symbols or, in other words, a way of *measuring* these quantities[1]. Of course, a physical theory is a single entity. Its mathematical and measurement ingredients are by no means independent of each other, yet, one cannot directly substitute one for the other. For this reason, in all physical (and in general practical) applications of the mathematical probability theory, the latter should be supplemented by at least one (realizable and concrete) *measuring technique* for the respective quantity P, the probability. It is only natural that we should turn to the relative frequency when looking for this technique or measurement axiom.

We assume that the probability of an event has been *measured* (approximately, as in any measurement) in terms of the *relative frequency* of its occurrence in a sufficiently long series of tests performed under certain unchanged conditions, relying on a sufficiently large ensemble of homogeneous systems, i.e., in a statistical experiment. Since it is an independent postulate that does not belong to the axioms of the mathematical theory of probability, this assumption is by no means the only one and does not a priori ensure success. Of course, a basic requirement is that it be consistent with the mathematical theory. As we shall see below (Sect. 4.6), the "frequency" measurement axiom

[1] This, at any rate, is the approach to measurement in classical (pre-quantum) physics. But the equations of quantum mechanics also contain quantities whose direct measurement is not required. (See Ref. [2.4], *Lectures in Quantum Mechanics*).

of probability meets this requirement. But whether or not the physical theory of random events, resulting from the incorporation of this axiom, will be in agreement with experiment is a *further* question, which can only be answered by statistical experiment. Moreover, it is important here to stress that without *some* means of measuring the probability, it is impossible not only to answer this question, but even to formulate it.

In adopting the proposed way of measuring probability, we by no means identify it with relative frequency, as is the case with the *frequency concept* put forward in 1928 by *Mizes* [2.5]. He suggested to define probability as the limit of the relative frequency n_i/n as $n \to \infty$ in a "statistical collective", i.e., an ensemble or series of tests satisfying definite requirements. However, since nobody knows what the limit of an empirical quantity is, this definition of probability could not be considered to be at the foundation of the mathematical theory. In the probabilistic context (Sect. 4.6), such a limit would again lead us into a vicious circle, since we would then be attempting to define probability in terms of probability.

Even now one sometimes encounters in the literature the so-called *intuitive concept* of probability, according to which probability is a measure of our ignorance. For instance, we do not know which face of a die will show up and *therefore* $P = 1/6$. If we were consistent in adhering to this point of view, we would witness a miracle: this very ignorance would bring knowledge, i.e., some positive statement. This, in essence, amounts to saying that if a distribution is unknown, then it is uniform. Of course, we could test this hypothesis too, to assess its correctness from its consequences. But why should this particular hypothesis be regarded as a necessary consequence of our ignorance? In the case of the false die we also do not know which face will appear, but our ignorance brings us no further in our quest for the true values of the probabilities for the faces of this die.

Referring to the mathematical theory, it simply does not require the distribution of possible outcomes to be uniform. Roughly speaking, the mathematical theory of probability based on certain axioms tell us how to determine, from known distributions of some random events or variables, the distribution of other events or variables somehow related to the former. For example, by the additivity axiom, the probability for an even number of points will be

$$p^{\text{even}} = p_2 + p_4 + p_6,$$

regardless of whether or not the probabilities p_2, p_4, and p_6 of faces 2, 4, and 6, respectively, are similar or even known at all.

A discussion of the question of where the initial distributions come from is beyond the scope of the mathematics. Algebra, for example, tells us how the roots x_1 and x_2 of a quadratic equation can be derived, i.e., how x_1 and x_2 can be expressed through the known coefficients a, b, and c. Algebra does not concern itself with the origin of these coefficients. It is this non-mathematical question of the source of the initial distributions that was at the roots of different concepts of probability and the theory of probability in its preaxiomatic period, i.e., before it was once and for all relieved of the duty to answer this question. The answer is given by measurement, i.e., by statistical experiment.

2.2 Distribution Laws for Random Variables

It is clear from the above why a user of the mathematical theory of probability, when speaking about probability, always means the relative frequency in a fairly large statistical ensemble. He would treat a probability distribution law for a random variable ξ, i.e., for a numerical characteristic of the outcome of a test in exactly the same way. If the possible values x of the quantity ξ belong to the interval $a<x\leq b$ (in particular, to the infinite interval $-\infty<x<\infty$), the distribution law is given by the (*integral*) *distribution function*

$$W(x) = P\{a<\xi\leq x\} \quad ,$$

i.e., by a non-negative, non-decreasing function of x, describing the probability that ξ belongs to the subinterval $a<\xi\leq x$ (random event). Here $W(a) = 0$ and, by Axiom II, $W(b) = P\{a<\xi\leq b\} = 1$.

If $W(x)$ grows with x by increments p_i at points $x = x_i$ and is constant between these points (a step function), we have the special case of discrete possible values of ξ, so that p_i is the probability of the event $\xi = x_i$:

$$p_i = P\{\xi = x_i\}.$$

Here p_i are similar to the intensities of discrete lines in a line spectrum, or to masses concentrated at points x_i on the x axis, so that the total intensity or total mass is unity

$$\sum_i p_i = 1.$$

If the distribution function $W(x)$ is differentiable, i.e., if a *probability density* $w(x) = dW(x)/dx$ exists, this is equivalent to saying that a continuous set of values x is possible, such that

$$w(x)dx = P\{x<\xi\leq x + dx\} \quad ,$$

$$\int_a^b w(x)dx = W(b) = 1.$$

The non-negative function $w(x)$ is analogous to the intensity density in a continuous spectrum, or the density of a mass spread continuously over the interval (a, b) on the x axis.

As indicated above, for the physicist the numbers p_i and $w(x)dx$ are relative frequencies in a statistical ensemble. In other words, they are those proportions of realizations of a random phenomenon in a large statistical experiment which have given $\xi = x_i$ for discrete possible values, or ξ falling within the interval $(x, x + dx)$ for continuous possible values.

Suppose that at time $t = 0$ a droplet of paint is added to a vessel of water, every precaution being taken to prevent agitation of the water. The subsequent behavior of the droplet will be determined by the diffusion of the paint. The concentration of paint $c(\mathbf{r}, t)$ at any point $\mathbf{r}$ and at any time $t>0$ can be uniquely determined by solving the diffusion equation for specified initial and boundary conditions. We have thus a completely deterministic phenomenon governed by a certain dynamical law. But if we turn to the behavior of individual particles or molecules of paint, we will be dealing with a statistical experiment. Even a minute drop of paint with a volume of, say, $10^{-3}\mathrm{mm}^3$ contains more than 10^{15} molecules. All of them are subject to virtually the same conditions and in the water undergo random Brownian motion, which is responsible for the diffusive spreading of the droplet. Consequently, here we have a fairly large ensemble of paint molecules such that $c(\mathbf{r}, t)d^3r$ is the number of molecules which happened to be at time t within the volume element $d^3\mathbf{r} = dx\,dy\,dz$. For the physicist the relative concentration $c(\mathbf{r}, t)$ [normalized so that for the whole volume of the vessel $\int c(\mathbf{r}, t)d^3r = 1$] is precisely the probability density of finding an individual molecule within $d^3\mathbf{r}$ at time t, if it started from point $\mathbf{r} = 0$ at $t = 0$. The deterministic behavior of $c(\mathbf{r}, t)$ is a manifestation of stability of the relative frequencies for a fairly large (10^{15}) ensemble.

More generally, the probability distribution may be mixed, i.e., discrete-continuous similar to the superposition of discrete and continuous spectra. To obtain a simple expression containing both the discrete and the continuous component, it is convenient to use the Stieltjes integral with respect to $W(x)$, such that the increment of the distribution function $dW(x)$ in the interval $(x, x + dx]$ may be both infinitesimal [continuous spectrum; $dW(x) = w(x)dx$] and finite [discrete spectrum; $dW(x) = W(x + 0) - W(x) = p$]. Using the Stieltjes integral, we can, for instance, write the mean of the deterministic function f of a random variable ξ as follows[2]:

$$\langle f(\xi)\rangle = \int_a^b f(x)dW(x), \tag{2.1}$$

no matter whether the function $W(x)$ is differentiable or not. Physicists, however, prefer another approach which, although less rigorous, is more practical. It is assumed that the density $w(x)$ is meaningful everyhwere, thus

$$\langle f(\xi)\rangle = \int_a^b f(x)w(x)dx. \tag{2.2}$$

[2] The statistical average, i.e., the one over a probability distribution (or, as a physicist would put it, the *ensemble average*) will be denoted by angle brackets, or when appropriate by a bar over the averaged quantity.

In the discrete case, instead of $w(x)$ we have the sum of delta-functions at points $x = x_i$ with weights p_i:

$$w(x) = \sum_i p_i \delta(x - x_i). \tag{2.3}$$

Integral (2.2) then becomes the discrete sum

$$\langle f(\xi) \rangle = \sum_i f(x_i) p_i.$$

The expression (2.3) holds not only for strictly discrete, but also continuous values, provided the probability density $w(x)$ has *sufficiently sharp* peaks. The width of these peaks should be small as compared with the scale of a specific problem. It should, however, be remembered that an approximation of these peaks by delta-pulses, assumed at the outset of the argument, would not always yield the result obtained by passing in the final relations to the limit (2.3). Given this reservation, the non-random (deterministic) quantity ξ, which can be said with certainty to take on the value x_0, can nearly always be viewed as a random variable with the extremely sharp probability density $w(x) = \delta(x - x_0)$. Any dynamical theory then becomes a special case of a statistical theory with extremely sharp distributions in the respective variables. This reasoning does not lead to a paradox, since quantum mechanics has long shown that the most fundamental laws of physics are statistical in nature.

In practice the interpretation of probability as the relative frequency of a sufficiently large ensemble also covers, of course, multivariate random functions, i.e., collections of random variables. In the above example of paint diffusion the random variable is the radius vector of each molecule of the droplet, i.e, the collection of three random coordinates of a molecule. In addition, in this example we are, in fact, dealing not with random variables, but with a random function: the probability density (concentration) c also varies with time t. In what follows we will focus on this generalization of the classical theory of probability. At this point it is important to stress that this generalization, too, does not preclude an interpretation of probability distribution laws in terms of relative frequencies.

Now we turn to one special, discrete distribution law – the binomial law – that will not only reveal some interesting facts and concepts, but also take us a step further, namely, from random variables to random functions.

2.3 The Binomial Distribution Law

If a *finite* number of tests can be arranged in a given time, e.g., by assigning the time $t_\nu = \nu\tau$ to the νth test (or in some other way), we will have a *model for a random process*, for which the concepts and methods of the classical probability theory suffice. By way of example we consider a well-known problem in the theory of probability – the Bernoulli problem (1713). It deals with a sequence of *independent* tests and is one of the most spectacular illustrations of how one and the same mathematical scheme (dynamic or statistic) encompasses a multitude of widely different phenomena.

We now list a number of specific issues that correspond directly to the mathematical scheme of the Bernoulli problem. Some of these issues do not require any localization of the tests in time, whereas others allow, or even presuppose, such a localization, i.e., they are concerned with processes unfolding in time.

The Urn Problem. An urn contains black and white balls, so that the probability of drawing a white ball (event A) is $P(A) = p$; and a black ball (event $\overline{A}$), $P(\overline{A}) = q = 1 - p$. Altogether, N tests are performed, so that in each case the ball drawn is replaced and the balls are shuffled (thereby ensuring the independence of the tests). What is the probability $p_N(n)$ that in N tests a white ball will be drawn n times?

Coin Tossing. What is the probability that in N tosses a head will appear n times, if the probability of showing a head in one toss is $p = 1/2$? Or what is the probability that of N babies born n will be boys, if the probability of a boy being born is $p = 0.51$?

Density Fluctuations. A volume V contains N gas molecules. The probability of a molecule being in element v of the volume V (event A) is p. What is the probability that there are n molecules in v?

Intensity Fluctuations. Let N oscillations given by $a_i \cos \omega t$ with $a_i = \pm a$ be superposed, with probabilities p and q, respectively. What is the probability that the intensity of the resulting oscillation is

$$J(N, n) = [n(+a) + (N - n)(-a)]^2 = (2n - N)^2 a^2,$$

i.e., that the amplitude $+a$ will enter n times?

Telephone Calls. A telephone operator is on duty for a time T. The probability that in a very short interval τ there will be a call is p. What is the probability of n calls during the shift?

The Shot Effect in a Vacuum. Electrons are ejected from a hot filament so that the probability of one being ejected in a rather short time τ is p. What is the probability that n electrons, i.e., charge ne, will reach the anode in time T? If the anode circuit incorporates an integrating device that sums up the charge transferred in time T, then the question should be rephrased as follows: what will be the average reading of the instrument and what will be its fluctuation?

The Random Walk. We will use the following model of the Brownian motion of a particle: The particle jumps by an amount a to the right or left with probabilities p and q, respectively. What is the probability that after N jumps the particle will travel a distance $s(N, n) = (2n - N)a$, i.e., will make n steps to the right? If here $p = q = 1/2$, this is equivalent to the one-dimensional motion of an "absolutely drunken man". If A is a step to the right, $\overline{A}$ is "no step", then it is the motion of a "wary man".

The last three problems concern processes occuring in time. They are most naturally associated with our ideas of tests carried out in a temporal sequence.

The general formulation of the problem therefore runs as follows: N *independent* tests are performed, each of which results in one of the opposite events A or $\bar{A}$. What is the probability that in N tests the event A will occur n times? The solution to the Bernoulli problem, though well known, is given below.

Thus,

$$P(A) = p, \quad P(\bar{A}) = q = 1 - p.$$

What is the probability that N independent tests will yield the sequence $A\bar{A}AA\bar{A}\ldots A$, i.e., *first* A, *then* $\bar{A}$, etc., with A occuring n times all in all? The answer is obvious:

$$pqppq\ldots p = p^n q^{N-n} \quad .$$

We are interested in the event B that A occurs n times in *any* n tests out of N. In *definite* tests, events where A occurs n times are all special cases of B that mutually exclude each other. By the additivity axiom we have

$$P_N(n) = \sum p^n q^{N-n} \quad ,$$

where the summation is to be performed over all the special cases of B. But the number of special cases of B is the number of ways in which n elements may be chosen from N, i.e., $C_N^n = N!/n!(N-n)!$. Thus,

$$P_N(n) = C_N^n p^n q^{N-n} \tag{2.4}$$

is the distribution law termed *binomial* since $P_N(n)$ is the nth term of the binomial series

$$(p+q)^N = \sum_{n=0}^{N} C_N^n p^n q^{N-n} = \sum_{n=0}^{N} P_N(n).$$

The terms should add up to unity, as the sum is the probability of *any one* of the possible values of n. This is really so because $p + q = 1$ [3].

More generally, when any test may yield one of k mutually exclusive events $A_1, A_2, \ldots, A_k$ with probabilities $p_1, p_2, \ldots, p_k (\sum_{i=1}^k p_i = 1)$, the probability that in N tests the event A_i will occur n_i times $(\sum_{i=1}^k n_i = N)$, is given by

[3] Note the solution to the so-called de Méré problem: what must the number of tests N be for an event A to occur *at least once*? The probability of A occuring once, twice, thrice, etc., in N tests is

$$P_N = \sum_{n=1}^{N} P_N(n) = \sum_{n=0}^{N} P_N(n) - P_N(0) = 1 - q^N = 1 - (1-p)^N.$$

With increasing N it tends to unity. Accepting one or another of P_N as a value that guarantees certainty, we can readily compute the corresponding value of N.

the respective term of the power series $(p_1 + p_2 + \ldots + p_k)^N$, i.e.,

$$P_N(n_1, n_2, \ldots, n_k) = \frac{N!}{n_1!n_2!\ldots n_k!} p_1^{n_1} p_2^{n_2} \ldots p_k^{n_k}.$$

We now consider certain properties and consequences of the binomial distribution law.

If we denote by n_m the value of n at which the probability $P_N(n)$ is maximal, i.e., the *most probable* value of n, then from the condition

$$P_N(n_m - 1) \le P_N(n_m) \ge P_N(n_m + 1)$$

it is clear that

$$(N+1)p - 1 \le n_m \le (n+1)p.$$

We thus have either one or two optimal values. If $Np \gg 1$, then

$$n_m \approx Np. \tag{2.5}$$

Now we find the mean $\bar{n}$ and *variance* $D[n] \equiv \sigma_n^2 \equiv \overline{(\Delta n)^2} \equiv \overline{n^2} - \bar{n}^2$. The kth moment is given by

$$\begin{aligned} \overline{n^k} &= \sum_{n=0}^{N} n^k P_N(n) = \sum_{n=0}^{N} C_N^n n^k p^n q^{N-n} \\ &= \underbrace{p\frac{\partial}{\partial p} p\frac{\partial}{\partial p} \ldots}_{k \text{ times}} \sum_{n=0}^{N} C_N^n p^n q^{N-n} = \left(p\frac{\partial}{\partial p}\right)^k (p+q)^N. \end{aligned}$$

After the operator $p(\partial/\partial p)$ has been applied k times, we should take into account in the final result that $p + q = 1$. Thus,

$$\bar{n} = p\frac{\partial}{\partial p}(p+q)^N = pN(p+q)^{N-1} = pN, \tag{2.6}$$

$$\begin{aligned} \overline{n^2} &= p\frac{\partial}{\partial p} p\frac{\partial}{\partial p}(p+q)^N = p\frac{\partial}{\partial p} pN(p+q)^{N-1} \\ &= pN[(p+q)^{N-1} + p(N-1)(p+q)^{N-2}] \\ &= pN[1 + p(N-1)] \quad , \end{aligned} \tag{2.7}$$

hence

$$\sigma_n^2 = \overline{n^2} - \bar{n}^2 = Npq. \tag{2.8}$$

It should be noted that by virtue of (2.5) the mean $\bar{n}$ coincides for large N with the most probable value. The variance σ_n^2 grows with the number of tests N in exactly the same way as the mean, i.e., in proportion to N. Therefore, the mean *relative* deviation

$$\sigma_n/\overline{n} = \sqrt{q/pN}$$

decreases with N. Thus, as N increases, fluctuations grow, but the relative fluctuations decrease.

2.4 Examples of Applications of the Binomial Law

To clarify the physical meaning of the results obtained, it is instructive to consider some of the specific problems formulated above.

When considering a gas in a vessel (N molecules in a volume V), we will assume that the probability of a molecule being in a volume v is $p = v/V$. The gas density in v is $\varrho_v = n/v$. Consequently,

$$\overline{\varrho}_v = \frac{\overline{n}}{v} = \frac{pN}{v} = \frac{N}{V} \equiv \overline{\varrho} = \text{const},$$

i.e., on average the density is uniform.

The variance of n determines the intensity of density fluctuations,

$$D[\varrho_v] = D\left[\frac{n}{v}\right] = \frac{1}{v^2}D[n] = \frac{Npq}{v^2} = \frac{N}{vV}\left(1 - \frac{v}{V}\right),$$

where the known relation $D[ax+b] = a^2 D[x]$ has been used, or

$$D[\varrho_v] \equiv \overline{(\Delta\varrho_v)^2} = \overline{\varrho}\left(\frac{1}{v} - \frac{1}{V}\right).$$

The absolute fluctuations of density thus grow with decreasing volume v and increasing mean density, i.e., with an increasing total number of molecules N in V. Typically, the case of $v \ll V$, where the term $1/V$ can be ignored, is of interest.

The relative fluctuations are given by

$$\frac{\overline{(\Delta\varrho_v)^2}}{\overline{\varrho}^2} = \frac{1}{\overline{p}}\left(\frac{1}{v} - \frac{1}{V}\right) \approx \frac{1}{\overline{\varrho}_v},$$

i.e., they also increase with decreasing v, but decrease with increasing ϱ.

We now return to the particular condition that ensures the independence of tests. We supposed that the probability of a molecule being in the volume v is $p = v/V$, regardless of where, and in what number, other molecules are present. In other words, the gas is taken to be ideal. If the finite volume of molecules and their interaction were taken into account, the above assumption would not be valid: the probability p in a given test (the probability of a *given* molecule being in the volume v) would depend on the outcome of the other tests.

Next we take into account the addition of oscillations. The binomial law (2.4)is the probability that n of N superposed oscillations have the amplitude $+a$. It follows that $P_N(n)$ is the probability of the resulting intensity

$$J \equiv J(N,n) = (2n - N)^2 a^2.$$

Thus, the mean intesity is

$$\overline{J} = \overline{(2n - N)}^2 a^2 = (4\overline{n}^2 - 4N\overline{n} + N^2)a^2 = N^2 a^2 \left[1 + 4pq\left(\frac{1}{N} - 1\right)\right] \quad .$$

If $p = q = 1/2$, then $\overline{J} = Na^2$, i.e., the intensities add up (complete incoherence). For $p = 1$ or $q = 1$, we have $\overline{J} = N^2a^2$ (complete coherence).

The variance of J is

$$D[J] = \overline{J^2} - \overline{J}^2 = \overline{(2n - N)^4 a^4} - \overline{J}^2.$$

Thus, the computation of *intensity fluctuations* – a quantity quadratic in n – requires a knowledge of the higher-order moments of n : $\overline{n^3}$ and $\overline{n^4}$. We have already outlined above the procedure for calculating $\overline{n^k}$ for the Bernoulli distribution: the application of the operator $p(\partial/\partial\varrho)$ to the binomial $(p+q)^N$ k times. Omitting the details of the calculations, we provide below the final result for the case of complete incoherence, where $p = q = 1/2$, and accordingly, $\overline{J} = Na^2$. In this case the variance is

$$D[J] \equiv \overline{(\Delta J)^2} = 2N(N-1)a^4,$$

and the relative intensity fluctuations are

$$\sqrt{\overline{(\Delta J)^2}/\overline{J}^2} = \sqrt{2(1 - 1/N)}.$$

The relative fluctuations do not decrease with N at all; on the contrary, they increase and tend to $\sqrt{2}$, i.e., they do not level off. With the addition of two oscillations the relative fluctuations of the intensity will be smaller than when many oscillations are superposed.

The familiar idea of levelling off derives from the *second* moment. If the noise depends on higher-order moments, then its contribution may grow with N.

In this connection, it is instructive to consider what is to be regarded as noise in the case of, say, a quadratic device, which measures not n, but n^2. The instrument measuring the quantity $\xi = n^2$ indicates $\overline{\xi} = \overline{n^2}$. It detects no ξ-noise, though $D[n] = \overline{n^2} - \overline{n}^2 \neq 0$, i.e., n itself fluctuates. The noise for the "ξ-meter" is determined by the quantity $D[\xi] = \overline{\xi^2} - \overline{\xi}^2 = \overline{n^4} - (\overline{n^2})^2$ and is thus related to the fluctuations of the fluctuations of n.

A phenomenon may evidently be classified as fluctuating or non-fluctuating depending on the quantity used to characterize it, as noted by *Leontovich* [Ref. 2.6, Sect. 34; see also Ref. 2.7]. Consider, for instance, the energy U of thermal radiation confined within a volume V. It fluctuates about its mean $\overline{U}$, so that the variance

$$D[U] = \overline{U^2} - \overline{U}^2$$

is a measure of the intensity of fluctuations of U. In electrodynamics the energy U is equal to the integral of the sum of squared strengths of electric and magnetic fields

$$U = \frac{1}{8\pi} \int_v (E^2 + H^2) dV,$$

hence,

$$\overline{U} = \frac{1}{8\pi} \int_v (\overline{E^2} + \overline{H^2}) dV.$$

Now we note that the means of the fields for a thermal radiation are zero ($\overline{\boldsymbol{E}} = \overline{\boldsymbol{H}} = 0$), consequently, $\overline{E^2} = D[\boldsymbol{E}]$, $\overline{H^2} = D[\boldsymbol{H}]$ and

$$\overline{U} = \frac{1}{8\pi} \int_v (D[\boldsymbol{E}] + D[\boldsymbol{H}]) dV,$$

i.e., the mean energy $\overline{U}$ as such is a measure of the intensity of fluctuations of $\boldsymbol{E}$ and $\boldsymbol{H}$. With these two approaches we thus have the following discrepancy in terminology:

Energy approach	*Electrodynamic approach*
Quantity U with mean $\overline{U}$.	*Quantities* $\boldsymbol{E}$ and $\boldsymbol{H}$ with zero means.
Fluctuations $\overline{(\Delta U)^2}$.	Fluctuations of fields $\overline{U}$.
	"Fluctuations of fluctuations" $\overline{(\Delta U)^2}$.

Turning now to problems where the process of "tests" occurs in time, we take as a first example the model of Brownian motion mentioned previously. The N-step displacement of a particle to the right by the amount

$$s \equiv s(N, n) = (2n - N)a$$

has the probability $P_N(n)$. Therefore,

$$\overline{s} = (2\overline{n} - N)a = N(p - q)a,$$

i.e., in the isotropic case ($p = q = 1/2$) the mean displacement is $\overline{s} = 0$.

The measure of scattering about $\overline{s}$, defined by the variance, is

$$D[s] = \overline{s^2} - \overline{s}^2 = D[2na - Na] = D[2na] = 4a^2 D[n] = 4a^2 Npq.$$

In the isotropic case, i.e., where $\overline{s} = 0$ and $D[s] = \overline{s^2}$, we obtain

$$\overline{s^2} = Na^2.$$

If discontinuities occur at equal time intervals τ, so that $t = N\tau$ is the total time, then

$$\overline{s^2} = \frac{a^2}{\tau} t.$$

This is the so-called *diffusion* time-dependence of the path traced by a particle. A detailed consideration of the connection between the statistical scheme under investigation and diffusion is deferred until Sects. 5.6, 9. We shall see that this simple model is also of interest in radiophysics, as it is directly connected to the problem of phase fluctuations in a self-oscillatory system.

2.5 Shot Effect. The Poisson Distribution

Now we consider the shot effect, which is quite similar to the telephone-call problem.

We assume that in each test the probability of event A (here it is the emission of an electron from the cathode surface during a short period of time τ) has the same value p. This value is independent of the outcome of other tests and of the number of tests in question, i.e., of the emission of other electrons and the presence of space charge. If we take into account the space charge responsible for the so-called depression of the shot effect, the problem becomes very complicated.

The arrival of an electron at the anode causes a pulse of anode current, whose duration is compatible with the electron transit time (in conventional vacuum tubes it is about 10^{-9}s), and the integral value is equal to the electron charge e. If the mean anode current amounts to, say, 5 mA, then about $3 \cdot 10^{16}$ electrons per second arrive at the anode. This suggests that the anode-current pulses overlap densely. If they followed one another without interruptions and overlapping, the transit of this number of electrons would take about a year. But here we are concerned neither with the degree of pulse overlap nor with the pulse form. By the event A we will understand here an instantaneous act of electron ejection from the cathode, i.e., the *onset* of an anode-current pulse.

We are interested in a charge coming to the anode in a time $T = N\tau$, where each time interval τ is a test (one of N tests) during which we register either an ejection (with probability p) or no ejection. The question now arises of how to deal with those intervals τ which contain two or more ejections. Intuitively, the probability of an ejection p must decay with decreasing τ. Then with the division of T into smaller and smaller subintervals $\tau = T/N$, the probabilities of two, three, etc., ejections in τ will decrease as quantities of second, third, etc. respectively order, with respect to τ. We can therefore anticipate that for sufficiently large N (without any limitations), for most of τ we only have two possibilities: either one ejection (probability p), or no ejection (probability $q = 1 - p$). This reduces the problem to the Bernoulli scheme, so that the probability of n ejections in T (for N tests) is $P_N(n)$. The current averaged over the time T is[4]

[4] In the gas-density example we assumed $\varrho_v = n/v$, i.e., we introduced a quantity averaged over v. Then it was a region of space and here we are dealing with a period of time T.

$$I_T = ne/T.$$

The mean of I_T is

$$\overline{I}_T = \overline{n}e/T = epN/T \equiv \overline{I}, \tag{2.9}$$

the intensity of fluctuations of I_T (variance) is

$$D[I_T] \equiv \overline{(\Delta I_T)^2} = D[ne/T] = e^2/T^2 D[n] = e^2 Npq/T^2,$$

and the relative mean-square deviation is given by

$$\overline{(\Delta I_T)^2}/\overline{I}_T^2 = q/NP.$$

But written in this way the results obtained cannot claim to be of any physical meaning, as they depend on the *arbitrary* number of tests N. This naturally suggests passing to the limit $N \to \infty$, subject to the condition that the average current $\overline{I}$ is kept constant, i.e.,

$$\overline{I} = n_1 e = \text{const},$$

where n_1 is the average number of electrons ejected per second. According to (2.9), this means that the probability p of an electron ejection in τ is related to N by

$$p = n_1 T/N = n_1 \tau,$$

i.e., p tends to zero together with τ. Inserting $p = n_1 T/N$ and $q = 1 - n_1 T/N$ into the relation for $D[I_T]$ and taking the limit $N \to \infty$ gives

$$\overline{(\Delta I_T)^2} = e\overline{I}/T \quad \text{and} \quad \overline{(\Delta I_T)^2 \overline{I}^2} = e/\overline{I}T. \tag{2.10}$$

For longer averaging times T both absolute and relative current fluctuations decay. As the average current $\overline{I}$ increases the fluctuations increase in absolute terms, but in relative terms they level off. Relations (2.10) are the principal equations for the shot effect without space charge, i.e., without depression. They suggest explicitly that the current fluctuations are governed by the discreteness of the electric charge.

The shot effect is an example of a problem where it pays to assume that the number of tests N in time T grows without limit, but the average number of occurrences of event A in unit time is n_1, and hence their mean in T, i.e., $\overline{n} = n_1 T$, is finite. Therefore, the probability of A in a single test

$$p = n_1 T/N = \overline{n}/N$$

tends to zero as $N \to \infty$, i.e., in the overwhelming majority of cases there is no event A.

All problems of this type (telephone calls, scintillations, waiting for a tram, intermolecular collisions, collisions of molecules with walls, etc.) involve the transition from a general formula for $P_N(n)$ to the limit as $N \to \infty$ and $p \to 0$, with $\bar{n}$ fixed.

Substituting $\bar{n}/N$ for p in (2.4), we obtain

$$\begin{aligned} P_N(n) &= \frac{N(N-1)\ldots[N-(n-1)]}{n!}\left(\frac{\bar{n}}{N}\right)^n\left(1-\frac{\bar{n}}{N}\right)^{N-n} \\ &= \frac{\bar{n}^n}{n!}\cdot 1\cdot\left(1-\frac{1}{n}\right)\ldots\left(1-\frac{n-1}{N}\right)\left(1-\frac{\bar{n}}{N}\right)^{N-n} \end{aligned} .$$

In the limit as $N \to \infty$, this relation yields for the probability n the Poisson distribution law

$$P(n) = \frac{\bar{n}^{-n}}{n!}e^{-\bar{n}} \quad . \tag{2.11}$$

The only parameter $\bar{n}$ figuring in this law is the mean of n, i.e.,

$$\sum_{in=0}^{\infty} nP(n) = \bar{n}.$$

All the higher-order moments are also expressed in terms of $\bar{n}$. So, we have

$$\overline{n^2} = \bar{n}^2 + \bar{n}, \quad \overline{n^3} = \bar{n}^3 + 3\bar{n}^2 + \bar{n}.$$

Note that if from the very beginning we consider that the probability of A in dt is $n_1 dt$ (dt instead of τ), then the Poisson law is obtained rigorously to second order of smallness in dt.

It is easily seen that

$$\sum_{n=0}^{\infty} P(n) = 1,$$

as expected. Of course, the mean $\bar{n}$ and the standard deviation σ_n derived from the Poisson distribution can also be deduced, using a limiting process, from the appropriate expressions given by the binomial law

$$\bar{n} = pN = n_1 T, \quad \sigma_n = \sqrt{Npq} = \sqrt{\bar{n}(1-p)} \to \sqrt{\bar{n}}.$$

Uniquely related to the uniform probability distribution for event A ($p = n_1 dt$, with $n_1 = \text{const}$) is not only the Poisson law, which gives the probability of n occurrences of A in time T, but also a quite definite (exponential) distribution law for the time period t from an arbitrary moment to the *first* occurrence of A. Let us treat this problem, proceeding again from the Bernoulli scheme.

The probability $w(t)\Delta t$ of the two independent events - the interval $(0, t)$ being empty and the interval $(t, t + \Delta t)$ containing at least one event A - is the product of the probabilities of these events. Let us divide the t axis into small subintervals of length τ, such that $(0, t)$ contains N such subintervals ($t = N\tau$); and Δt contains ΔN subintervals [$\Delta N = \Delta T/\tau = (\Delta t/t)N$]. Thus the probability that the interval $(0, t)$ will be empty is

$$q^N = (1-p)^N = (1-n_1\tau)^N = \left(1 - \frac{n_1 t}{N}\right)^N,$$

and the probability of *at least one* occurrence of A within Δt is (see Footnote 3)

$$1-(1-p)^{\Delta N} = 1 - \left(1 - \frac{n_1 t}{N}\right)^{(\Delta t/t)N} .$$

In the limit as $N \to \infty$, the former probability tends to $\exp(n_1 t)$ and the latter to $1 - \exp(-n_1 \Delta t)$, i.e., to $n_1 \Delta t$ in a first approximation. As a result, we have[5]

$$W(t) = n_1 \exp(-n_1 t). \tag{2.12}$$

According to (2.12), the average expectation time is

$$\bar{t} = \int_0^\infty t w(t) dt = \frac{1}{n_1},$$

which could be written simply from the definition of n_1. It is worth noting that the origin of time is not singled out. So, the origin may be the instant of the occurrence of an event A, and so t will be the time interval between two occurrences of A (molecular collisions, telephone calls, electron ejections, etc.)[6].

2.6 The De Moivre-Laplace Limit Theorem

Another limiting case of the binomial law for $N \to \infty$ arises when p does not tend to zero but assumes a definite value other than zero and unity. Thus, $\bar{n} = pN$ also grows with N. This is the content of the so-called *local limit theorem* proven by De Moivre (1730) for $p = q = 1/2$, and then by Laplace, for any p and q (naturally, $p + q = 1$). Here we will not consider the general limiting case for $P_N(n_1, n_2, \ldots, n_k)$, but will restrict our discussion to the conditions of the Laplace theorem.

[5] In this limiting process we are no longer interested in the serial numbers of tests, concentrating instead on a fixed time interval Δt within which the total number and the series numbers of tests grow indefinitely with N. This change of approach can be illustrated by an example from hydrodynamics, when the so-called "material" equations are replaced by the "local" ones. Another example of this change is the frequently used transition from Fourier series to Fourier integrals, when fixed numbers (or frequencies) of discrete harmonics are replaced by a fixed frequency interval $\Delta\omega$.

[6] At first glance, this appears to be strange. It would seem natural that, if we arrived at a tram stop just when a tram was leaving, we would have to wait longer than if we arrived during the time gap between two trams. But without any fixed time schedule for the tram traffic, this is not so (see [2.8] where the time between molecular collisions is considered).

Now we formulate first a physical problem for which the above-mentioned limiting process appears to be natural. For this purpose the Brownian-motion model discussed above is quite adequate. In N steps a particle has travelled from the initial position through a distance $s(N,n) = (2n - N)a$ with the probability $P_N(n)$. If a step takes a time τ, the distance $s(N,n)$ is covered in $t = N\tau$. Since τ is small, real observations deal with time intervals t involving an enormous number N of elementary displacements $\pm a$. But the respective probabilities p and $q = 1 - p$ of displacements $+a$ and $-a$ characterize the system in question and are independent of the observed number of steps N. Thus, for $t \gg \tau$, fairly large N and fixed p, it is expedient to apply the asymptotic expression for $P_N(n)$.

The De Moivre-Laplace local theorem states: *If the probability p of an event is independent of the test serial number and is different from 0 and 1, then the probability $P_N(n)$ at large N is given by the asymptotic expression*

$$L_N(n) = \frac{\exp(-x_n^2/2)}{\sqrt{2\pi N pq}} \quad \text{with} \quad x_n = \frac{n - pN}{\sqrt{Npq}}, \tag{2.13}$$

and $L_N(n)/P_N(n) \to 1$ as $N \to \infty$ uniformly for all n, for which x_n lies in a finite interval, i.e., $|x_n| < M$, where M is a positive number.

It is easily verified that the *Laplace distribution* (2.13) yields the same mean $\bar{n} = pN$ and variance $\sigma^2 = Npq$, as the binomial law. Therefore, $L_N(n)$ may be written as

$$L_N(n) = \frac{\exp[-(n - \bar{n})^2/2\sigma^2]}{\sqrt{2\pi}\sigma}. \tag{2.14}$$

As N increases, $L_N(n)$ spreads in absolute terms, but in relative terms it becomes narrower. If we fix the mean, and assume that σ varies independently, then, as $\sigma \to 0$, the distribution becomes sharper and eventually tends to a delta-function.

Note that the distribution $P_N(n)$ approximates $L_N(n)$ ever more precisely and faster with N, the closer p is to $1/2$. If we set $\eta = P_N(n)/L_N(n)$, then at $p = 1/2$ we have, for example, the numbers given in Table 2.1.

Figure 2.1 illustrates the special case of $n = \bar{n}$, showing the regions in the plane ($\bar{n}$, $1/p$) that are suitable for the approximation obtained. Within the region with the hatching inclined to the right, $P_N(\bar{n})$ *exceeds* $P(\bar{n})$ by no more than 1%, and in the region with the hatching inclined to the left $P_N(\bar{n})$

Table 2.1. For different values of N the related n and y are shown

N	n	η
25	15	1.0065
100	55	1.0030
1150	595	1.0001

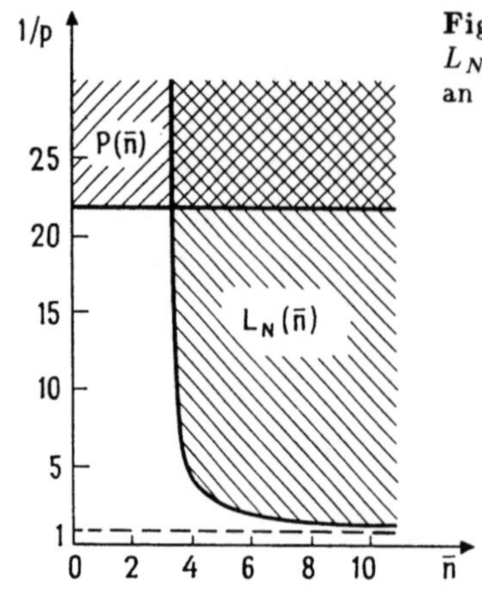

Fig. 2.1. The regions in the $(1/p, \overline{n})$ plane where the Laplace $L_N(n)$ and the Poisson $P(\overline{n})$ distributions are applicable (with an error of less than 1%)

is *less* than $L_N(\overline{n})$, also by no more than 1%. It is worth noting that, since we have chosen $n = \overline{n}$, the boundary curves, strictly speaking, are only suitable for integral $\overline{n}$.

In the upper right corner of the figure both asymptotic relations are suitable, but the Laplace formula is more convenient because it contains no factorials. In this region (of small p and not very small $\overline{n}$) we have $\sigma^2 = Npq \to Np = \overline{n}$ as $p \to 0$. Hence the Laplace distribution, too, depends here only on one parameter $\overline{n}$

$$L_N(n) = \frac{\exp\left[-(n-\overline{n})^2/2\overline{n}\right]}{\sqrt{2\pi\overline{n}}}. \tag{2.15}$$

2.7 Normal or Gaussian Distribution Law

In the region where the Laplace distribution can be applied, we are no longer interested in the number of tests N. It is known a priori that in a given time t a Brownian particle has undergone an exceedingly large number N of very small displacements. What is important is the distribution of the resulting displacement s (or the resulting amplitude when osicllations with random amplitudes $\pm a$ are added). Therefore, it is also advisable to use the local description (see Footnote 5), i.e., the continuous distribution $w(s)ds$, which gives the probability of s being in the interval $(s, s+ds)$.

What is the probability that x_n lies in the interval $(x, x+\Delta x)$? This is the interval of n ranging from $n = \overline{n} + \sigma x$ to $n = \overline{n} + \sigma(x + \Delta x)$, i.e., the interval contains $\sigma \Delta x = \sqrt{Npq}\Delta x$ values of n. This number grows as $\sqrt{N}$, whereas the relative width of the interval decreases as

$$\frac{\sigma \Delta x}{\overline{n}} \sim \frac{1}{\sqrt{N}}.$$

All values of $L_N(n)$, for n belonging to the interval of width $\sigma\Delta x$, are essentially equal. The realization of any of them is one of a number of mutually exclusive special cases of an event that is of interest to us. Consequently, by the additivity axiom, we have

$$P\{x<x_n\leq x + \Delta x\} = \sum_{n+\sigma x}^{\bar{n}+\sigma(x+\Delta x)} L_N(n) = L_N(n)\sigma\Delta x$$

$$= \frac{\exp(-x^2/2)}{\sqrt{2\pi}}\Delta x, \qquad (2.16)$$

which is the *normal*, or *Gaussian*, distribution law.

Thus, the De Moivre-Laplace theorem can also be formulated as follows: when $0<p<1$ and $N\to\infty$ the binomial law goes over into the normal law for x, this change being uniform for any finite x.

The law (2.16) is derived for the so-called *normalized* random variable

$$x_n = \frac{n-\bar{n}}{\sqrt{D[n]}},$$

i.e., a variable whose mean is zero ($\bar{x}_n = 0$) and whose variance is unity ($D[x_n] = 1$). If the Gaussian distribution is rewritten for a non-normalized variable y with mean $\bar{y}$ and variance $D[y] = \sigma^2$, then, since $x = (y-\bar{y})/\sigma$, we obtain

$$w(y)dy = \frac{\exp[-(y-\bar{y})^2/2\sigma^2]}{\sqrt{2\pi}\sigma}dy. \qquad (2.17)$$

In particular, for the displacement s during time t we have found that in the case of isotropic motion of a Brownian particle $\bar{s} = 0, D[s] = s^2 = Bt$ holds (see Sect. 2.4), where $B = a^2/\tau$ is the diffusion coefficient. Hence

$$w(s)ds = \frac{\exp(-s^2/2Bt)}{\sqrt{2\pi Bt}}ds. \qquad (2.18)$$

At $t = 0$ the density $w(s)$ becomes $\delta(s)$ (the particle is with certainty at the point $s = 0$), and then, with increasing t, the distribution flattens indefinitely, as shown in Fig. 2.2.

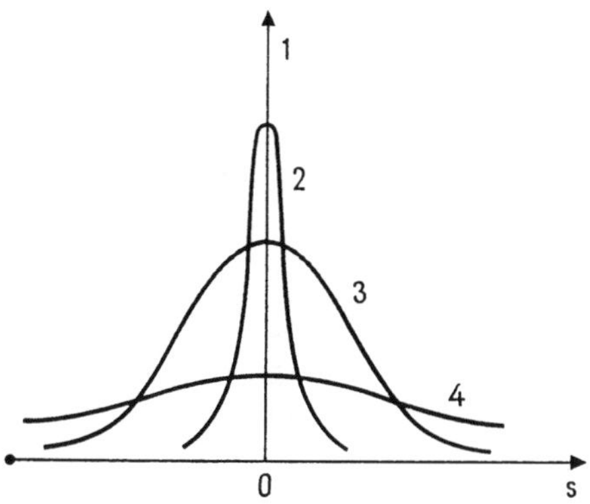

Fig. 2.2. Smearing out of $w(s)$ with time

It can readily be shown that, for a random variable obeying the normal distribution law (2.17), the central moments are

$$\gamma_k = \overline{(y - \bar{y})^k} = \begin{cases} 0 & \text{for odd } k, \\ 1.3\ldots(k-1)\sigma^k & \text{for even } k. \end{cases}$$

We have already indicated that the Gauss distribution is of general importance. The Bernoulli problem at large N and $\bar{n}$ is just one of the many that reduce to this distribution. Below, we shall see why the normal law holds a unique position among distributions. But for the time being the following remarks will suffice.

The probability of state X (X being a collection of generalized coordinates q and momenta p) for a system in a thermostat with (energy) temperature $\Theta = kT$ ($k = 1.38 \cdot 10^{-16}$erg/K is the Boltzmann constant) is given by the so-called canonical Gibbs distribution

$$w(X)dX = \exp\left[\frac{\Psi - H(X)}{\Theta}\right]dX, \tag{2.19}$$

where Ψ is the free energy, and $H(X)$ the Hamiltonian of the system. If H is a non-negative bilinear form, we have the (multidimensional) Gauss distribution. The kinetic energy, which gives the velocity distribution, is always a quadratic form. So, for one particle with kinetic energy $m(v_x^2 + v_y^2 + v_z^2)/2$, (2.19) gives a normal distribution for v_x, v_y, v_z, i.e., the Maxwell distribution

$$w(v_x, v_y, v_z)dv_x dv_y dv_z = A \exp\left[-\frac{m}{2\Theta}\left(v_x^2 + v_y^2 + v_z^2\right)\right]dv_x dv_y dv_z.$$

Hence, the magnitude of the velocity v has the distribution

$$w(v)dv = B \exp\left(-\frac{mv^2}{2\Theta}\right)v^2 dv, \tag{2.20}$$

which is no longer normal.

At times one is surprised by the following paradox. According to (2.20), arbitrarily large v have the finite probability

$$P\{v > v_o\} = B \int_{v_o}^{\infty} \exp\left(-\frac{mv^2}{2\Theta}\right)v^2 dv.$$

Therefore, a finite probability can also be assigned to the impossible event that one molecule has a kinetic energy exceeding the energy of the entire gas (of the whole thermostat). This occurred when we went over from very large but finite N to $N \to \infty$. The fact is that in the binomial law, n only assumes values from 0 to N, whereas the Poisson and Laplace distributions admit any values of n. In passing from $L_N(n)$ to the continuous normal distribution, there is no upper bound for x, either.

If the physical system in question is close to a stable equilibrium, undergoing only small deviations from it (fluctuations), then in many cases the potential energy $U(q)$ can also be thought of as a non-negative quadratic form, i.e., the distribution of coordinates is Gaussian, too.

When discussing the normal distribution we gave the Gibbs and Maxwell distributions as examples of distributions describing a *system* of random variables. This natural generaliaztion – a transition from one random variable to ensembles of such variables, and thus from a one-dimensional to a multidimensional problem – is of significance for applications of the theory, and also as a next step in the development of the theory of random processes. Suffice it to say that not infrequently a random function is represented as a *deterministic* function of t and a system of random parameters $\alpha_1, \alpha_2, \ldots, \alpha_n$

$$\xi(t) = f(t; \alpha_1, \alpha_2, \ldots, \alpha_n).$$

The substitution of a system of such variables for one random variable is also of special importance because it involves the notion of *statistical dependence* between random variables. We now recall the meaning of statistical independence.

Let x_k be possible values of a random variable ξ_k. The variables $\xi_1, \xi_2, \xi_3, \ldots, \xi_n$ are called *mutually* independent if their joint distribution function can be represented as the product of their distribution functions, i.e.,

$$W(x_1, x_2, \ldots, x_n) = W_{\xi 1}(x_1) \cdot W_{\xi 2}(x_2) \ldots W_{\xi n}(x_n).$$

Evidently, this form is also valid for the probability densities

$$w(x_1, x_2, \ldots, x_n) = w_{\xi 1}(x_1) \cdot w_{\xi 2}(x_2) \ldots w_{\xi n}(x_n),$$

and any mixed moment of the order $m = m_1 + m_2 + \ldots + m_n$ reduces to the product of moments

$$\langle \xi_1^m 1 \xi_2^m 2 \ldots \xi_n^m n \rangle = \langle \xi_1^m 1 \rangle \langle \xi_2^m 2 \rangle \ldots \langle \xi_n^m n \rangle.$$

In particular, for two independent random variables ξ and η we have

$$\overline{\xi\eta} = \overline{\xi}\overline{n}.$$

But if $\overline{\xi\eta} - \overline{\xi} \cdot \overline{n} = 0$, this does not suggest that ξ and η are statistically independent. The statistical connection between ξ and η, the measure of which is the difference $\overline{\xi\eta} - \overline{\xi} \cdot \overline{\eta}$, is called the *correlation* between the random variables ξ and η. It is often more convenient to use the symmetrical and dimensionless measure of correlation, namely *the correlation coefficient*, which is defined as

$$K_{\xi\eta} = \frac{\overline{\xi\eta} - \overline{\xi} \cdot \overline{\eta}}{\sqrt{D[\xi] \cdot D[\eta]}} = \frac{\overline{\xi\eta} - \overline{\xi}\overline{\eta}}{\sqrt{(\overline{\xi^2} - \overline{\xi}^2)(\overline{\eta^2} - \overline{\eta}^2)}} \quad . \tag{2.21}$$

It is readily seen that for a linear relation $\eta = a\xi + b$, we have $K_{\xi\eta} = 1$. It can also be shown quite simply that $|K_{\xi\eta}| \leq 1$ always. When $K_{\xi\eta} = 0$, the variables ξ and η are said to be uncorrelated; for $K_{\xi\eta} < 0$ the correlation is negative. The case of $K_{\xi\eta} = -1$ is sometimes (not quite appropriately) referred to as *anticorrelation*. The statistical independence of ξ and η indicates that they are uncorrelated, but the converse is not generally true.

The mean of the sum of random variables is at all times equal to the sum of their means, but the variance of the sum of random variables is

$$D\Big[\sum_i \xi_i\Big] = \Big\langle\Big[\sum_i(\xi_i - \bar{\xi}_i)\Big]^2\Big\rangle = \sum_i D[\xi_i] + 2\sum_{i<k}\sqrt{D[\xi_i]D[\xi_k]}K\xi_i\xi_k. \tag{2.22}$$

Thus the variance of a sum is equal to the sum of variances, subject to the sufficient condition that all the random variables are pairwise uncorrelated (which, as mentioned above, is by no means indicative of their mutual independence).

A system of random variables $\xi_1, \xi_2, \ldots, \xi_n$ may be thought of as a random point or vector ($\boldsymbol{\varrho}$) in an n-dimensional space. Therefore, this system is often referred to as an n-dimensional random variable.

By way of example, we consider the general form of the normal, or Gaussian distribution law for an n-dimensional random variable $\boldsymbol{\varrho}$ (here the variable $\boldsymbol{\varrho}$ is said to be normal or Gaussian). The general expression for this law is

$$w(\boldsymbol{r})d^n r = Ce^{-H(\boldsymbol{r})}d^n r, \quad d^n r = dx_1 dx_2 \ldots dx_n, \tag{2.23}$$

where $H(\boldsymbol{r})$ is the non-negative, symmetric and quadratic form

$$H(\boldsymbol{r}) = \sum_{i,k} a_{ik}x_i x_k, \quad a_{ik} = a_{ki} \quad , \tag{2.24}$$

and the constant C is represented through the elements of the matrix $A = (a_{ik})$ in accordance with the normalization condition

$$\int_{-\infty}^{+\infty} w(r)d^n r = 1.$$

Thus, equal-probability-density surfaces are a family of concentric, similar n-

[7] For the distribution (2.23) the means ξ_i vanish; the same is true of all odd moments in general. If $\xi_i \neq 0$, it simply indicates that the center of the ellipsoids is displaced from the origin to the point $\bar{\boldsymbol{\varrho}}\{\bar{\xi}_1, \bar{\xi}_2, \ldots, \bar{\xi}_n\}$. In this case, in (2.24) H is a quadratic form of the components of the difference $\boldsymbol{r} - \bar{\boldsymbol{\varrho}}$, i.e.,

$$H(\boldsymbol{r}) = \sum_{i,k} a_{ik}(x_i - \bar{\xi}_i)(x_k - \bar{\xi}_k).$$

dimensional ellipsoids $H(\mathbf{r}) = \text{const}$. The probability density is maximal at the origin of coordinates, $\mathbf{r} = 0$, where it is equal to C^{7}.

The normal distribution is generally written in a canonical form, i.e., through the so-called *covariance matrix* $B = (b_{ik})$, where the b_{ik} are the second order moments,

$$b_{ik} = \overline{\xi_i \xi_k} = \int_{-\infty}^{+\infty} x_i x_k w(\mathbf{r}) d^n r. \tag{2.25}$$

If we introduce standard deviations σ_k for the components ξ_k of the n-dimensional random variable $\mathbf{r}$ under consideration ($b_{ii} = \overline{\xi_i^2} = \sigma_i^2$), then we can represent b_{ik} through the correlation coefficients K_{ik}

$$b_{ik} = \sigma_i \sigma_i K_{ik} \quad .$$

Clearly, the matrix B is symmetric, as is the initial matrix A. We have therefore $n(n+1)/2$ equations (2.25), which express all the independent b_{ik} through independent a_{ik}. If these equations are solved for a_{ik} and the resulting expressions relating a_{ik} to b_{ik} inserted into (2.23,24), the distribution (2.23) assumes the canonical form

$$w(\mathbf{r}) d^n r = \frac{1}{(2\pi)^{n/2} |B|^{1/2}} \exp\left(-\frac{1}{2} \sum_{i,k} B_{ik}^{-1} x_i x_k\right) d^n r. \tag{2.26}$$

Here $|B| = Det(B)$, and B_{ik}^{-1} are the elements of the matrix B^{-1}, which is the inverse of B.

So, for a two-dimensional normal random variable the distribution law is

$$w(x_1, x_2) dx_1 dx_2 = \frac{1}{2\pi \sqrt{b_{11} b_{22} - b_{12}^2}} \times \exp\left[-\frac{b_{22} x_1^2 + b_{11} x_2^2 - 2 b_{12} x_1 x_2}{2(b_{11} b_{22} - b_{12}^2)}\right] dx_1 dx_2, \tag{2.27}$$

or, in terms of standard deviations and correlation coefficients K ($b_{11} = \sigma_1^2$, $b_{22} = \sigma_2^2$, $b_{12} = b_{21} = \sigma_1 \sigma_2 K$),

$$w(x_1, x_2) dx_1 dx_2 = \frac{dx_1 dx_2}{2\pi \sigma_1 \sigma_2 \sqrt{1 - K^2}} \times \exp\left[-\frac{1}{2(1 - K^2)} \left(\frac{x_1^2}{\sigma_1^2} + \frac{x_2^2}{\sigma_2^2} - 2K \frac{x_1 x_2}{\sigma_1 \sigma_2}\right)\right]. \tag{2.28}$$

Another feature of the normal distribution is that the absence of correlation between ξ_i implies their statistical independence. This is clearly seen in the two-dimensional case (2.28), since at $K = 0$ the probability density $w(x_1, x_2)$

becomes the product of two one-dimensional normal densities. The same can also be shown to hold true for the general n-dimensional case. In fact, ξ_i being pairwise uncorrelated means that the matrix B is diagonal (only $b_{ii} = \sigma_i^2$ are nonzero). Consequently, the inverse matrix B^{-1} is also diagonal, so that

$$B_{ii}^{-1} = \frac{1}{\sigma_i^2}, \quad |B|^{1/2} = \sigma_1\sigma_2\ldots\sigma_n.$$

It follows from (2.26) that

$$w(\boldsymbol{r}) = \prod_i \frac{1}{\sqrt{2\pi}\sigma_i} \exp\left(-x_i^2/2\sigma_i^2\right),$$

i.e., $w(\boldsymbol{r})$ is the product of n one-dimensional normal probability densities for each of ξ_i.

Thus, for *normal* random variables *the absence of correlation is equivalent to statistical independence.*

Finally, we would like to point out that it is customary in physics to designate a random variable by the same symbol as its possible values. This we have already done; for example, in considering the Brownian-motion model where we denoted by n both the *random number* of steps to the right, and the *possible values* $n = 1, 2, \ldots, N$ of this random number. This convention reduces the number of symbols used and in most cases does not result in confusion. From now on, we will follow this convention. Different symbols for a random variable and its possible values will only be used when this is likely to lead to some confusion.

2.8 Exercises

2.8.1. Write the joint probability density for the random variables $x = \sin\theta$ and $y = \cos\theta$, given the probability density $w(\theta)$ in the interval $(-\pi, \pi)$.
Solution. We express $w_2(x, y)$ in terms of the conditional probability density of y

$$w_2(x, y) = w_1(x)v(y|x).$$

Since y is a deterministic function of x, i.e.,

$$y = \begin{cases} +\sqrt{1-x^2} & \text{for} \quad |\theta| < \pi/2 \\ -\sqrt{1-x^2} & \text{for} \quad \pi/2 < |\theta| < \pi, \end{cases}$$

the conditional probability density is

$$v(y|x) = \delta(y \mp \sqrt{1-x^2}).$$

The one-dimensional distribution of x follows from the given distribution of θ:

$$w(\theta)d\theta = w(arc\ \sin x)\frac{dx}{\sqrt{1-x^2}} = w_1(x)dx.$$

Thus

$$w_2(x,y) = \frac{w(arc\ \sin x)}{\sqrt{1-x^2}}\delta(y \mp \sqrt{1-x^2})$$

with the minus sign for $|arc\ \sin x| < \pi/2$, and the plus sign for $\pi/2 < |arc\ \sin x| < \pi$. Of course, a similar expression can be derived in terms of $v(x|y)$.

2.8.2. Which distributions of a random variable x on the interval $(-a, a)$ ensure that x and any even deterministic function $f(x)$ are uncorrelated?
Solution. A sufficient condition for x and $f(x)$ to be uncorrelated is that $w(x)$ is an even function, because the moments $\langle xf(x)\rangle$ and $\langle x\rangle$ vanish.

2.8.3. Consider oscillations $\xi = \sin(\omega t + \theta)$ and $\eta = \cos(\omega t + \theta)$ where θ is a random phase. What conditions must the probability density of θ meet for ξ and η to be uncorrelated on the interval $(-\pi, \pi)$?
Solution. The correlation coefficient for ξ and η varies as the difference

$$\begin{aligned}\overline{\xi\eta} - \overline{\xi}\cdot\overline{\eta} = {} & [\langle \sin\theta \cos\theta\rangle - \langle \sin\theta\rangle \cdot \langle \cos\theta\rangle] \cos 2\omega t \\ & + \tfrac{1}{2}[\langle \cos^2\theta\rangle - \langle \cos\theta\rangle^2 - \langle \sin^2\theta\rangle + \langle \sin\theta\rangle^2] \sin 2\omega t.\end{aligned} \tag{2.29}$$

Let us represent the probability density $w(\theta)$ as

$$\omega(\theta) = \frac{1}{2\pi} + \sum_{k=1}^{\infty}(a_k \sin k\theta + b_k \cos k\theta).$$

For this distribution we now compute the moments that enter into the coefficients of sin $2\omega t$ in (2.29), and equate them to zero to obtain the following two conditions:

$$a_2 = 2\pi a_1 b_1 \quad \text{and} \quad b_2 = \pi(b_1^2 - a_1^2).$$

If we introduce the amplitudes and phases of the first and second harmonics of $w(\theta)$

$$a_i = A_i \sin a_i \quad \text{and} \quad b_i = A_i \cos a_i \quad \text{with} \quad i = 1, 2,$$

these two conditions become

$$A_2 = \pi A_1^2, \quad a_2 = 2a_1.$$

The oscillations ξ and η will thus be uncorrelated if $w(\theta)$ does not contain the first and second harmonics ($A_1 = 0$, $A_2 = 0$) and, all the more so, if it contains no harmonics at all, i.e., when we have the uniform distribution $w(\theta) = 1/2\pi$.

2.8.4. Find the variance of the sum of a *random* number n of independent random variables x_i with the same variances σ^2.
Solution. The joint distribution function $W(y,n)$ of the random variables $y = \sum_{i=1}^{n} x_i$ and n can be written in terms of the conditional probability $V(y|n)$ times the probability p_n of the value n

$$W(y,n) = p_n V(y|n).$$

Hence

$$D[y] = \sum_n p_n D[y|n] = \sum_n p_n n\sigma^2 = \bar{n}\sigma^2.$$

2.8.5. Show that for n normal variables x_i, n linear mutually independent combinations $y_k = \sum_i \sigma_{ki} x_i$ always exists.
Solution. The validity of this statement follows from the fact that there is always an orthogonal transformation that reduces the quadratic form of $H(\boldsymbol{r})$ to the sum of squares. Geometrically, this corresponds to the rotation of axes in the n-dimensional space x_i, such that the new axes would coincide with the principal axes of the ellipsoids $H = \text{const}$.

2.8.6. Consider a normal variable x with mean $\bar{x} = 0$ and variance $\overline{x^2} = \sigma^2$. Show that for any deterministic differentiable function $f(x)$ that increases, as $|x| \to \infty$, slower than $\exp(x^2/2\sigma^2)$ the following formula holds:

$$\langle x f(x)\rangle = \sigma^2 \langle f'(x)\rangle.$$

Solution. Integration by parts gives

$$\begin{aligned}\langle x f(x)\rangle &= \int_{-\infty}^{+\infty} x f(x) \exp\left(-x^2/2\sigma^2\right) \frac{dx}{\sqrt{2\pi}\sigma} \\ &= -\sigma^2 \left[\frac{f(x)}{\sqrt{2\pi}\sigma} \exp\left(-x^2/2\sigma^2\right) \Big|_{-\infty}^{+\infty} \right. \\ &\quad \left. - \int_{-\infty}^{+\infty} f'(x) \exp\left(-x^2/2\sigma^2\right) \frac{dx}{\sqrt{2\pi}\sigma} \right] \\ &= \sigma^2 \langle f'(x)\rangle. \end{aligned} \tag{2.30}$$

Generalization of this relation to functionals of random Gaussian fields is used in the theory of random fields.

If $\bar{x} \neq 0$, (2.30) takes the form

$$\langle x f(x)\rangle = \bar{x}\langle f(x)\rangle + \sigma^2 \langle f'(x)\rangle,$$

where $\sigma^2 = \overline{x^2} - \bar{x}^2$.

2.8.7. Consider an n-dimensional random variable $\boldsymbol{\xi} = (\xi_1, \ldots, \xi_n)$, whose n-dimensional probability density $w_{\xi}(x_1, \ldots, x_n)$ is known, and where x_k are the possible values of ξ_k. The variable $\boldsymbol{\eta}(\eta_1, \ldots, \eta_n)$ is related to $\boldsymbol{\xi}$ by the functional dependence $\boldsymbol{\eta} = \boldsymbol{F}(\boldsymbol{\xi})$, where $\boldsymbol{F}$ is a deterministic one-dimensional vector function with the single-valued inverse function $\boldsymbol{\xi} = \boldsymbol{F}^{-1}(\boldsymbol{\eta})$. Express the probability density $w_{\eta}(\boldsymbol{y})$ in terms of $w_{\xi}(\boldsymbol{x})$.

Solution. The single-valued relation $\boldsymbol{\xi} = \boldsymbol{F}^{-1}(\boldsymbol{\eta})$ may be regarded as a coordinate transformation in the n-dimensional space of possible values

$$x_i = F_i^{-1}(y_1, \ldots, y_n), \quad i = 1, 2, \ldots, n.$$

The volume element $d^n x = dx_1 \ldots dx_n$ is known to be transformed using the formula $d^n x = Q(y) d^n y$, where $Q(y)$ is the Jacobian of the transformation

$$Q(\boldsymbol{y}) = \operatorname{Det}\left|\frac{\partial x_i}{\partial y_k}\right| = \operatorname{Det}\left|\frac{\partial F_i^{-1}(\boldsymbol{y})}{\partial y_k}\right|.$$

Since this is a one-to-one transformation, the probability of $\boldsymbol{\eta}$ being within the interval $(\boldsymbol{y}, \boldsymbol{y} + d\boldsymbol{y})$ is the same as that of $\boldsymbol{\xi}$ being in $(\boldsymbol{x}, \boldsymbol{x} + d\boldsymbol{x})$, i.e.,

$$w_{\eta}(\boldsymbol{y}) d^n y = w_{\xi}(\boldsymbol{x}) d^n x.$$

Substituting $\boldsymbol{x} = \boldsymbol{F}^{-1}(\boldsymbol{y})$ and $d^n x = Q(\boldsymbol{y}) d^n y$ on the right-hand side gives

$$w_{\eta}(\boldsymbol{y}) = w_{\xi}[\boldsymbol{F}^{-1}(y)] \cdot Q(\boldsymbol{y}).$$

3. Random Pulses

This chapter deals already with random processes occuring in time t although it is still introductory as far as methodology is concerned. This is because a random function of t may in certain cases be viewed as a deterministic function of t and a finite number of random variables. This enables us to use the tools of the classical probability theory, without having to employ its generalization to random functions. The simplified model of the pulsed Poisson process is a problem of this type (Sect. 3.1).

From the computational point of view, the above problem can be more successfully tackled by looking for a characteristic function uniquely related to the probability distribution, rather than trying to find the distribution itself. Therefore, Sect. 3.2 recapitulates the definition and properties of the characteristic function which is then (Sect. 3.3) obtained for the above-mentioned model of the random process. A similar approach enables us to obtain not only the mean and variance of a process, but also its covariance (Sect. 3.4).

Section 3.5 gives some insight into how complex the problem of random pulses becomes if we forgo the assumption that they are statistically independent, or if we allow the (averaged) pulse repetition rate to vary with time. Section 3.6 reminds the reader of the contents of the central limit theorem of the theory of probability and clarifies the role of that theorem as applied to problems involving random pulses.

3.1 Statement of the Problem

We now turn to a special type of the random process, the so-called Poisson pulse process. This specific class of particular processes is of special interest since many important practical results can be obtained using classical probability theory. A brief explanation is given below.

As has already been indicated, a random function of time t can sometimes be defined as a deterministic function f of t and a *finite* number of random variables $\alpha_1, \ldots, \alpha_n$. All the statistical properties of the function $\xi(t) = f(t; \alpha_1, \ldots, \alpha_n)$ are then completely determined by the n-dimensional random variable $\alpha_1, \ldots, \alpha_n$, i.e., by its distribution $w(\alpha_1, \ldots, \alpha_n) d\alpha_1 \ldots d\alpha_n$.

An example of such a definition is the sum of trigonometric functions weighed with random coefficients a_k, b_k

$$\xi(t) = f[t; (a_k), (b_k)] = \sum_{k=0}^{n} (a_k \cos kt + b_k \sin kt),$$

defining the random function $\xi(t)$ that is periodic with period 2π.

As long as the number n of random parameters is finite, we may use the concepts and methods of classical probability theory. However, the transition to a countable infinite set of random parameters (to the Fourier series in the present example) requires a more general approach. Pulse processes can, at least in the simplest cases, be treated as deterministic functions of time and a finite number of random parameters. This enables us to consider such processes without resorting to the general theory of random functions.

We now look at a process that represents a superposition of pulses of identical shape. Let the pulse shape be described by a deterministic function $F(t)$, with the pulses differing in amplitude. We can then write

$$\xi(t) = \sum_{\nu} a_\nu F(t - t_\nu), \tag{3.1}$$

where t_ν marks the onset of the νth pulse, and a_ν is its amplitude. The choice of t_ν is arbitrary in so far as we do not suppose $F(t - t_\nu) = 0$ for $t<t_\nu$, i.e., it may well be that the function $F(t)$ has no initial value before which it is zero. We only assume that $F(t)$ tends to zero rather fast as $|t|\to\infty$. The times t_ν may be associated with any characteristic points, e.g., with one of the extrema of $F(t)$ or with zero points, if any, and so forth.

If the parameters a_ν and t_ν are random variables, then $\xi(t)$ is a random time function also; i.e., at any t the value of $\xi(t)$ is a random variable. Thus, we have an example of the above definition of a random function as a deterministic function of a collection of random parameters (in this case, a_ν and t_ν).

We now list some physical problems involving random processes of the form (3.1). One such process is the current or voltage at the output of a quadrupole whose input is essentially described by (3.1), but which consists of very short bursts, i.e., in the limit of delta-functions

$$I_{\text{inp}}(t) = \sum_{\nu} a_\nu \delta(t - t_\nu). \tag{3.2}$$

Here "very short" means much shorter than the time constant of the quadrupole. Given that transit time of the electron in a vacuum tube meets this condition, the anode current may be written as (3.2) with $a_\nu = e = \text{const}$. If we are to consider the transit time, and thereby the shape of anode-current pulses produced by individual electrons, the same anode current may be represented by (3.1) with $a_\nu = e$, provided that

$$\int_{-\infty}^{+\infty} F(t)dt = 1.$$

Another example of a random process described by (3.1) is the reflection of a pulse traveling through a medium that contains random inhomogeneities with random reflection or scattering coefficients. The pulses reflected from the inhomogeneities will, in the absence of dispersion, have the same shape as the initial pulse, but they will return with random delays and amplitudes.

The problem can be formulated still more generally. We shall now assume that the random variable is the *shape of the pulses* themselves, i.e, the pulse is described by a deterministic function $F(t-t_\nu)$ and an ensemble of a finite number m of random parameters or, in other words, of an m-dimensional random variable $\boldsymbol{a}_\nu = \{a_{\nu 1}, \ldots, a_{\nu m}\}$

$$\xi(t) = \sum_\nu F(t - t_\nu, \boldsymbol{a}_\nu). \tag{3.3}$$

Here again we define the random function F as a deterministic function of a finite number of random parameters. The components of $\boldsymbol{a}$ may also be statistically dependent on each other.

Referring to stochastic properties of the random parameters $\boldsymbol{a}_\nu$ and t_ν governing the statistical characteristics of the process $\xi(t)$, we adopt the simple assumptions:

1) All the $\boldsymbol{a}_\nu$ and t_ν are statistically independent, and their distributions are independent of the pulse index ν. Then, for a system consisting of any number $n(m+1)$ of parameters $\boldsymbol{a}_\nu$ and t_ν, it is sufficient to define the distribution functions $w_a(\boldsymbol{a})d\boldsymbol{a}$ and $w_t(t)dt$, since the distribution function of the entire system is presented in the factorized form

$$\begin{aligned} &w(\boldsymbol{a}_1, \ldots, \boldsymbol{a}_n; t_1, \ldots, t_n) d\boldsymbol{a}_1 \ldots d\boldsymbol{a}_n dt_1 \ldots dt_n \\ &\quad = \prod_{\nu=1}^{n} w_a(\boldsymbol{a}_\nu) d\boldsymbol{a}_\nu w_t(t_\nu) dt_\nu . \end{aligned} \tag{3.4a}$$

Here $d\boldsymbol{a}_\nu \equiv d^m a_\nu = da_{\nu 1} \ldots, da_{\nu m}$.

2) The probability of occurrence of a pulse in the time interval from t to $t+dt$ is independent of t and proportional to dt, i.e.,

$$w_t(t)dt = n_1 dt, \quad n_1 = \text{const.} \tag{3.4b}$$

As indicated in Sect. 2.5, this assumption is equivalent to saying that the probability of occurrence of n pulses in the time interval T is given by the Poisson distribution

$$P(n) = \frac{\bar{n}^n e^{-\bar{n}}}{n!}, \quad \bar{n} = n_1 T. \tag{3.5}$$

Bearing these assumptions in mind, we now compute the distribution function for $\xi(t)$.

In order to avoid dealing with the infinite sequences of random parameters $\boldsymbol{a}_\nu$ and t_ν, we proceed as follows. We take a time interval $(-T/2,\ T/2)$, so large compared with the length of an individual pulse and with the average gap $1/n_1$ between pulses, that we may ignore boundary effects. In other words, the pulses that contain the boundaries of the interval $(-T/2,\, T/2)$ are suppressed. This omission is permissible since $F(t)$ has to tend to zero sufficiently fast as $|t|\to\infty$. At the moment we do not concern ourselves with the question of how fast $F(t)$ vanishes. We only mention that the boundary effects may be neglected even if the time interval $F(t)\neq 0$ is not finite.

We should like to find the distribution of the random variable $\xi(t)$ (which is itself a random variable for any fixed t), i.e., the probability that $\xi(t)$ will occur in the interval $(x,\, x+dx)$, say event B,

$$P\{B\} = P\{x<\xi(t)\leq x+dx\} = w(x)dx.$$

Event B can occur if any of the events $A_0, A_1, \ldots, A_n, \ldots,$ occurs. A_n is the occurrence of n pulses in the interval $(-T/2,\, T/2)$. Events A_n being mutually exclusive, we have according to the formula for the total probability

$$P\{B\} = \sum_{n=0}^{\infty} P\{A_n\}P\{B|A_n\}.$$

Obviously, $P\{A_n\}$ is nothing other than the probability $P(n)$ given by the Poisson law. Given that n pulses occur in $(-T/2,\, T/2)$ we now introduce the density of the conditional probability $P\{B|A_n\}$, i.e., of the probability of event $B = \{x<\xi(t)\leq x+dx\}$:

$$P\{B|A_n\} = v(x|n)dx. \tag{3.6}$$

Then

$$w(x)dx = \sum_{n=0}^{\infty} P(n)v(x|n)dx \tag{3.7}$$

and, consequently, the mean of any function $f(\xi)$ can be obtained in two steps:

$$\begin{aligned}\overline{f(\xi)} &= \int_{-\infty}^{+\infty} f(x)w(x)dx \\ &= \sum_{n=0}^{\infty} P(n)\int_{-\infty}^{+\infty} f(x)v(x|n)dx = \sum_{n=0}^{\infty} P(n)\overline{{}^{n}f(\xi)},\end{aligned} \tag{3.8}$$

where $\overline{{}^{n}f(\xi)}$ is the conditional mean, provided that there are n pulses in $(-T/2,\, T/2)$.

Thus, according to (3.7), we can find $w(x)$ if we know the conditional probability density $v(x|n)$. This $v(x|n)$ may be said to be the probability density of the random function

$$\xi_n(t) = \sum_{\nu=1}^{n} F(t-t_\nu, \boldsymbol{a}_\nu), \tag{3.9}$$

which depends on a finite number of parameters $\boldsymbol{a}_\nu$ and t_ν related to those n

pulses that occur in the interval $(-T/2, T/2)$. To find $v(x|n)$, it is expedient to utilize the effective tool of the so-called *characteristic functions.*

3.2 Characteristic Functions

Characteristic functions are useful in proving many important theorems of probability theory and in solving specific problems. A characteristic function is *uniquely* related to the distribution function $W(x)$ and has a significant advantage over $W(x)$. Namely, if we take the sum of *independent* random variables, then the distribution function of the sum is computed using the convolution relation (composition or symbolic multiplication of distribution functions). In contrast, the characteristic function of the sum is obtained simply by multiplying the characteristic functions of summands. This is of special importance for the central limit theorem of probability theory, which deals precisely with the distribution of a sum of *independent* random variables (Sect. 3.6).

The characteristic function of a random variable x (as indicated, we shall frequently denote the random variable and its possible values by the same letter) is the mean (mathematical expectation) of the function $\exp(iux)$, where u is a real parameter:

$$\varphi_x(u) = \langle e^{iux}\rangle = \int_{-\infty}^{\infty} e^{iux} dW(x), \tag{3.10}$$

and $\overline{a + ib} = \bar{a} + i\bar{b}$. The Fourier-Stieltjes integral (3.10) always converges since

$$|\varphi_x(u)| \leq \int_{-\infty}^{+\infty} |e^{iux}| dW(x) = 1.$$

Further, we see that $\varphi_x(0) = 1$ always and $\varphi_x(-u) = \overline{\varphi_x(u)}$. Consider two examples of the characteristic function:

For the *normal distribution*

$$dW(x) = \frac{1}{\sqrt{2\pi}\sigma} \exp[-(x-a)^2/2\sigma^2]dx$$

we have

$$\varphi_x(u) = \frac{1}{\sqrt{2\pi}\sigma} \int_{-\infty}^{+\infty} \exp[iux - (x-a)^2/2\sigma^2]dx = \exp[iau - \sigma^2 u^2/2].$$

The narrower the distribution (the smaller σ), the wider the characteristic function. Of course, this feature is not only characteristic of the normal distribution, but is a well-known common property of a pair of functions connected by the Fourier integral. If $\sigma \to 0$, i.e., $w(x) \to \delta(x-a)$, then $\varphi_x(u) \to \exp(iau)$, in particular, at $a = 0$ we have $\varphi_x(u) \to 1$.

For the *Poisson distribution*

$$P(n) = \frac{\overline{n}^n}{n!}e^{-\overline{n}}$$

we obtain

$$\varphi_n(u) = \sum_{n=0}^{\infty} e^{iun}P(n) = e^{-\overline{n}} \sum_{n=0}^{\infty} \frac{(\overline{n}\, e^{iu})^n}{n!} = \exp\left[\overline{n}(e^{iu} - 1)\right] \quad .$$

If for any $k>0$ the absolute moment $\beta_k = \overline{|x|^k}$ is finite, then the expression $\varphi_x(u)$ can be differentiated k times with respect to u. Since

$$e^{iux} = 1 + \frac{iux}{1!} + \ldots + \frac{(iux)^k}{k!} + R_k(iux), \quad \text{we get}$$

$$\varphi_x(u) = \sum_{m=0}^{k} \frac{\alpha_m(iu)^m}{m!} + o(u^k), \quad \alpha_m = \overline{x^m}. \tag{3.11}$$

If *all* β_k are finite, an expansion into an infinite series is reasonable, i.e.,

$$\varphi_x(u) = \sum_{m=0}^{\infty} \frac{\alpha_m(iu)^m}{m!}. \tag{3.12}$$

It should be noted that the mth moment, if finite, is computed using the characteristic function. Its mth derivative with respect to iu at $u = 0$ is

$$\alpha_m = \left.\frac{\partial^m \varphi_x(u)}{\partial(iu)^m}\right|_{u=0}. \tag{3.13}$$

The moments (for integral m) are thus obtained by differentiation, that is to say by a simpler operation than integration as required when using the distribution function.

Hardly any explanation is needed for the fact that the characteristic function for a multidimensional random variable $\boldsymbol{r} = \{x_1, x_2, \ldots, x_n\}$ is determined in exactly the same way, i.e., as the Fourier conjugate function

$$\varphi_r(\boldsymbol{u}) = \int_{-\infty}^{+\infty} \exp(i\boldsymbol{ur})dW(\boldsymbol{r}), \tag{3.14}$$

where $\boldsymbol{u}$ is a vector with the components $u_1, u_2, \ldots, u_n$.

For the probability density of the variable $z = x + y$, where x and y are *independent* random variables, we have the convolution equation

$$w(z) = w_1 * w_2 = \int_{-\infty}^{+\infty} w_1(x)w_2(z-x)dx = \int_{-\infty}^{+\infty} w_1(z-y)w_2(y)dy.$$

There is a fairly complicated functional relationship between w_1, w_2, and w. Since x and y are independent, the characteristic function of z is given by

$$\varphi_z(u) = \overline{e^{iuz}} = \overline{e^{iux}}\,\overline{e^{iuy}} = \varphi_x(u)\varphi_y(u).$$

Thus, the convolution of w and W, i.e., the symbolic multiplication of distribution functions, corresponds to the *conventional multiplication* of the characteristic functions.

It can easily be seen that if $y = \sum_i x_i$ and each term is independent of the sum of the preceding terms, then

$$\varphi_y(u) = \prod_i \varphi_{x_i}(u). \tag{3.15}$$

There is, of course, even greater justification for (3.15) in the case of mutually independent x_i.

Another important problem is how to find the distribution function $W(x)$ from the characteristic function $\varphi_x(u)$. If the distribution is differentiable, i.e., $dW(x) = w(x)dx$, (3.10) becomes a conventional Fourier integral

$$\varphi_x(u) = \int_{-\infty}^{+\infty} e^{iux} w(x)dx \tag{3.16}$$

whose inversion immediately gives

$$w(x) = \frac{1}{2\pi}\int_{-\infty}^{+\infty} e^{-iux}\varphi_x(u)du, \tag{3.17}$$

and hence

$$W(x) - W(0) = \int_0^x w(x)dx = \frac{1}{2\pi}\int_{-\infty}^{+\infty} \frac{1 - e^{-iux}}{iu}\varphi_x(u)du. \tag{3.18}$$

If the distribution is discrete, then using a delta-function we may apply the same inversion formulas (3.16,17). Indeed, for

$$w(x) = \sum_k p_k \delta(x - a_k)$$

we have by (3.16),

$$\varphi_x(u) = \int_{-\infty}^{+\infty} e^{iux} \sum_k p_k \delta(x - a_k)dx = \sum_k \exp(iua_k)p_k.$$

The inversion equation (3.17) leads us back to the correct expression for $w(x)$, i.e.,

$$w(x) = \frac{1}{2\pi}\int_{-\infty}^{+\infty} e^{iux}\sum_k p_k \exp(iua_k)du$$

$$= \sum_k p_k \frac{1}{2\pi}\int_{-\infty}^{+\infty} \exp[iu(a_k - x)]du = \sum_k p_k \delta(x - a_k).$$

It is this unique relation between the distribution and the characteristic functions that is widely used both in the general theory and in solving specific problems. It transpires that in some cases it is easier to find the distribution function directly, and in others the characteristic function. In this connection we would like to refer to the Levy theorem: if a sequence of distribution functions $W_k(x)$ converges to the distribution function $W(x)$ as $k \to \infty$ [at points where $W(x)$ is continuous], then the sequence of corresponding characteristic functions $\varphi_k(u)$ also converges to the characteristic function $\varphi(u)$ corresponding to $W(x)$, and vice versa.

If we take $\ln\varphi_x(u)$ and expand it into a power series with respect to u [we should take that branch of the logarithm which also vanishes with u, and expand it, according to (3.11), up to the term of the order of u^k], then we get

$$\ln\varphi_x(u) = \sum_{m=1}^{k} \frac{\lambda_m}{m!}(iu)^m + o(u^k).$$

The quantities λ_m are called the *cumulants,* or *semi-invariants*, of a distribution. The mth cumulant is a rational algebraic function of the first m moments α_m, or of the central moments γ_m. In particular,

$$\lambda_1 = \alpha_1, \ \lambda_2 = \alpha_2 - \alpha_1^2 = \gamma_2, \ \lambda_3 = \alpha_3 - 3\alpha_2\alpha_1 + 2\alpha_1^3 = \gamma_3,$$

$$\lambda_4 = \alpha_4 - 3\alpha_2^2 - 4\alpha_3\alpha_1 + 12\alpha_2\alpha_1^2 - 6\alpha_1^4 = \gamma_4 - 3\gamma_2^2.$$

The additivity of cumulants in composing the distributions of independent variables should be noted. In the convolution, φ_{x_k} are multiplied, $\ln\varphi_{x_k}$ add up, and hence the cumulants add up too:

$$\lambda_m = \sum_{k=1}^{n} \lambda_m^{(k)} \quad , \quad x = \sum_{k=1}^{n} x_k.$$

It is a fairly straightforward matter to establish with the aid of the characteristic functions, whether or not some distributions remain true to type in convolution. If, for instance, the independent random variables n_k have a Poisson distribution, then

$$\varphi_{n_k}(u) = \exp[\overline{n}_k(e^{iu} - 1)] \quad .$$

Thus, for $n = \sum_k n_k$ we have

$$\varphi_n(u) = \exp\left[\bar{n}(e^{iu} - 1)\right], \quad \bar{n} = \sum_k \bar{n}_k,$$

i.e., the Poisson distribution again. If the sum of independent random variables with a similar distribution represents the same distribution as the summands, this distribution is referred to as *stable*. We have thus shown that the Poisson distribution is stable. It can readily be seen that the normal distribution is stable, too.

Let independent terms be normally distributed with means $\bar{x}_k = a_k$ and variances $D[x_k] = \sigma_k^2$, i.e.,

$$W_k(x) = \Phi\left(\frac{x - a_k}{\sigma_k}\right), \quad \text{where} \quad \Phi(x) = \frac{1}{\sqrt{2\pi}} \int_{-\infty}^{x} \exp\left(-x^2/2\right) dx.$$

The characteristic function for x_k is

$$\varphi_k(u) = \exp\left(iua_k - \sigma_k^2 u^2/2\right).$$

This implies that for $\sum_{k=1}^{n} x_k$ the characteristic function will be

$$\varphi(u) = \prod_{k=1}^{n} \varphi_k(u) = \exp\left(iuA_n - B_n^2 u^2/2\right), \quad \text{where}$$

$$A_n = \sum_{k=1}^{n} a_k, \quad B_n^2 = \sum_{k=1}^{n} \sigma_k^2, \qquad \text{and thus}$$

$$W_1(x) * W_2(x) * \ldots * W_n(x) = \Phi\left(\frac{x - a_1}{\sigma_1}\right) * \ldots * \Phi\left(\frac{x - a_n}{\sigma_n}\right) = \Phi\left(\frac{x - A_n}{B_n}\right).$$

If all $W_k(x)$ are identical, i.e., $a_k = a$, $\sigma_k = \sigma$, then $A_n = na$, $B_n = \sqrt{n}\sigma$. Denoting the composition of similar normal distribution functions $W(x)$ by $[W(x)]^{n*}$, we have

$$[W(x)]^{n*} = \left[\Phi\left(\frac{x - a}{\sigma}\right)\right]^{n*} = \Phi\left(\frac{x - na}{\sqrt{n}\sigma}\right).$$

In particular, at $a = 0$ and $\sigma = 1/\sqrt{n}$

$$[\Phi(x\sqrt{n})]^{n*} = \Phi(x).$$

This signifies that if the x_k have the same distribution function $\Phi(x)$, then the sum

$$S_n = \frac{1}{\sqrt{n}} \sum_{k=1}^{n} x_k$$

also has the distribution function $\Phi(x)$.

At times it appears more convenient to use not the characteristic function, i.e., the mean of $\exp(iux)$, but rather the means of the quantities $\exp(\theta x)$

$$F_x(\theta) = \overline{e^{\theta x}} = \int e^{\theta x} dW(x),$$

or of u^x

$$F_x(u) = \overline{u^x} = \int u^x dW(x),$$

subject, of course, to the limitations that ensure the existence of these means. Such functions are known as *moment-generating* or simply *generating* functions of the distribution $W(x)$. If $F_x(\theta)$ can be differentiated m times with respect to θ, then by putting $\theta = 0$ we obtain $\alpha_m = \overline{x^m}$. Similarly, we may derive α_m from $F_x(u)$, using m times the operation $u(\partial/\partial u)$ to set then $u = 1$.

3.3 Distribution Function for a Poisson Pulse Process

The characteristic function is equivalent to the distribution function. Nevertheless, for the random variable (3.9), i.e., for

$$\xi_n(t) = \sum_{\nu=1}^{n} x_\nu, \quad x_\nu = F(t - t_\nu, \boldsymbol{a}_\nu),$$

the characteristic function is derived far more simply than the distribution function. This is due to the statistical *independence* of the terms x_ν. Therefore, the characteristic function (3.9), i.e., a *conditional* function, is given by

$$\varphi_\xi(u|n) = \prod_{\nu=1}^{n} \varphi_{x_\nu}(u),$$

provided there are n pulses in the interval $(-T/2, T/2)$. In turn, the characteristic function for x_ν (that is independent of the pulse subscript ν) is

$$\varphi_{x_\nu}(u) = \overline{\exp(iux_\nu)} = \int w_a(\boldsymbol{a}) d\boldsymbol{a} \int \exp[iuF(t-\tau, \boldsymbol{a})] w_t(\tau) d\tau.$$

Here we did not indicate the limits in the integral with respect to $\boldsymbol{a}$. This is to imply that the integration is over the whole region of possible values (finite or infinite) of the components of this m-dimensional random variable. For the distribution of pulse-onset times $w_t(\tau)d\tau$, we cannot use (3.4). This is because pulses do not originate arbitrarily on the t axis, but only within the interval $(-T/2, T/2)$. This means that the uniform distribution of τ must be normalized to unity over this interval

$$w_t(\tau)d\tau = \frac{n_1 d\tau}{\int_{-T/2}^{T/2} n_1 d\tau} = \frac{d\tau}{T}. \tag{3.19}$$

Thus

$$\varphi_\xi(u|n) = \left\{ \int w_a(\boldsymbol{a})d\boldsymbol{a} \int\limits_{-T/2}^{T/2} \exp\left[iuF(t-\tau, \boldsymbol{a})\right] \frac{d\tau}{T} \right\}^n . \tag{3.20}$$

Fourier transforming $w(x)dx$ cf. (3.7) [or inserting $f(x) = \exp(iux)$ into (3.8)] we get the following relationship between the unconditional characteristic function and the conditional one:

$$\varphi_\xi(u) = \sum_{n=0}^{\infty} P(n)\varphi_\xi(u|n). \tag{3.21}$$

Substituting (3.5) for $P(n)$ and (3.20) for $\varphi_\xi(u|n)$ into (3.21) gives

$$\begin{aligned} \varphi_\xi(u) &= \sum_{n=0}^{\infty} \frac{\exp(-n_1 T)}{n!} \left[n_1 \int w_a(\boldsymbol{a})d\boldsymbol{a} \int\limits_{-T/2}^{T/2} e^{iuF(t-\tau,\boldsymbol{a})} d\tau \right]^n \\ &= \exp\left[n_1 \int w_a(\boldsymbol{a})d\boldsymbol{a} \int\limits_{-T/2}^{T/2} e^{iuF(t-\tau,\boldsymbol{a})} d\tau - n_1 T \right] , \end{aligned}$$

which can be rewritten as

$$\begin{aligned} \varphi_\xi(u) &= \exp\left\{ n_1 \int w_a(\boldsymbol{a})d\boldsymbol{a} \int\limits_{-T/2}^{T/2} [e^{iuF(t-\tau,\boldsymbol{a})} - 1]d\tau \right\} \\ &= \exp\left\{ n_1 \int w_a(\boldsymbol{a})d\boldsymbol{a} \int\limits_{t-T/2}^{t+T/2} [e^{iuF(\theta,\boldsymbol{a})} - 1]d\theta \right\} . \end{aligned}$$

In the latter expression a new integration variable $\theta = t - \tau$ is used.

We notice that the term in the integrand with respect to θ is only non-zero at those values of θ for which $F(\theta, \boldsymbol{a}) \neq 0$, i.e., within the confines of a pulse. Therefore, for all t separated from the interval boundaries $(-T/2, T/2)$ by no less than the pulse duration ϑ (ignoring the boundary effects), we may extend the limits of integration with respect to θ up to $\pm\infty$. In other words, we may take T to infinity. We shall in future make this transition without mentioning it specifically on each occasion. But, it should be borne in mind that it is only possible for $F(\theta, \boldsymbol{a})$ decaying sufficiently fast as $|\theta| \to \infty$. This gives

$$\varphi_\xi(u) = \exp\left\{ n_1 \int w_a(\boldsymbol{a})d\boldsymbol{a} \int\limits_{-\infty}^{+\infty} [e^{iuF(\theta,\boldsymbol{a})} - 1]d\theta \right\}. \tag{3.22}$$

Thus, we have the characteristic function $\xi(t)$. The exponent in (3.22) is the logarithm of this characteristic function. It has to be recalled that the coefficients at $(iu)^m/m!$ in the power series expansion of $\ln \varphi_\xi(u)$ with respect to iu, are the cumulants of the distribution $\xi(t)$. By expanding the logarithm of (3.22) into a power series with respect to iu we derive the mth cumulant

$$\lambda_m = n_1 \int w_a(\boldsymbol{a})d\boldsymbol{a} \int_{-\infty}^{+\infty} F^m(\theta, \boldsymbol{a})d\theta. \tag{3.23}$$

So, for the first and second cumulants, i.e., for the mean and variance of $\xi(t)$, we obtain

$$\lambda_1 = \overline{\xi(t)} = n_1 \int w_a(\boldsymbol{a})d\boldsymbol{a} \int_{-\infty}^{+\infty} F(\theta, \boldsymbol{a})d\theta, \tag{3.24a}$$

$$\lambda_2 = D[\xi(t)] = n_1 \int w_a(\boldsymbol{a})d\boldsymbol{a} \int_{-\infty}^{+\infty} F^2(\theta, \boldsymbol{a})d\theta. \tag{3.24b}$$

In the case where $\boldsymbol{a}_\nu$ is a one-dimensional random variable representing the amplitude of a pulse in a process of the type (3.1), (3.24) takes the form

$$\overline{\xi(t)} = n_1\bar{a} \int_{-\infty}^{+\infty} F(\theta)d\theta, \quad D[\xi(t)] = n_1\overline{a^2} \int_{-\infty}^{+\infty} F^2(\theta)d\theta. \tag{3.25}$$

Note that $\bar{\xi} = 0$ when either $\bar{a}$ or the pulse area $\int_{-\infty}^{+\infty} F(\theta)d\theta$ vanishes. Finally, if the amplitude is fixed, say $a_\nu = 1$, [in other words, the probability density a_ν is $w_a(a_\nu) = \delta(a_\nu - 1)$], then we arrive at relations originally derived by *Campbell* [3.1]

$$\overline{\xi(t)} = n_1 \int_{-\infty}^{+\infty} F(\theta)d\theta, \quad D[\xi(t)] = n_1 \int_{-\infty}^{+\infty} F^2(\theta)d\theta. \tag{3.26}$$

Both the characteristic function (3.22) and the cumulants (3.23) contain as the only characteristic of the density of pulses the average number of pulses per unit time n_1. In addition, all these quantities do not vary with time. This is a manifestation of the stationary nature of the process under consideration, a feature which will be discussed in Sect. 4.3.

Clearly, the above relationships cover the elementary theory of the shot effect (Sect. 2.5). Let the individual pulses be identical ($a = e =$ const) and rectangular in shape, of length ϑ and height $1/\vartheta$ (Fig. 3.1). Thus

$$\int_{-\infty}^{+\infty} F(\theta)d\theta = 1, \quad \int_{-\infty}^{+\infty} F^2(\theta)d\theta = \frac{1}{\vartheta}.$$

According to (3.25), for the pulse current $I(t) = e\sum_\nu F(t - t_\nu)$ we get its mean and variance

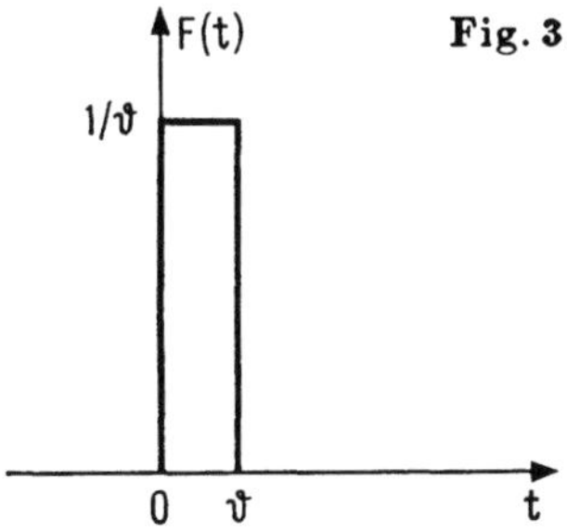

Fig. 3.1. Sketch of a rectangular pulse of unit area

$$\bar{I} = n_1 e \quad \text{and} \quad D[I] = \frac{n_1 e^2}{\vartheta} = \frac{e\bar{I}}{\vartheta}, \tag{3.27}$$

respectively. At first glance, the expression for $D[I]$ disagrees with the result deduced earlier, see (2.10), since the denominator now contains the pulse duration ϑ, i.e., a quantity of the order of the electron's transit time, whereas the denominator of (2.10) contained the observation time $T \gg \vartheta$. However, there is in fact no contradiction, since (3.27) gives the variance of the *instantaneous* current $I(t)$, whereas (2.10) is the variance of a current that is already averaged over the time T, i.e., the variance of the quantity

$$I_T = \frac{1}{T}\int_0^T I(t)dt. \tag{3.28}$$

Hence, the fluctuations of I_T, in comparison with those of $I(t)$, are averaged out and this all the more so the larger T is. It is clear that for $D[I_T]$ we get the previous expression (2.10) (see Exercise 3.7.6).

Assuming that $a_\nu = e = \text{const}$, we obtained a model of the shot effect. If the random parameter a_ν is a multiple of the electron charge, then the process

$$\xi(t) = \sum_\nu e m_\nu \delta(t - t_\nu)$$

may, with an appropriate probability distribution for m_ν, describe the current of secondary electron emission and its fluctuations: The νth electron of the primary current $I(t) = e\sum_\nu \delta(t - t_\nu)$ gives rise to a random number $m_\nu = 0, 1, 2, \ldots$ of secondary electrons. Instead of (3.27) the moments of secondary emission current $\xi(t) \equiv I_{\text{s.e.}}$ will be

$$\bar{I}_{\text{s.e.}} = n_1 \bar{m} e,$$

$$\overline{(\Delta I_{\text{s.e.}})^2} = \frac{n_1 \overline{m^2} e^2}{\vartheta} = \frac{e\bar{I}_{\text{s.e.}}}{\vartheta}\frac{\overline{m^2}}{\bar{m}}.$$

Knowing the characteristic function (3.22), we in principle also know its Fourier transform, i.e., the probability density $w(t)$ of the process $\xi(t)$. But the appropriate Fourier integral in its closed form is only obtainable in a few simple cases. If, for example, the process consists of identical square pulses of fixed amplitude a and duration ϑ,

$$F(\theta,a) = aF(\theta), \quad F(\theta) = \begin{cases} 1 & \text{in the interval } (0,\vartheta), \\ 0 & \text{otherwise} \end{cases}, \tag{3.29}$$

then (3.22) gives

$$\varphi_\xi(u) = \exp\left[n_1 \int_0^\vartheta (e^{iua}-1)d\theta\right] = \exp\left[n_1\vartheta(e^{iua}-1)\right] \quad .$$

But this is the characteristic function of the Poisson distribution with parameter $\overline{n} = n_1\vartheta$ and discrete possible values na, with $n = 0,1,2,\ldots$:

$$P\{\xi(t) = na\} = \frac{(n_1\vartheta)^n}{n!}\exp(-n_1\vartheta). \tag{3.30}$$

In general, it is advantageous in the above problem to look not for the distribution function but the characteristic function.

Under certain conditions, however, both (3.22) and the corresponding distribution may take on a universal form. We shall now discuss this form and the respective conditions. Taking the Fourier transform of the characteristic function

$$\varphi_\xi(u) = \exp\left(\sum^{\infty}\lambda_m\frac{(iu)^m}{m!}\right)$$

we find the probability density

$$\begin{aligned} w(\xi) &= \frac{1}{2\pi}\int_{-\infty}^{+\infty} e^{-iu\xi}\varphi_\xi(u)du \\ &= \frac{1}{2\pi}\int_{-\infty}^{+\infty}\exp\left(-iu\xi + \sum_{m=1}^{\infty}\lambda_m\frac{(iu)^m}{m!}\right)du. \end{aligned}$$

If we introduce the notation

$$\lambda_1 = \overline{\xi}, \ \lambda_2 = \sigma_\xi^2 = D[\xi,t)], \ x = \frac{\xi-\lambda_1}{\sqrt{\lambda_2}}$$

$$\Phi^{(n)}(x) = \frac{1}{\sqrt{2\pi}}\frac{d^n e^{-x^2/2}}{dx^n}$$

and, taking $m = 3$ as our starting point, expand the exponential function in the integrand in a power series of iu, then term-by-term integration gives the representation of $w(\xi)$ as a series with respect to the derivatives $\Phi^{(n)}(x)$. Collecting in this series terms of the same order with respect to n_1, we arrive at the so-called Edgeworth series [Ref. 3.2, Chap. 17]

$$w(\xi) = \frac{\Phi^{(0)}(x)}{\sigma_\xi} - \frac{\lambda_3\Phi^{(3)}(x)}{3!\sigma_\xi^4} + \left(\frac{\lambda_4\Phi^{(4)}(x)}{4!\sigma_\xi^5} + \frac{\lambda_3^2\Phi^{(6)}(x)}{72\sigma_\xi^7} + \ldots\right). \tag{3.31}$$

Here the first term is of order $n_1^{-1/2}$, the second of order n_1^{-1}, and the third of order $n_1^{-3/2}$, etc. All the terms are proportional to $\exp(-x^2/2)$, the first being simply the normal distribution for x. Clearly, as n_1 increases, (i.e., as the density of pulses grows), the first term will increasingly dominate, so that $w(\xi)$ approaches the normal distribution.

We now try to ascertain for which values of n_1, we can restrict ourselves to the first term of the series (3.31). The two first terms of the series are

$$w(\xi) = \frac{e^{-x^2/2}}{\sqrt{2\pi}\sigma_\xi}\left[1 - \frac{\lambda_3}{6\sigma_\xi^3}x(3 - x^2)\right] \quad .$$

The presence of the exponential factor allows us to ignore large values of x for which $w(\xi)$ is already very small. For $|x| \leq 2$, we have $|x(3 - x^2)| \leq 2$. Hence, the condition under which the second term is small is

$$\frac{|\lambda_3|}{3\sigma_\xi^2} \leq \frac{n_1\overline{|a|^3}\int\limits_{-\infty}^{+\infty}|F(\theta)|^3 d\theta}{3\left[n_1\overline{a^2}\int\limits_{-\infty}^{+\infty}F^2(\theta)d\theta\right]^{3/2}} \ll 1$$

or, if we discard a numerical factor of the order of unity,

$$\frac{\int\limits_{-\infty}^{+\infty}|F(\theta)|^3 d\theta}{n_1^{1/2}\left[\int\limits_{-\infty}^{+\infty}F^2(\theta)d\theta\right]} \ll 1.$$

If $\int_{-\infty}^{\infty} F(\theta)d\theta = 1$ and the pulse has a duration of about ϑ, the integral in the numerator is of order $1/\vartheta^2$, and that in the denominator of order $1/\vartheta$. Hence, the above condition reduces to

$$n_1\vartheta \gg 1. \tag{3.32}$$

This means that the *number of pulses generated during the time interval in which the given pulse has non-zero amplitude, should be sufficiently large.* In other words, at the instant t, the variable $\xi(t)$ is composed of pulses produced within the interval from $t - \vartheta$ to t, since at the time t the pulses generated earlier have already decayed. Consequently, $n_1\vartheta$ is the average number of terms x_ν constituting $\xi(t)$ at each given instant of time. The condition (3.32) thus tells us that the closer the distribution $\xi(t)$ is to the normal one, the greater the number of pulses superimposed at each time. If the impacts are rare, the distribution will be sensitive to the shape of the individual impacts, and no normal distribution will ensue. This is only obtained either by making the impacts more frequent, or by increasing the duration of the pulses. In both cases the number of impacts during one pulse will increase.

This result is quite a special case of the so-called *central limit theorem* in probability theory. In fact, under the specific conditions of the problem in hand the above reasoning is a proof of the theorem. We shall see later (Sect. 3.6), that the result obtained and the condition (3.32) directly stem from much more general concepts.

For the general case – i.e., without passing to the limit of dense noise ($n_1\vartheta \to \infty$) – *Gilbert* and *Pollak* [3.3] obtained the distribution of the pulse process (3.1) for some pulse shapes $F(t)$ in an explicit form. First, *Gilbert* and *Pollak* indicated that the process (3.1), and even the more general process

$$\xi(t) = \sum_{\nu} F(t - t_\nu, a_\nu), \tag{3.33}$$

can always be reduced to an equivalent noise of the form

$$\xi(t) = \sum_{\nu} F_{eq}(t - t_\nu) \tag{3.34}$$

with a pulse shape $F_{eq}(t)$ such that the distribution

$$W(\xi) = \int_{-\infty}^{\xi} w(x)dx$$

will be the same for (3.33) and (3.34). Second, for $W(\xi)$ they derived (by two different techniques) the integral equation

$$\int_{-\infty}^{\xi} x\,dW(x) = n_1 \int_{-\infty}^{+\infty} W[\xi - F(\theta)]F(\theta)d\theta,$$

and succeeded in solving it for some pulse shapes $F(t)$.

For rectangular pulses of height a and duration ϑ, this equation is clearly satisfied by the Poisson distribution (3.30). We recall, that (3.30) turns into the Laplace distribution with increasing $n_1\vartheta$, i.e., as the height of the steps a levels off it turns into the normal distribution.

If the pulse shape is specified somewhat artificially in the following way (Fig. 3.2, where γ denotes the Euler constant)

$$F(t) = 0 \qquad \text{for} \quad t \leq 0,$$

$$t = \int_{F(t)}^{\infty} \frac{e^{-y}dy}{y} \qquad \text{for} \quad t > 0,$$

then the distribution of ξ becomes a power-exponential (continuous Poisson) one

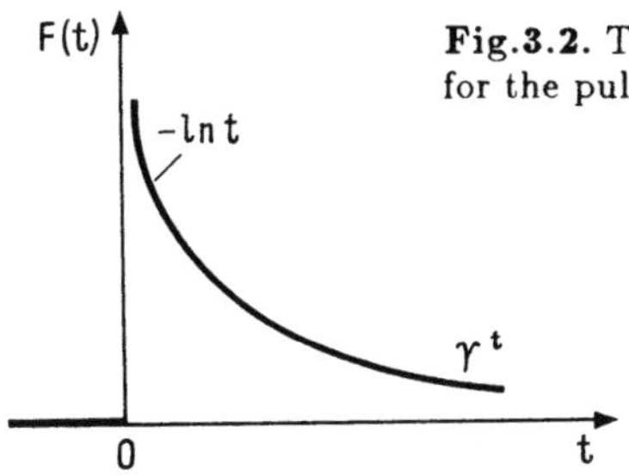

Fig.3.2. The pulse shape resulting in the probability density (3.17) for the pulse Poisson process (3.16)

$$w(x) = \frac{x^{\alpha-1}e^{-x}}{\Gamma(\alpha)} \quad (0<x<\infty) \tag{3.35}$$

where $\alpha = \overline{\xi}$. It also becomes normal with increasing α.

Unfortunately, for a pulse shape as important as the damped oscillation (the impulse function of a resonant circuit), the equation for W can only be solved numerically, but even this is to be regarded as a success.

Note that (3.35) is a special case of the so-called *gamma-distribution*

$$w(x) = \frac{\kappa^\alpha x^{\alpha-1}}{\Gamma(\alpha)} e^{-\kappa x} \quad (0<x<\infty), \tag{3.36}$$

whose characteristic function is

$$\vartheta(u) = \left(\frac{\kappa}{\kappa - iu}\right)^\alpha, \tag{3.37}$$

the mean and variance being expressed in terms of the positive parameters α and κ:

$$\overline{x} = \frac{\alpha}{\kappa}, \quad D[x] \equiv \sigma_x^2 = \frac{\alpha}{\kappa^2} \quad .$$

Clearly, (3.35) follows from (3.36) if the parameter κ (scale parameter) is made equal to unity. For an integer α ($\alpha = n$) it follows from (3.36) that

$$w(x) = \frac{\kappa^n}{(n-1)!} x^{n-1} e^{-\kappa x} \quad .$$

This distribution is commonly written in terms of the parameter $\mu = 1/\overline{x}$ (so that $\kappa = n\mu$ and $\sigma_x^2 = 1/n\mu^2$)

$$w(x) = \frac{(n\mu)^n}{(n-1)!} x^{n-1} e^{-n\mu x} \quad . \tag{3.38}$$

3.4 Covariance

Treating $\xi(t)$ as a random variable, we find the characteristic function (3.22) and thereby the cumulants of $\xi(t)$, in particular the mean and variance, see (3.24). But to compute mixed moments of two random variables $\xi_1 = \xi(t_1)$

and $\xi_2 = \xi(t_2)$, (i.e., the values of the random process $\xi(t)$ at two *different* instants of time), we need the joint (bivariate) distribution function for ξ_1 and ξ_2 or the corresponding bivariate characteristic function. It is not our aim to solve this problem here. We shall only calculate the mixed moment $\overline{\xi_1\xi_2 - \overline{\xi}_1 \cdot \overline{\xi}_2}$, which is one of the most important characteristics of random processes since it defines the correlation between the variables ξ_1 and ξ_2. This moment depends in general on t_1 and t_2 or (if we set $t_1 = t$, $t_2 = t + \tau$) on t and τ. This moment is known as the *covariance*, or *correlation function*, of the random process $\xi(t)$

$$\psi_\xi(t,\tau) = \overline{\xi(t)\xi(t+\tau)} - \overline{\xi(t)} \cdot \overline{\xi(t+\tau)} \quad . \tag{3.39}$$

As the cumulant $\lambda_1 = \overline{\xi(t)}$ appears to be independent of t (like all λ_m), the second term on the right-hand side of (3.39) is λ_1^2. Thus, we only need the joint moment

$$B_\xi(t,\tau) = \overline{\xi(t)\xi(t+\tau)}. \tag{3.40}$$

The scheme of calculation here is like that in the univariate case.

Assuming that the time interval $(-T/2, T/2)$ contains n pulses we now express the bivariate probability density for the random variables

$$\begin{aligned} \xi &\equiv \xi(t) = \sum_\nu F(t - t_\nu, \boldsymbol{a}_\nu), \\ \xi_\tau &\equiv \xi(t+\tau) = \sum_\mu F(t + \tau - t_\mu, \boldsymbol{a}_\mu), \end{aligned} \tag{3.41}$$

in terms of conditional probability densities,

$$w(x,y) = \sum_{n=0}^{\infty} P(n) v(x,y|n). \tag{3.42}$$

It follows that

$$\overline{\xi\xi_\tau} = \sum_{n=0}^{\infty} P(n)\, {}^n\overline{\xi\xi_\tau},$$

holds where

$${}^n\overline{\xi\xi_\tau} = \int_{-\infty}^{+\infty} xyv(x,y|n)dx\,dy$$

is the *conditional* moment, i.e., the mean of the product of the variables (3.41), but with the summation over only those n pulses which fall into $(-T/2, T/2)$. This leads to

$${}^n\overline{\xi\xi_\tau} = \sum_{\nu,\mu=1}^{n} \langle F(t - t_\nu, \boldsymbol{a}_\nu) F(t + \tau - t_\mu, \boldsymbol{a}_\mu) \rangle$$

where the averaging is performed over all the random parameters $t_\nu, t_\mu, \boldsymbol{a}_\nu, \boldsymbol{a}_\mu$ ($\nu, \mu = 1, 2, \ldots, n$). As the variables with $\nu \neq \mu$ are assumed to be independent, the double sum can be divided into two parts

$$
\begin{aligned}
{}^n\overline{\xi\xi_\tau} = & \sum_{\nu=1}^{n} \langle F(t - t_\nu, \boldsymbol{a}_\nu) F(t + \tau - t_\nu, \boldsymbol{a}_\nu) \rangle \\
& + \sum_{\nu \neq \mu} \langle F(t - t_\nu, \boldsymbol{a}_\nu) \rangle \langle F(t + \tau - t_\mu, \boldsymbol{a}_\mu) \rangle .
\end{aligned}
$$

However, due to the initial assumptions, the mean values do not depend on the subscripts ν and μ. Therefore, the first sum contains n identical terms, and the second $(n^2 - n)$:

$$
\begin{aligned}
{}^n\overline{\xi\xi_\tau} = & \, n \langle F(t - \chi, \boldsymbol{a}) F(t + \tau - \chi, \boldsymbol{a}) \rangle \\
& + (n^2 - n) \langle F(t - \chi, \boldsymbol{a}) \rangle \langle F(t + \tau - \chi', \boldsymbol{a}') \rangle ,
\end{aligned}
$$

where

$$
\begin{aligned}
& \langle F(t - \chi, \boldsymbol{a}) F(t + \tau - \chi, \boldsymbol{a}) \rangle \\
& \quad = \int w_a(\boldsymbol{a}) d\boldsymbol{a}) \int_{-T/2}^{T/2} F(t - \chi, \boldsymbol{a}) F(t + \tau - \chi, \boldsymbol{a}) \frac{d\chi}{T} ,
\end{aligned}
$$

$$
\langle F(t - \chi, \boldsymbol{a}) \rangle = \int w_a(\boldsymbol{a}) d\boldsymbol{a} \int_{-T/2}^{T/2} F(t - \chi, \boldsymbol{a}) \frac{d\chi}{T} ,
$$

$$
\langle F(t + \tau - \chi', \boldsymbol{a}) \rangle = \int w_a(\boldsymbol{a}) d\boldsymbol{a} \int_{-T/2}^{T/2} F(t - \tau - \chi', \boldsymbol{a}) \frac{d\chi'}{T} .
$$

In the first two expressions we change the integration variable to $t - \chi = \boldsymbol{\theta}$, in the third to $t + \tau - \chi' = \boldsymbol{\theta}$, and extend the integration limits with respect to $\boldsymbol{\theta}$ to $\pm\infty$. We thus have

$$
\begin{aligned}
{}^n\overline{\xi\xi_\tau} = & \frac{T}{n} \int w_a(\boldsymbol{a}) d\boldsymbol{a} \int_{-\infty}^{+\infty} F(\theta, \boldsymbol{a}) F(\theta + \tau, \boldsymbol{a}) d\theta \\
& + \frac{n^2 - n}{T} \left[\int w_a(\boldsymbol{a}) d\boldsymbol{a} \int_{-\infty}^{+\infty} F(\theta, \boldsymbol{a}) d\theta \right]^2 .
\end{aligned}
$$

It remains only to average this conditional mean over the Poisson distribution for which, as we now know, $\overline{n} = n_1 T$ and $D[n] = \overline{n^2} - \overline{n}^2 = \overline{n} = n_1 T$). Using (3.42) we obtain for the second moment

$$B_\xi(\tau) = \overline{\xi\xi_\tau} = n_1 \int w_a(\boldsymbol{a})d\boldsymbol{a} \int_{-\infty}^{+\infty} F(\theta,\boldsymbol{a})F(\theta+\tau,\boldsymbol{a})d\theta + \lambda_1^2,$$

where we have used (3.24a). Hence, the covariance (3.39) is

$$\psi_\xi(\tau) = n_1 \int w_a(\boldsymbol{a})d\boldsymbol{a} \int_{-\infty}^{+\infty} F(\theta,\boldsymbol{a})F(\theta+\tau,\boldsymbol{a})d\theta. \tag{3.43}$$

B_ξ and ψ_ξ depend only on the time *difference* τ and not on the time t. Again, this is due to the stationary nature of the random process already mentioned. For $\tau = 0$, (3.43) reduces to (3.24b) giving the variance of $\xi(t)$, i.e.,

$$\psi_\xi(0) = D[\xi(t)] = \lambda_2,$$

which is to be expected according to the definition (3.39) of the covariance.

It is evident that $\psi_\xi(\tau)$ is an even function of τ. If $F(\theta)$ decays for $|\theta|\to\infty$ faster than $1/\sqrt{\theta}$, then $\psi_\xi(\tau)$ tends to zero as $|\tau|\to\infty$. Clearly, as $|\tau|$ increases, the factors in the integrand in (3.43) behave in such a way that for large $|\theta|$ the product approaches zero faster than $1/|\theta|$.

In the special case where $\boldsymbol{a}$ is a one-dimensional random variable or, more specifically, the amplitude of a pulse $[F(\theta,\boldsymbol{a}) = aF(\theta)]$, we arrive at

$$\psi_\xi(\tau) = n_1\overline{a^2} \int_{-\infty}^{+\infty} F(\theta)F(\theta+\tau)d\theta. \tag{3.44}$$

The random duration of the νth pulse ϑ_ν may also be one of the components of an n-dimensional random variable $\boldsymbol{a}$. Let ϑ_ν be independent of the other components of $\boldsymbol{a}_\nu$, so that the distribution $w_{\boldsymbol{a}}(\boldsymbol{a}_\nu)d\boldsymbol{a}_\nu$ must be replaced by $w_a(\boldsymbol{a}_\nu)d\boldsymbol{a}_\nu \cdot w_\vartheta(\vartheta_\nu)d\vartheta_\nu$. If we wish to take into account explicitly the random variation of the pulse duration due to ϑ_ν, then it is advantageous to represent the deterministic function F describing the pulse shape in the form $F[(t-t_\nu)/\vartheta_\nu,\, a_\nu]$. Here the derivation of the characteristic function, the cumulants, and the covariance is the same as above. However, when averaging over the random times of pulse generation t_ν it is convenient to go over to the new integration variable $\theta = (t-t_\nu)/\vartheta_\nu$. Then the distribution dt_ν/T in $(-T/2, T/2)$ becomes $\vartheta_\nu d\theta/T$ in the region from $(t-T/2)/\vartheta_\nu$ to $(t+T/2)/\vartheta_\nu$. After extending the limits to $\pm\infty$, the parameter ϑ_ν will be retained only as a factor in the integrand. (In deducing the covariance, the ϑ_ν-scale enters also as scale of the time shift τ, i.e., instead of τ we have τ/ϑ_ν). Consequently, the cumulants λ_m are given by

$$\lambda_m = n_1\bar{\vartheta} \int_{-\infty}^{+\infty} \langle F^m(\theta,\boldsymbol{a})\rangle d\theta, \tag{3.45}$$

instead of (3.23), and for the covariance by

$$\psi_\xi(\tau) = n_1 \int\limits_0^{\infty} \vartheta w_\vartheta(\vartheta) d\vartheta \int\limits_{-\infty}^{+\infty} \left\langle F(\theta, \boldsymbol{a}) F\left(\theta + \frac{\tau}{\vartheta}, \boldsymbol{a}\right) \right\rangle d\theta, \tag{3.46}$$

instead of (3.43). Here $\langle \ldots \rangle$ is a short-hand notation for the averaging over the distribution of $\boldsymbol{a}$.

Consider the following example. Let $\boldsymbol{a}_\nu$ be a univariate random amplitude, such that

$$\xi(t) = \sum_\nu a_\nu F\left(\frac{t - t_\nu}{\vartheta_\nu}\right). \tag{3.47}$$

Further, let $F(\theta)$ be unity in the interval (0,1), and zero otherwise, and the durations ϑ being distributed according to

$$w(\vartheta) d\vartheta = e^{-\vartheta/\bar{\vartheta}} \frac{d\vartheta}{\bar{\vartheta}},$$

then, for $\tau > 0$, we have

$$\int\limits_{-\infty}^{+\infty} F(\theta) F\left(\theta + \frac{\tau}{\vartheta}\right) d\theta = \begin{cases} 0 & \text{for} \quad \vartheta < \tau, \\ 1 - \dfrac{\tau}{\vartheta} & \text{for} \quad \vartheta > \tau, \end{cases} \qquad \text{and hence}$$

$$\psi(\tau) = n_1 \overline{a^2} \int\limits_\tau^{\infty} (\vartheta - \tau) e^{-\vartheta/\bar{\vartheta}} \frac{d\vartheta}{\bar{\vartheta}} = n_1 \overline{a^2} \bar{\vartheta} e^{-\tau/\bar{\vartheta}} \quad .$$

The covariance is even in τ, so that at any sign of τ the result may be represented as

$$\psi(\tau) = n_1 \overline{a^2} \bar{\vartheta} e^{-|\tau|/\bar{\vartheta}} \quad . \tag{3.48}$$

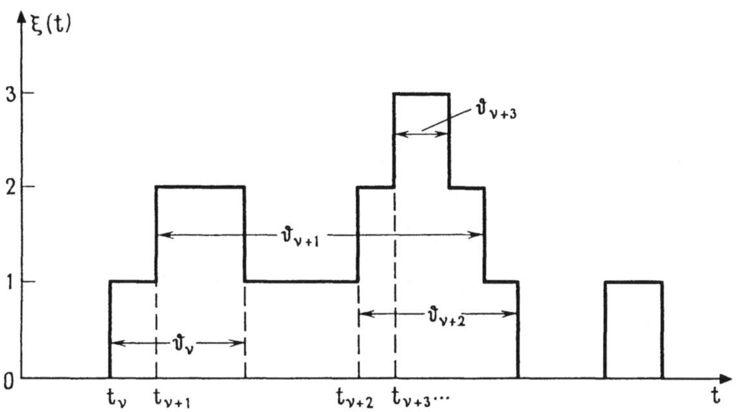

Fig. 3.3. A realization of the Poisson process consisting of rectangular pulses of height one and random length ϑ_ν

Figure 3.3 provides some idea of the character of the process in the special case of fixed $a_\nu = 1$. It has been assumed that this process may serve as a model for the flicker effect, i.e., the relatively slow fluctuations of emission current due to local variations in the emissivity of the cathode surface [3.4]. If this variation is accounted for by the presence of impurity atoms on the surface (in particular of residual gas molecules that have settled on the cathode) then $\xi(t)$ is the random instantaneous number of these atoms, and the ϑ_ν are their residence times on the surface. But such a model fails to agree with the form of the spectrum for the flicker effect which, within a wide range of ω, is fairly close to ω^{-1} [Ref. 3.5, Sect. 4.6]. Along with attempts to improve the theory by introducing a *collection* of the mean residence times, $\bar{\vartheta}$, with certain weights or by other similar assumptions, it has been suggested that the desired spectrum may be connected with the process (3.1). These attempts invoke the notion that the pulses $F(t)$ decay very slowly with t, i.e., that they decrease for sufficiently large t as $t^{\alpha/2-1}$ with $0<\alpha<2$. However, for $\alpha \geq 1$ the process is no longer stationary. See [3.6], where a review of the literature is also given. We shall return to this problem later on [3.5] when discussing the spectrum analysis of random processes.

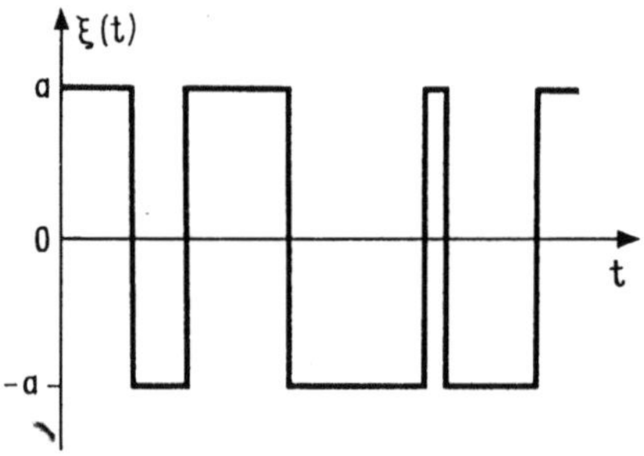

Fig. 3.4. An illustration of a generalized telegraph signal

Let us consider one more example of a pulse process associated with a Poisson process. This is the so-called generalized telegraph signal: a function $\xi(t)$ alternating between the values $-a$ and $+a$, with the instants t_ν of the jumps being subject to a Poisson distribution (Fig. 3.4). For simplicity we take the density of jumps n_1 to be constant. The addition of a bias $\xi_0 > a$ to $\xi(t)$ would produce single-polarity pulses, so that the process in question could be an oversimplified model of conductivity fluctuations in a granular resistor, which is essentially an agglomerate (mostly a film) of well-conducting grains. Being of no importance for the statistical properties of the process, the bias $\bar{\xi} = \xi_0$ is assumed to be zero. The Poisson distribution of times t of sign inversions of ξ implies that the durations of pulses have the exponential distribution

$$n_1 \exp(-n_1\vartheta)d\vartheta .$$

It seems that by considering the *derivative* of $\xi(t)$ to be composed of the pulses $\pm 2a\delta(t - t_\nu)$, we come back to the process described by (3.1). However, it is evident that this is not the case, as the jumps here must show a sign *alternation*, whereas in (3.1) the values of the a_ν are mutually independent.

If in the interval $(t, t+\tau)$ the sign of $\xi(t)$ changes an even number of times, then $\xi(t+\tau)$ will have the same sign as $\xi(t)$, i.e., $\xi(t+\tau)\xi(t) = a^2$. Otherwise, the product is equal to $-a^2$. Hence, for $\tau>0$

$$\psi(\tau) = a^2 P_{\text{ev}} - a^2 P_{\text{odd}} \quad ,$$

where P_{ev} and P_{odd} are the probabilities of even and odd numbers of jumps, respectively, in τ. By the Poisson law

$$P(n) = \frac{\bar{n}^n}{n!} e^{-\bar{n}} \quad \text{with} \quad \bar{n} = n_1 \tau,$$

we get

$$P_{\text{ev}} = P(0) + P(2) + \ldots = e^{-\bar{n}}\left(1 + \frac{\bar{n}^2}{2!} + \frac{\bar{n}^4}{4!} + \ldots\right) = e^{-\bar{n}} \cosh \bar{n},$$

$$P_{\text{odd}} = P(1) + P(3) + \ldots = e^{-\bar{n}}\left(\bar{n} + \frac{\bar{n}^3}{3!} + \frac{\bar{n}^5}{5!} + \ldots\right) = e^{-\bar{n}} \sinh \bar{n}.$$

Consequently, for $\tau>0$

$$\psi(\tau) = a^2 e^{-\bar{n}}(\cosh \bar{n} - \sinh \bar{n}) = a^2 e^{-2\bar{n}} = a^2 \exp(-2n_1\tau),$$

holds and for any sign of τ

$$\psi(\tau) = a^2 \exp(-2n_1|\tau|). \tag{3.49}$$

Note that the covariance (3.49) is similar to (3.48), in spite of the disparate forms of the random processes (Figs. 3.3 and 3.4). Also, a similar covariance is found for a process (3.1) consisting of exponential pulses,

$$F(t) = \begin{cases} 0 & \text{for} \quad t<0, \\ e^{-\alpha t} & \text{for} \quad t>0. \end{cases}$$

Straightforward calculation, using (3.44), then yields

$$\psi(\tau) = \frac{n_1 \overline{a^2}}{2\alpha} e^{-\alpha|\tau|} \quad . \tag{3.50}$$

Figure 3.5 conveys some idea of the realization of such a process, that may be the response of an RC network to a random sequence of delta pulses.

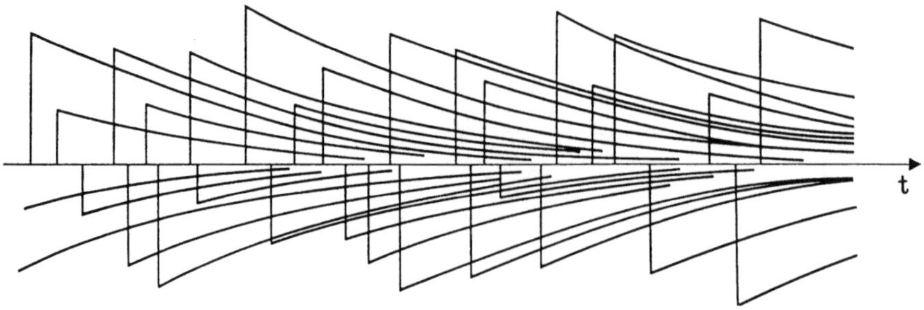

Fig. 3.5. A realization of the Poisson process consisting of exponentially decaying pulses with random amplitudes

The covariance of the form $A \exp(-\alpha|\tau|)$ is commonly used as an approximation to those cases where the real $\psi(\tau)$ decreases monotonously without changing its sign. Such an approximation is most convenient (along with the Gaussian curve) for various theoretical estimates and approximate derivations.

3.5 Some Generalization of the Pulse Problem

The conclusions reached above have been based on the assumption that the probability density for a pulse in the interval $(t, t + dt)$ is constant, and hence independent of both time t and the presence and position of other pulses. Elimination of any one of these assumptions leads to more general problems.

If the probability of a pulse being within $(t, t + dt)$ depends on the differences between t and the times at which a certain number of *preceding pulses* emerges, then we have the *conditional* probability. We shall first consider the problem which has been solved by *Domb* [3.7]. We restrict ourselves to the case where the probability of the νth pulse is only influenced by the $(\nu - t)$th pulse, i.e., the probability varies with the time ϑ elapsed since the occurrence of the $(\nu - 1)$th pulse,

$$P\left\{\begin{array}{c|c} \nu\text{th pulse} & \text{if interval } \vartheta \\ \text{in}(\vartheta, \vartheta + d\vartheta) & \text{is empty} \end{array}\right\} = \varphi(\vartheta)d\vartheta. \tag{3.51}$$

It is only natural to expect the influence of the $(\nu - 1)$th pulse to weaken with increasing ϑ, i.e., $\varphi(\vartheta) \rightarrow \text{const}$. Previously we had $\varphi(\vartheta) = n_1^0 = \text{const}$ for any ϑ, i.e., the distribution of (3.51) was unconditional. (The superscript 0 at n_1 is introduced to distinguish this from the expressions for the uncorrelated pulses.) This lead to the Poisson distribution for the number of pulses n in the interval $(0, T)$. It is obvious that now the distribution of n will not be Poissonian, and below we shall consider the respective variations of the moments $\overline{n}$ and $\overline{n^2}$.

What is the relationship of $\varphi(\vartheta)$ to the probability $W(\vartheta)$ that the time span ϑ after the $(\nu - 1)$th pulse will be empty? Evidently, the probability $W(\vartheta + d\vartheta)$ for an empty time span $\vartheta + d\vartheta$ is the product of the probability $W(\vartheta)$ of the empty span ϑ and the conditional probability that given an empty span ϑ, the interval $(\vartheta, \vartheta + d\vartheta)$ is empty, too. If the interval $d\vartheta$ is not empty, then it may not only contain the νth pulse, but in addition the $(\nu+1)$th, $(\nu+2)$th, and so forth. However, we have already seen that the probability of finding m pulses in an infinitesimal interval $d\vartheta$ is a quantity of order $(d\vartheta)^m$. Therefore, the conditional probability of an empty $d\vartheta$ is to first order in $d\vartheta$ given by $1 - \varphi(\vartheta)d\vartheta$. Thus

$$W(\vartheta + d\vartheta) = W(\vartheta)[1 - \varphi(\vartheta)d\vartheta] \quad .$$

Restricting the left-hand side also to terms of first order in $d\vartheta$, we have, after the integration,

$$W(\vartheta) = \exp\left[-\int_0^\vartheta \varphi(\theta)d\theta\right]. \tag{3.52}$$

Further, we introduce the probability $g(\vartheta)d\vartheta$ that between two successive pulses, the time span from ϑ to $\vartheta + d\vartheta$ will elapse. This probability is equal to the product of the probability $W(\vartheta)$ of an empty ϑ and the conditional probability (3.51) of finding a pulse in $(\vartheta, \vartheta + d\vartheta)$ for the empty ϑ

$$g(\vartheta)d\vartheta = W(\vartheta)\varphi(\vartheta)d\vartheta = \exp\left[-\int_0^\vartheta \varphi(\theta)d\theta\right]\varphi(\vartheta)d\vartheta. \tag{3.53}$$

We now focus our attention on the desired probability $P_n(T)$ of n pulses being found in the interval $(0, T)$. The event "n pulses in $(0, T)$" may be broken up into mutually exclusive special n cases. The nth event describes the occurrence of the last (nth) pulse within the time interval $(\eta, \eta + d\eta)$ located at the end of $(0, T)$. Denoting the probability of this special case by $p_n(T, \eta)d\eta$, we arrive at

$$P_n(T) = \int_0^T p_n(T, \eta)d\eta.$$

But $p_n(T, \eta)d\eta$ can be written as the product of the probability $W(\eta)$ that the interval η will be empty (i.e., the nth pulse is the least) and the probability that the nth pulse will occur at the end of the interval $(0, T - \eta)$ that is equal to $p_n(T - \eta, 0)d\eta$, namely

$$p_n(T, \eta) = W(\eta)p_n(T - \eta, 0).$$

Consequently,

$$P_n(T) = \int_0^T W(\eta)p_n(T - \eta, 0)d\eta, \tag{3.54}$$

whereas the probabilities $p_n(t, 0) \equiv w_n(t)$ satisfy the following integro-difference equation

$$w_n(T) = \int_0^T w_{n-1}(T-\eta)g(\eta)d\eta. \tag{3.55}$$

In fact, the integrand is the product of the probabilities that the $(n-1)$th pulse will occur at the end of the interval $(0, T-\eta)$, and that the time η will elapse between the $(n-1)$th and nth pulses, i.e., that the nth pulse will be at the end of the interval $(0, T)$. Different values of η yield mutually exclusive special cases for the event of "the nth pulse being at the end of $(0,T)$". Its total probability is represented by (3.55).

By fixing $g(\eta)$ and $w_1(t)$ and solving (3.55), we obtain $w_n(t) \equiv p_n(t, 0)$. This enables us to compute $P_n(T)$ using (3.54). However, the direct solution of this problem is rather complicated.

In [3.7] the problem is treated asymptotically (for sufficiently large T) using the Laplace transform and the moment-generating function

$$f(T,\theta) = \sum_{n=1}^{\infty} P_n(T)e^{n\theta} \quad .$$

At large T this function is dependent only on $g(t)$, not on $w_1(t)$, and may be represented by a series in terms of powers of θ with coefficients expressed through (asymptotic) values of $\overline{n}, \overline{n^2}, \ldots$. This leads in particular, to

$$\overline{n} = \frac{T}{\overline{\vartheta}}, \quad \overline{n^2} - \overline{n}^2 = \frac{\overline{\vartheta^2} - \overline{\vartheta}^2}{\overline{\vartheta}^3}T, \quad \text{with} \quad \overline{\vartheta^m} = \int_0^{\infty} \vartheta^m g(\vartheta)d\vartheta. \tag{3.56}$$

For *uncorrelated* pulses, (3.51) takes the form (Sect. 3.1)

$$\varphi(\vartheta)d\vartheta = n_1^0 d\vartheta,$$

where n_1^0 is in this case the average number of pulses per unit time. Accordingly, (3.52,53) yield the following expressions for the empty time span ϑ after the pulse:

$$W(\vartheta) = \exp(-n_1^0\vartheta),$$

and for the probability in the interval $(\vartheta, \vartheta + d\vartheta)$ between the subsequent pulses:

$$g(\vartheta)d\vartheta = \exp(-n_1^0\vartheta)n_1^0 d\vartheta. \tag{3.57}$$

If for large ϑ we identify $g(\vartheta)d\vartheta$ with $w_1(\vartheta)d\vartheta$, i.e., with the probability that the *first* pulse (being also the last) will occur at the end of the empty time span $(0, \vartheta)$, then direct calculation utilizing (3.55) gives

$$w_n(t) = \frac{\exp(-n_1^0 t)(n_1^0)^n t^{n-1}}{(n-1)!}$$

and from (3.54) we get the Poisson distribution

$$P_n(T) = \frac{\exp(-n_1^0 T)(n_1^0 T)^n}{n!}, \qquad \text{with}$$

$$\bar{n} = n_1^0 T, \quad \overline{n^2} - \bar{n}^2 = n_1^0 T.$$

This same result is also obtained from (3.56) for

$$g(\vartheta) = \exp(-n_1^0 \vartheta) n_1^0.$$

We now consider examples which offer other choices for the function $g(\vartheta)$.

(1) Let $g(\vartheta)$ be the density of the gamma distribution

$$g(\vartheta) = \frac{\kappa^\alpha \vartheta^{\alpha-1}}{\Gamma(\alpha)} e^{-\kappa\vartheta} \quad .$$

If $0<\alpha<1$, then the probability of the time span ϑ occuring between pulses decreases faster with increasing ϑ than if $\alpha = 1$, (which corresponds to the absence of correlations, see (3.57) with $n_1^0 = \kappa$). Thus, for $0<\alpha<1$ we have an "attractive" correlation: a pulse increases the probability of the next pulse. On the other hand, for $\alpha>1$ the correlation is in this sense "repulsive". Using (3.56) we have

$$\bar{n} = \frac{\kappa T}{\alpha}, \quad \overline{n^2} - \bar{n}^2 = \frac{\kappa T}{\alpha^2} = \frac{\bar{n}}{\alpha}.$$

It is seen from the expression for $\bar{n}$ that now the average number of pulses per unit time is

$$n_1 = \frac{\kappa}{\alpha} = \frac{n_1^0}{\alpha}.$$

If we introduce into the general expressions (3.56) the density of pulses κ, corresponding to the absence of correlations, (3.56) may be rewritten as

$$\bar{n} = \gamma\kappa\Gamma, \quad \overline{n^2} - \bar{n}^2 = \Gamma^2 \bar{n}, \tag{3.58}$$

where

$$\gamma = \frac{1}{\kappa\bar{\vartheta}}, \quad \Gamma^2 = \frac{\overline{\vartheta^2}}{\bar{\vartheta}^2} - 1. \tag{3.59}$$

In the example under consideration this reduces to

$$\gamma = \frac{1}{\alpha}, \quad \Gamma^2 = \frac{1}{\alpha}.$$

Expressions similar to (3.58) are common in calculations of shot noise when the contribution of space charge is taken into account. The quantity Γ^2 is known as the *coefficient of depression* when (as in the case of $\Gamma^2<1$) it describes the depression of the shot noise as compared with the case without space charge. Calculations of Γ^2 for various physically interesting conditions are readily available in the literature. The theory, due to *Domb*, indicates that γ and Γ^2 are generally expressed in terms of the correlation between events, so that a physical theory for the depression of the shot effect may be reduced to the problem of finding the form of the distribution of $g(\vartheta)$ [or $\varphi(\vartheta)$ or $W(\vartheta)$][1]. The depression occuring in this effect implies that the space charge produces a repulsive correlation between anode-current pulses.

(2) A particle counter saturates after the passage of a particle and recovers within a constant time interval β. During this interval the probability of registering a particle is zero, and thereafter the distribution $\varphi(\vartheta)d\vartheta = \lambda d\vartheta$ is uniform. Thus

$$g(\vartheta) = \begin{cases} 0 & \text{for } \vartheta<\beta, \\ \lambda e^{-\lambda(\vartheta-\beta)} & \text{for } \varphi>\beta. \end{cases}$$

This gives

$$\overline{\vartheta} = \beta + \frac{1}{\lambda}, \quad \overline{\vartheta^2} - \overline{\vartheta}^2 = \frac{1}{\lambda^2},$$

therefore, by (3.59), we have

$$\gamma = \frac{1}{1+\lambda\beta}, \quad \Gamma^2 = \frac{1}{(1+\lambda\beta)^2},$$

i.e., always $\Gamma^2<1$, as is to be expected.

(3) Let

$$g(\vartheta) = \beta\delta(\vartheta) + (1-\beta)\lambda e^{-\lambda\vartheta} \quad .$$

The first term suggests that there are *finite* probabilities $\beta, \beta^2, \ldots$ of the joint occurrence of two, three, etc., events. In this case

$$\overline{\vartheta} = \frac{1-\beta}{\lambda}, \quad \overline{\vartheta^2} - \overline{\vartheta}^2 = \frac{1-\beta^2}{\lambda^2}$$

is obtained and hence

$$\gamma = \frac{1}{1-\beta}, \quad \Gamma^2 = \frac{1+\beta}{1-\beta}.$$

Here, $\Gamma^2>1$ always.

[1] If the probability of a pulse depends on the time of occurrence of not one, but r preceding pulses, the relations (3.58) are still valid, but the derivation of γ and Γ^2 requires both means $\overline{\vartheta}_1$ and $\overline{\vartheta_1^2}$, and the joint moments for the interval ϑ_1 and r succeeding intervals $\overline{\vartheta_1\vartheta_i}$ $(i = 2, 3, \ldots, r)$. For $i>r$ the interval lengths become uncorrelated, and thus $\overline{\vartheta_1\vartheta_i} = \overline{\vartheta}_1^2$.

Now we turn our attention to the occurrence of pulses at *independent* times but with a nonuniform distribution of t_ν. The probability $n_1(t)dt$ of the occurrence of a pulse in the interval $(t, t+dt)$ is then still independent of whether any pulses (and how many) were available before time t, but it does depend on this time. It should be noted that t is measured from some fixed origin $t = 0$. This clearly introduces a certain degree of order: the density of impacts depends explicitly on t, and thereby the process becomes nonstationary. How will the expressions for the mean, the mean-square values, and the covariance of the pulse process $\xi(t)$ change?

It may be shown that in this case the probability of occurrence of n pulses within the interval $(0, T)$ still obeys the Poisson law with the parameter

$$\bar{n} = \int_0^T n_1(t)dt$$

(see Exercise 3.7.10). However, the probability $w_t(t)dt$ of a pulse occuring with certainty in the interval $(0, T)$, and lying within $(t, t+dt)$ is,

$$w_t(t)dt = \frac{n_1(t)dt}{\bar{n}}.$$

(At $n_1(t) = n_1^0 = \text{const}$ we come back to the distribution dt/T.) With these deviations we should now repeat the derivations previously made for constant n_1 in Sects. 3.3 and 3.4. So, for example, we now have

$$\langle F(t-t_\nu)F(t+\tau-t_\mu)\rangle$$
$$= \begin{cases} \dfrac{1}{\bar{n}^2}\displaystyle\int_{-\infty}^{+\infty} F(\theta)n_1(t-\theta)d\theta \int_{-\infty}^{+\infty} F(\theta'+\tau)n_1(t-\theta')d\theta' & (\nu \neq \mu), \\ \dfrac{1}{\bar{n}}\displaystyle\int_{-\infty}^{+\infty} F(\theta)F(\theta+\tau)n_1(t-\theta)d\theta & (\nu = \mu). \end{cases}$$

This gives

$$\overline{\xi(t)} = \int_{-\infty}^{+\infty} n_1(t-\theta)\langle F(t, \boldsymbol{a})\rangle d\theta, \tag{3.60}$$

$$\begin{aligned} \psi(t,\tau) &= \overline{\xi(t)\xi(t+\tau)} - \overline{\xi(t)}\cdot\overline{\xi(t+\tau)} \\ &= \int_{-\infty}^{+\infty} n_1(t-\theta)\langle F(\theta, \boldsymbol{a})F(\theta+\tau, \boldsymbol{a})\rangle d\theta. \end{aligned} \tag{3.61}$$

Hence we have, in particular

$$D[\xi(t)] = \psi(t,0) = \overline{\xi^2(t)} - \overline{\xi(t)}^2 = \int_{-\infty}^{+\infty} n_1(t-\theta)\langle F^2(\theta,\boldsymbol{a})\rangle d\theta. \tag{3.62}$$

According to (3.60,62) the mean and variance here are not constant but vary with t, which is indicative of the nonstationary character of the process. The covariance (3.61) that characterizes the statistical connection between $\xi(t)$ and $\xi(t+\tau)$ now depends not only on the time shift τ, but also on the time t. Clearly, if we let $n_1(t) = n_1 = \text{const}$, the formulas reduce to those derived above. The generalization to nonstationary pulse processes has been treated extensively in the literature (e.g., [3.8]).

Suppose that $n_1(t)$ is an almost periodic function of time (periodic being a special case). Then the same almost periodic (or periodic) character will also manifest itself in the moments (3.60–62) of the nonstationary process under discussion.

If during one pulse, i.e, in period ϑ, the function $n_1(t)$ oscillates many times, i.e., $n_1(t)$ varies rapidly compared with $F(t)$, then only the mean $n_1(t-\theta)$ which is equal to $\overline{n}/T = \tilde{n}_1$, contributes to the integrals. We then have

$$\overline{\xi(t)} = \tilde{n}_1 \int_{-\infty}^{+\infty} \langle F(\theta,\boldsymbol{a})\rangle d\theta, \qquad \psi(\tau) = \tilde{n}_1 \int_{-\infty}^{+\infty} \langle F(\theta,\boldsymbol{a})F(\theta+\tau,\boldsymbol{a})\rangle d\theta,$$

i.e., the same expression as for the stationary process with the time-averaged value of the number of pulses per unit time, $\tilde{n}_1$.

Such a situation occurs, for instance, in the intermediate-frequency (i.f.) amplifier. Heterodyning changes the parameters of the noisy mixer with high frequency ω_0 and therefore with this frequency the density $n_1(t)$ of shot-noise pulses varies, too. The pulse duration ϑ at the output of the i.f. amplifier is known to be determined by its bandwith $\Delta\omega \ll \omega_0$ $(\vartheta \sim 1/\Delta\omega)$. Thus, $\vartheta \gg 1/\omega_0$ and the noise at the i.f. amplifier output may be regarded as stationary.

By contrast, if $n_1(t)$ varies very slowly, i.e., during the pulse duration ϑ the function $n_1(t)$ remains essentially unchanged, then in the integrals we may assume $n_1(t-\theta) = n_1(t)$ and factor it out. We then obtain the *quasi-stationary* variation of statistical characteristics of the process

$$\bar{\xi}(t) = n_1(t) \int_{-\infty}^{+\infty} \langle F(\theta,\boldsymbol{a})\rangle d\theta,$$

$$\psi(t,\tau) = n_1(t) \int_{-\infty}^{+\infty} \langle F(\theta,\boldsymbol{a})F(\theta+\tau,\boldsymbol{a})\rangle d\theta.$$

At all times t the process behaves as if it were stationary with instantaneous values of the pulse density at respective times. This is the case, for example, when noise is transmitted through a linear system with slowly varying parameters.

In the remainder of this text we shall be dealing repeatedly with random processes constituting a random sequence of pulses, but calling upon the techniques of the theory of random functions. Before we go on to discuss these techniques, we shall look at the question of how the asymptotic result derived in Sect. 3.3 is related to the central limit theorem of probability theory.

3.6 Impulse Noise and the Central Limit Theorem

In probability theory the *central limit theorem* refers to a family of theorems of varying degrees of generality which are concerned with the limiting distribution of a sum of independent random variables as the number of terms increases indefinitely.

We have seen (Sect. 3.2) that the sum of an *arbitrary* number of normally distributed variables obeys Gauss' law. The central limit theorem leads us beyond this "normal family". Referring to the specific example of a random process with Poisson pulses (Sect. 3.3), we have seen that the distribution of the sum $\xi(t) = \sum_\nu x_\nu$ tends to a normal one as the density of noise grows, i.e., as the number of independent pulses $x_\nu = F(t - t_\nu, \boldsymbol{a}_\nu)$ that superimpose at each given instant of time t increases. The terms x_ν are distributed not according to a Gaussian, but to some special law which is the *same* for all the x_ν. This follows from the distributions of the random parameters $\boldsymbol{a}_\nu$ and t_ν.

Another example of the same type is provided by the De Moivre-Laplace theorem (Sect. 2.6). It may conveniently be written in the form of an arithmetic mean (i.e., as the sum of independent random variables, $\varepsilon_i = 1$ or 0):

$$\varsigma_N = \frac{1}{N}\sum_{i=1}^{N} \varepsilon_i = \frac{n}{N}.$$

According to the theorem, the distribution of the normalized sum ς_N tends to a normal distribution [when the number N of independent random variables ε_i increases without limit]. Here all the summands are distributed, as in the case of the pulse process, in the *same* manner, but now the distribution is a simple alternative: $\varepsilon_i = 1$ or 0 with probabilities p and $q = 1 - p$.

Will the normal distribution of the sum be a limiting one given other distributions of summands or even different distributions for each summand? If so, then how rigid must the limitations obeyed by the term distributions be? It is these questions that are answered by the central limit theorem.

These questions are of importance as they concern the statistics of those physical quantities which are influenced by many independent random factors, each of which causes only a slight random deviation. This situation is encountered in measurements, in fluctuations of macroscopic quantities due to the random character of locations and motions of an enormous number of microparticles (molecules, ions, electrons), and so on. The problem even attracted the attention of Gauss. However, Chebyshev, Markov, and Lyapunov were the first to solve it by showing that it suffices to impose a fairly general restriction on the random summands. This restriction is essentially that the individual terms should only have a small influence on the sum, their effect being the smaller, the larger N.

A rigorous proof of the theorem was in 1898 given by Markov, and in a more general form, in 1900, by Lyapunov. It is the Lyapunov theorem with its further generalizations that is known as the *central limit theorem* of probability theory.

We now illustrate the proof by referring to the very simple example of the *Lindeberg-Levy* theorem. This theorem (proved later than the Lyapunov theorem) contains an extremely strong constraint: it is assumed that *all the terms* x_ν have the same distribution. The theorem states: let x_ν be independent random variables with the same distribution function $W(x)$, with $\overline{x}_\nu = 0$, $\overline{x_\nu^2} = 1$. Then, the distribution of the sum

$$S_N = \frac{1}{\sqrt{N}} \sum_1^N x_\nu$$

tends for all x uniformly to

$$\Phi(x) = \frac{1}{\sqrt{2\pi}} \int_{-\infty}^{x} e^{-x^2/2} dx,$$

as $N \to \infty$, that is

$$[W(x\sqrt{N})]^{N*} \to \Phi(x),$$

and hence

$$\left[W\left(\frac{x-a}{\sigma}\right)\right]^{N*} \to \Phi\left(\frac{x-Na}{\sqrt{N}\sigma}\right)$$

(at any fixed a and σ). This is of course, a special case of the central limit theorem. However, unlike the Lyapunov theorem, x_ν here is not supposed to possess absolute moments of the order $k>2$ (see below).

Using characteristic functions the theorem is proven in a straightforward manner. Let the characteristic function $\varphi(u)$ correspond to $W(x)$. As stated previously $\alpha_1 = \overline{x} = 0$, $\alpha_2 = \overline{x^2} = 1$. Consequently

$$\varphi(u) = 1 - \frac{u^2}{2} + o(u^2), \quad \text{whence}$$

$$\varphi\left(\frac{u}{\sqrt{N}}\right) = 1 - \frac{u^2}{2N} + o(N^{-3/2}).$$

For the sum

$$S_N = \frac{1}{\sqrt{N}} \sum_1^N x_\nu$$

the characteristic function is

$$\left[\varphi\left(\frac{u}{\sqrt{N}}\right)\right]^N = \left[1 - \frac{u^2}{2N} + o(N^{-3/2})\right]^N .$$

As $N \to \infty$,this function tends in each finite interval of u uniformly to $\exp(-u^2/2)$, i.e., to the characteristic function for $\Phi(x)$. This is in line with the statement of Levy's theorem: if a sequence of characteristic functions tends to $\varphi(u)$, then the sequence of corresponding distribution functions tends to the $W(x)$ corresponding to $\varphi(u)$.

The Lyapunov theorem is more general as far as the sum of *differently distributed* independent random variables is concerned. If we assume (without loss of generality) that $\bar{x}_\nu = 0$ and the variances $\sigma_\nu^2 = \overline{x_\nu^2}$ are *finite*, then the distribution of the normalized sum

$$S_N = \frac{1}{B_N} \sum_1^N x_\nu$$

where $B_N^2 = \sum_1^N \sigma_\nu^2$ is the variance of the sum $\sum_1^N x_\nu$, tends under the appropriate conditions to $\Phi(x)$ as $N \to \infty$. It is just these conditions that are crucial.

Of special significance is the case in which for $N \to \infty$ we have $B_N \to \infty$, but with $\sigma_\nu / B_N \to 0$, i.e., each individual summand σ_ν^2 accounts for an infinitesimally small fraction of the total variance B_N^2 that grows infinitely with N. A *sufficient* condition here is the one used in the Lyapunov theorem. Denote by $[C_{kN}]^k$ the sum of the kth *absolute* moments x_ν

$$[C_{kN}]^k = \sum_1^N \overline{|x_\nu|^k} \; . \tag{3.63}$$

The Lyapunov condition can then be written as

$$\lim_{N \to \infty} \frac{C_{kN}}{B_N} = 0 \quad \text{at} \quad k>2, \tag{3.64}$$

i.e., it is necessary that the variables x_ν have absolute moments of an arbitrary order $k>2$ and the kth root of the sum of these moments grows for $N \to \infty$ more slowly than the square root of the sum of the variances. If the condition (3.64) is met, then the distribution $F_N(x)$ of the normalized sum S_N tends for $N \to \infty$ to the normal law $\Phi(x)$. Using condition (3.64) Lyapunov has not only proven his theorem, but also estimated the quantity $|F_N(x) - \Phi(x)|$, i.e., how fast it tends to zero with N (no slower than $C \ln N / \sqrt{N}$; Cramér has refined this estimate: at somewhat more rigid conditions the modulus of the difference $F_N(x)$ and $\Phi(x)$ is less than $C/\sqrt{N}$).

In the case of $B_N \to \infty$ and $\sigma_\nu / B_N \to 0$ for $N \to \infty$, the most general (*necessary* and *sufficient*) condition for the validity of the central limit theorem is the Lindeberg condition, see [Ref. 3.9, p. 73]. However, we will confine ourselves to the statement of the theorem subject to the Lyapunov condition since this will suffice for the majority of problems encountered in physics. Moreover, as a rule it is sufficient to use Lyapunov's condition at $k = 3$, i.e., the condition may be used in the form

$$C_{3N}/B_N \to 0 \quad \text{as} \quad N \to \infty. \tag{3.65}$$

For noise theory the Lyapunov condition implies that there are no large *rare pulses* (unless the pulses themselves are subject to normal distribution), i.e., at any moment of time many similar (in the sense of their relative smallness) random independent variables occur. Under these conditions, valid for a wide class of noises, the noise distribution is normal. But noises may differ substantially with regard to their other properties and in particular in their spectra, while having the same distribution. For noise resulting from the superposition of independent pulses, these differences depend on the individual terms and on the shapes of pulses to be summed up.

It should be noted that the above derivation of condition (3.32), where the distribution of random pulse superposition may be approximated by the normal distribution, is in fact an application of Lyapunov's condition (3.65). In this problem $n_1\vartheta$ just corresponds to the number N of random variables x_ν that define the instantaneous value of the sum $\sum_\nu x_\nu$.

3.7 Exercises

3.7.1. The logarithmic normal distribution (lognormal, for short) of a random variable $x \geq 0$ refers to the distribution

$$w(x)dx = \frac{dx}{\sqrt{2\pi}\sigma x} \exp\left\{-\frac{(\ln x - \lambda)^2}{2\sigma^2}\right\}, \tag{3.66}$$

where $\lambda = \langle \ln x \rangle$ and $\sigma^2 = D[\ln x]$. Equation (3.66) is thus a standard normal distribution of a normalized random variable $y = (\ln x - \lambda)/\sigma$, i.e.,

$$w(x)dx = \Phi(y)dy, \quad \Phi(y) = \frac{1}{\sqrt{2\pi}} e^{-y^2/2} \quad . \tag{3.67}$$

Use the transformation

$$x = e^{\sigma y + \lambda}$$

from x to y to find the moments $\alpha_k = \overline{x^k}$.

Solution. We have

$$\alpha_k = \overline{x^k} = \int_0^{\infty} x^k w(x)dx = \int_{-\infty}^{+\infty} e^{k(\sigma y+\lambda)}\Phi(y)dy = e^{k\lambda+k^2\sigma^2/2} \quad ,$$

or, as $\alpha_1 = \exp(\lambda + \sigma^2/2)$,

$$\alpha_k = e^{k(k-1)\sigma^2/2}\alpha_1^k.$$

For the variance of x we obtain

$$D[x] = \alpha_2 - \alpha_1^2 = (e^{\sigma^2} - 1)\alpha_1^2.$$

With $\lambda = \langle \ln x \rangle \equiv \ln x_0$ and $\exp(\sigma^2/2) = \varrho$, this yields

$$\alpha_1 = x_0\varrho, \ D[x] = (\varrho^2 - 1)\alpha_1^2, \ \alpha_k = \varrho^{k(k-1)}\alpha_1^k.$$

The lognormal distribution is the limiting distribution for the *product* of independent random variables z_k with an increasing number of terms, n. If $x = \prod_{k=1}^n z_k$, then $\ln x = \sum_{k=1}^n \ln z_k$, i.e., $\ln x$ is the sum of n independent random variables. If the conditions of the central limit theorem for $\ln x$ are satisfied (Sect. 3.6), the distribution of $\ln x$ must tend to a normal distribution.

This theorem also implies that the distribution of the *sum* of lognormal variables x_k, i.e., the quantity $X = \sum_{k=1}^n x_k$, tends with increasing n to the normal one. However, in the case of lognormal terms the distribution of the variable $Y = (X - \overline{X})/\sqrt{D[X]}$ tends very slowly to $\Phi(Y)$. Therefore, there is for large n a wide range of ϱ in which the distribution of Y is better approximated by the lognormal law than by the Gaussian, which it approaches as n increases still further. Appropriate estimates are examined in some detail in [3.10].

3.7.2. Express the characteristic function of the random variable $y = ax + b$ in terms of the characteristic function of the random variable x.
Solution. We have

$$\varphi_y(u) = \overline{e^{iuy}} = e^{ibu}\overline{e^{iaux}} = e^{ibu}\varphi_x(au).$$

In particular for the normalized quantity $y = (x - \overline{x})/\sigma_x$

$$\varphi_y(u) = \exp(-i\overline{x}u/\sigma_x)\varphi_x\left(\frac{u}{\sigma_x}\right).$$

3.7.3. Find the characteristic function of the Cauchy distribution

$$dW(x) = \frac{dx}{\pi(1 + x^2)}.$$

Solution. For this distribution, from equation (3.10), we get

$$\varphi_x(u) = \frac{1}{\pi}\int_{-\infty}^{+\infty} \frac{e^{iux}dx}{1+x^2} = \frac{2}{\pi}\int_0^{\infty} \frac{\cos(ux)}{1+x^2}dx = e^{-|u|} \quad .$$

A vertex at $u = 0$ makes it impossible to differentiate $\varphi_x(u)$ at zero, i.e., to calculate moments using (3.13).

3.7.4. Find the characteristic function of the random variable $\xi_n = n - pN$ for the binomial distribution of n and pass to the limit of characteristic functions of the Poisson and Laplace distributions.
Solution. We have

$$\begin{aligned}\varphi_{\xi_n}(u) &= \langle e^{iu(n-pN)}\rangle = e^{-iupN}\sum_{n=0}^{N} e^{iun}P_N(n)\\ &= e^{-iupN}\sum_{n=0}^{N} C_N^n(pe^{iu})^n q^{N-n} = e^{-iupN}(pe^{iu}+q)^N\\ &= (pe^{iqu}+qe^{-ipu})^N .\end{aligned}$$

For $p = \bar{n}/N$, and hence $q = 1 - \bar{n}/N$ we arrive at the characteristic function of the quantity $n - \bar{n}$ for the Poisson distribution of n, i.e.,

$$\begin{aligned}\lim_{N\to\infty}\varphi_{\xi_n}(u) &= \lim_{N\to\infty} e^{-i\bar{n}u}\left[1+\frac{\bar{n}}{N}(e^{iu}-1)\right]^N\\ &= e^{-i\bar{n}u}\exp\left[\bar{n}(e^{iu}-1)\right] \quad .\end{aligned}$$

For the normalized variable $x_n = (n-pN)/\sqrt{Npq} = \xi_n/\sqrt{Npq}$ the characteristic function is

$$\varphi_{x_n}(u) = \varphi_{\xi_n}\left(\frac{u}{\sqrt{Npq}}\right) = \left[p\,\exp\left(iu\sqrt{q/Np}\right) + q\,\exp\left(-iu\sqrt{p/Nq}\right)\right]^N .$$

Writing the exponential functions as a power series in u, it will be seen that as $N\to\infty$ we obtain $\varphi_{x_n}(u) = \exp(-u^2/2)$, i.e., the characteristic function of the normal distribution for the normalized random variable x_n.

3.7.5. Find the characteristic function for the n-variate normal distribution law.
Solution. By virtue of (2.26) and (3.14) we have

$$\varphi_r(\boldsymbol{u}) = \frac{1}{(2\pi)^{n/2}|B|^{1/2}}\int_{-\infty}^{+\infty} \exp\left\{i\boldsymbol{u}\cdot\boldsymbol{r} - \sum_{j,k} b_{jk}^{-1}x_j x_k\right\}dx_1\ldots dx_n,$$

where $\boldsymbol{u}\cdot\boldsymbol{r} = \sum_{k=1}^{n} u_k x_k$. The successive integration with respect to x_1, x_2, etc., gives

$$\varphi_r(u) = \exp\left\{-\frac{1}{2}\sum_{j,k} b_{jk}u_ju_k\right\} \quad ,$$

which may be written as

$$\sum_{j,k=1}^{n} b_{jk}u_ju_k = \sum_{j,k=1}^{n} \overline{x_jx_k}u_ju_k = \overline{\left(\sum_{k=1}^{n} x_ku_k\right)^2} = \overline{(\boldsymbol{u}\cdot\boldsymbol{r})^2}.$$

Consequently, for the normal distribution we arrive at

$$\varphi_r(\boldsymbol{u}) = \langle e^{i\boldsymbol{u}\cdot\boldsymbol{r}}\rangle = e^{-\langle(\boldsymbol{u}\cdot\boldsymbol{r})^2\rangle/2} \quad .$$

3.7.6. Random variables $x_k (k = 1, 2, \ldots)$ with $\overline{x}_k = 0$ are normally distributed. Use the characteristic function of the normal multivariate distribution

$$\varphi(\boldsymbol{u}) = e^{-S(\boldsymbol{u})} \quad , \quad S(\boldsymbol{u}) = \frac{1}{2}\sum_{j,k} b_{jk}u_ju_k, \; b_{jk} = b_{kj} = \overline{x_jx_k},$$

to show that odd moments are zero, i.e.,

$$\langle x_{i_1}x_{i_2}\ldots x_{i_{2n+1}}\rangle = 0, \tag{3.68}$$

whereas even moments obey the expression

$$\langle x_{i_1}x_{i_2}\ldots x_{i_{2n}}\rangle = \frac{1}{2^n n!}\sum b_{i_1i_2}b_{i_3i_4}\ldots b_{i_{2n-1},i_{2n}} \quad , \tag{3.69}$$

where i_k are positive integers, and the sum is taken over all the permutations of subscripts $i_1, \ldots, i_{2n}$.

Solution. Formulas (3.68) and (3.69) are derived using the general rule:

$$\left\langle x_{i_1}x_{i_2}\ldots x_{i_m}\right\rangle = \left.\frac{\partial^m\varphi(\boldsymbol{u})}{i^m\partial u_{i_1}\partial u_{i_2}\ldots\partial u_{i_m}}\right|_{n=0} \quad .$$

The number of permutations of subscripts in (3.69) is $(2n)!$, but some of the resulting terms coincide. This implies that the value of each term does not change under (the $n!$ possible) permutations of multipliers. Moreover, each multiplier remains unchanged under (the 2^n possible) permutations of its subscripts due to the symmetry of b_{jk}. The summands in (3.69) thus break up into groups with $2^n n!$ identical terms. The number of these groups is

$$\frac{(2n)!}{2^n n!} = 1\cdot 3\cdot 5\ldots(2n-1) \equiv (2n-1)!!$$

Therefore, (3.69) may be written as the sum of $(2n-1)!!$ *different* terms

$$\left\langle x_{i_1}x_{i_2}\ldots x_{i_{2n}}\right\rangle = \sum_{\text{d.p.}} b_{i_1i_2}b_{i_3i_4}\ldots b_{i_{2n-1},i_{2n}} \quad ,$$

where $\sum_{\text{d.p.}}$ is taken to be the sum over *different permutations* yielding different terms (considering the symmetry of b_{ij}). For example,

$$\langle x_1x_2x_3x_4\rangle = b_{12}b_{34} + b_{13}b_{24} + b_{14}b_{23} \quad ,$$

$$\begin{aligned}\langle x_1x_2x_3x_4x_5x_6\rangle = b_{12}&(b_{34}b_{56} + b_{35}b_{46} + b_{36}b_{45}) \\ + b_{13}&(b_{24}b_{56} + b_{25}b_{46} + b_{26}b_{45}) \\ + b_{14}&(b_{23}b_{56} + b_{25}b_{36} + b_{26}b_{35}) \\ + b_{15}&(b_{23}b_{46} + b_{24}b_{36} + b_{26}b_{34}) \\ + b_{16}&(b_{23}b_{45} + b_{24}b_{35} + b_{25}b_{34}).\end{aligned}$$

In particular, if $i_1 = i_2 = \ldots = i_{2n}$, we obtain

$$\langle x^{2n}\rangle = (2n-1)!!\langle x^2\rangle^n$$

If $\overline{x}_k \neq 0$, then the above expressions are the *central* moments

$$\langle (x_{i_1} - \overline{x_{i_1}})(x_{i_2} - \overline{x_{i_2}})\ldots(x_{i_m} - \overline{x_{i_m}})\rangle .$$

3.7.7. Show that for the deterministic function $f(x)$ of a Gaussian random variable x the formula

$$\langle f(x)\rangle = \sum_{n=0}^{\infty} f^{(2n)}(\overline{x})\frac{(\sigma^2/2)^n}{n!}, \tag{3.70}$$

with $\sigma^2 = D[x] = \overline{x^2} - \overline{x}^2$ is valid.
Solution. Equation (3.70) results from the averaging of the expansion of $f(x)$ in a Taylor series of $x - \overline{x}$, using the values of the central moments $\langle (x-\overline{x})^k\rangle$ of the Gaussian variable x (see the preceding exercise).

3.7.8. Let a random variable x be distributed as

$$w(x)dx = A(a)F(x)e^{af(x)}dx, \tag{3.71}$$

where $A(a)$ is a normalization constant. Show that for the cumulants of the random variable $f(x)$ the following expression holds:

$$\lambda_n(a) = \frac{d^{n-1}\lambda_1(a)}{da^{n-1}}. \tag{3.72}$$

Solution. The characteristic function for $f(x)$ is

$$\varphi_f(iu) = \langle e^{iuf(x)}\rangle = A(a)\int F(x)e^{(a+iu)f(x)}dx,$$

where the integration covers the whole range of possible values of x. Hence we obtain

$$\begin{aligned}\ln\varphi_f(iu) &= \ln A(a) + \ln\int F(x)e^{(a+iu)f(x)}dx \\ &= \ln A(a) + \mathcal{F}(a+iu).\end{aligned}$$

For the first cumulant $f(x)$ we have

$$\begin{aligned}\lambda_1(a) \equiv \langle f(x)\rangle &= \left.\frac{\partial \ln\varphi_f(iu)}{\partial iu}\right|_{u=0} = \left.\frac{\partial\mathcal{F}(a+iu)}{\partial iu}\right|_{u=0} \\ &= \left.\frac{\partial\mathcal{F}(a+iu)}{\partial a}\right|_{u=0} = \frac{d\mathcal{F}(a)}{da}.\end{aligned}$$

Thus, the nth cumulant is

$$\lambda_n(a) = \left.\frac{\partial^n \ln\varphi_f(iu)}{\partial(iu)^n}\right|_{u=0} = \left[\frac{\partial^{n-1}}{\partial a^{n-1}}\frac{\partial\mathcal{F}(a+iu)}{\partial iu}\right]_{u=0} = \frac{d^{n-1}\lambda_1(a)}{da^{n-1}}.$$

Note that (3.71) is, in particular, the Gibbs distribution for a system with the Hamiltonian $H(x) = H_0(x) - bf(x)$,

$$w(x)dx = A\,\exp\{-\beta[H_0(x) - bf(x)]\}dx \qquad (\beta = 1/kT),$$

i.e., in this case $F(x) = \exp[-\beta H_0(x)]$ and $a = \beta b$.

3.7.9. Find the characteristic function and distribution density of the random variable $z = f(\boldsymbol{x})$, where $\boldsymbol{x}$ is an n-dimensional random variable with the distribution $w_n(\boldsymbol{x})d\boldsymbol{x} = w_n(x_1,\ldots,x_n)\cdot dx_1\ldots dx_n$, and f is a deterministic function.

Solution. We have

$$\varphi_z(u) = \langle e^{iuz}\rangle = \langle e^{iuf(\boldsymbol{x})}\rangle = \int e^{iuf(\boldsymbol{x})}w_n(\boldsymbol{x})d\boldsymbol{x}$$

and consequently

$$\begin{aligned}w(z) &= \frac{1}{2\pi}\int_{-\infty}^{+\infty} e^{-iuz}\varphi_z(u)du = \left\langle\frac{1}{2\pi}\int_{-\infty}^{+\infty} e^{-iu[z-f(\boldsymbol{x})]}du\right\rangle \\ &= \langle\delta[z - f(\boldsymbol{x})]\rangle,\end{aligned}$$

where the latter becomes more transparent when written in the explicit form

$$w(z) = \int\delta[z - f(\boldsymbol{x})]w_n(\boldsymbol{x})d\boldsymbol{x}.$$

The delta-function retains from the set of possible values of $\boldsymbol{x}$ only those for which $f(\boldsymbol{x})$ has the value z. Thus $w(z)dz$ gives the probability measure of the interval $(z, z + dz)$ for z.

3.7.10. Show that for the time averaged shot current (3.28) the variance is given by (2.10), if the averaging time T is notably longer than the pulse duration ϑ.

Solution. According to (3.28)

$$D[I_T] = \overline{(I_T - \overline{I}_T)^2} = \frac{1}{T^2} \int\limits_0^T\!\!\int [\overline{I(t)I(t')} - \overline{I(t)} \cdot \overline{I(t')}]dt\,dt'$$

$$= \frac{1}{T^2} \int\!\!\int \psi_I(\tau)dt\,dt',$$

where $\tau = t - t'$. Since the covariance $\psi_I(\tau)$ is an even function of τ, we may write

$$D[I_T] = \frac{2}{T^2} \int\limits_0^T dt \int\limits_t^T dt' \psi_I(t - t') = \frac{2}{T^2} \int\limits_0^T dt \int\limits_0^{T-t} \psi_I(\tau)d\tau.$$

The covariance for the instantaneous current $I(t)$ is described by (3.44), if we put $a = e$ and assume the pulse area to be unity, i.e.,

$$\psi_I(\tau) = n_1 e^2 \int\limits_{-\infty}^{+\infty} F(\theta)F(\theta + \tau)d\theta, \qquad \int\limits_{-\infty}^{+\infty} F(\theta)d\theta = 1.$$

It suffices to compute $D[I_T]$ for some simple pulse shape, e.g., for a square pulse (Fig. 3.1). In this case we obtain the "triangular" covariance

$$\psi_I(\tau) = \begin{cases} \dfrac{e\overline{I}}{\vartheta}\left(1 - \dfrac{|\tau|}{\vartheta}\right) & \text{for} \quad |\tau| < \vartheta, \\ 0 & \text{for} \quad |\tau| < \vartheta. \end{cases}$$

Inserting this into the double integral for $D[I_T]$ we get

$$D[I_T] = \begin{cases} \dfrac{e\overline{I}}{T}\left(1 - \dfrac{\vartheta}{3T}\right) & \text{for} \quad T > \vartheta, \\ \dfrac{e\overline{I}}{\vartheta}\left(1 - \dfrac{T}{3\vartheta}\right) & \text{for} \quad T < \vartheta. \end{cases}$$

Thus the value of the variance (2.10) is attained as the averaging time T becomes longer. In practice, it is sufficient for T to be just a bit longer than the pulse duration ϑ.

3.7.11. Use the probabilities P_{ev} and P_{odd} derived in Sect. 3.4 to compute the bivariate characteristic function and probability density for a random telegraph signal.

Solution. If at a time t the values $\pm a$ have a probability of $1/2$, then (assuming $a = 1$) the characteristic function of a bivariate distribution is given by

$$\begin{aligned}\varphi_2(u_1,u_2,\tau) &= \exp\left[\overline{i(u_1\xi + u_2\xi_\tau)}\right] \\ &= \frac{P_{\text{ev}}}{2}\{\exp[i(u_1+u_2)] + \exp[-i(u_1+u_2)]\} \\ &+ \frac{P_{\text{odd}}}{2}\{\exp[i(u_1-u_2)] + \exp[-i(u_1-u_2)]\} \\ &= e^{-\bar{n}}[\cosh\bar{n}\cdot\cos(u_1+u_2) + \sinh\bar{n}\cdot\cos(u_1-u_2)] \quad .\end{aligned}$$

Using the inverse Fourier transform we obtain

$$\begin{aligned}w_2(\xi,\xi_\tau,\tau) &= \frac{1}{4\pi^2}\iint\limits_{-\infty}^{+\infty}\exp[-i(u_1\xi + u_2\xi_\tau)]\varphi_2(u_1,u_2,\tau)du_1\,du_2 \\ &= \frac{1+e^{-2\bar{n}}}{4}[\delta(\xi-1)\delta(\xi_\tau-1) + \delta(\xi+1)\delta(\xi_\tau+1)] \\ &+ \frac{1-e^{-2\bar{n}}}{4}[\delta(\xi-1)\delta(\xi_\tau+1) + \delta(\xi+1)\delta(\xi_\tau-1)] \quad ,\end{aligned}$$

where $\bar{n} = n_1\tau$.

3.7.12. In [3.11] a class of bivariate *elliptically symmetric* distributions has been described that generalizes the bivariate Gauss' law for variables x_1 and x_2 depicting the same variances. For $\sigma_1^2 = \sigma_2^2 = \sigma^2$ the Gaussian probability density (2.28) has the form

$$w_2(x_1,x_2) = \frac{1}{2\pi\sigma^2\sqrt{1-K^2}}\exp\left\{-\frac{r^2}{2\sigma^2(1-K^2)}\right\} \quad , \tag{3.73}$$

i.e., its dependence on x_1 and x_2 is owing to

$$r = \sqrt{x_1^2 + x_2^2 - 2Kx_1x_2}.$$

Elliptically symmetric distributions w_2, also depend solely on r :

$$w_2(x_1,x_2) = F(r);$$

but here F is *any* non-negative function normalized to unity. Such distributions, like their special case (3.73), are useful when dealing with a number of problems related to the transmission of the signals and noise in memoryless nonlinear systems.

Show that the bivariate characteristic function for elliptically symmetric distributions may be written as the modified Bessel transform of $F(r)$,

$$\varphi_2(u_1,u_2) \equiv \tilde{\varphi}(s) = \frac{2\pi}{\sqrt{1-K^2}}\int\limits_0^\infty F(r)J_0\left(\frac{sr}{\sqrt{1-K^2}}\right)r\,dr, \tag{3.74}$$

where J_0 is the zero-order Bessel function and

$$s = \sqrt{u_1^2 + u_2^2 + 2Ku_1u_2}.$$

Find the univariate characteristic function.

Solution. By definition we have

$$\varphi_2(u_1, u_2) = \iint_{-\infty}^{+\infty} F(r) \exp[i(u_1x_1 + u_2x_2)] dx_1\, dx_2.$$

The transformation in the (x_1, x_2) plane to the coordinates r and α

$$x_1 - Kx_2 = r \cos\alpha,$$

$$x_2\sqrt{1-K^2} = r \sin\alpha,$$

$$dx_1\, dx_2 = \frac{r\, dr\, d\alpha}{\sqrt{1-K^2}}, \qquad \text{gives}$$

$$\varphi_2(u_1, u_2) = \frac{1}{\sqrt{1-K^2}} \int_0^\infty F(r) r\, dr \int_0^{2\pi} d\alpha$$
$$\times \exp\left\{ir\left[u_1 \cos\alpha + \frac{Ku_1 + u_2}{\sqrt{1-K^2}} \sin\alpha\right]\right\} .$$

Using the well-known relation

$$\int_0^{2\pi} d\alpha \exp\{i[a \cos\alpha + b \sin\alpha]\} = 2\pi J_0\left(\sqrt{a^2 + b^2}\right),$$

we arrive at (3.74).

It follows from the definition of the characteristic function that the univariate function $\varphi_1(u_1)$ may be derived from the bivariate function $\varphi_2(u_1, u_2)$ if $u_2 = 0$ is used in the latter:

$$\varphi_1(u_1) = \varphi_2(u_1, 0).$$

For the elliptically symmetric distributions (3.74) then yields

$$\varphi_1(u_1) = \frac{2\pi}{\sqrt{1-K^2}} \int_0^\infty F(r) J_0\left(\frac{u_1 r}{\sqrt{1-K^2}}\right) r\, dr \equiv \tilde{\varphi}(u_1),$$

i.e., the univariate characteristic function varies with its argument in exactly the same way as the bivariate function varies with s :

$$\varphi_2(u_1, u_2) = \tilde{\varphi}(s) = \varphi_1(s).$$

Let us now specify an even univariate probability density $w_1(x)$ and find the corresponding even characteristic function in s

$$\varphi_1(s) = \int_{-\infty}^{+\infty} e^{isx} w_1(x)dx = 2\int_0^{\infty} w_1(x) \cos sx \cdot dx.$$

Using now the inverse transformation to (3.74), we may obtain the bivariate probability density $w_2(x_1, x_2) = F(r)$

$$F(r) = \frac{1}{2\pi\sqrt{1-K^2}} \int_0^{\infty} \varphi_1(s) J_0\Big(\frac{sr}{\sqrt{1-K^2}}\Big) s\, ds.$$

However, it should be borne in mind that not every even $w_1(x)$ is suitable, as not all of them result in a non-negative and integrable $w_2(x_1, x_2)$.

3.7.13. Show that the Poisson distribution (2.11) for the probability of n events in the fixed time interval $(0, T)$ follows from the exponential distribution (2.12) of the time before the occurrence of an event in a random flow of events.

Solution. Let $t_1, t_2, \ldots$ be the times of occurrence of events $1, 2, \ldots$ counted from $t = 0$. According to (2.12) the probability that event 1 will occur in $(t_1, t_1 + dt_1)$ is

$$u(t_1)dt_1 = \exp(-n_1 t_1) n_1 dt_1, \quad \text{where} \quad 0 \le t_1 < \infty.$$

The probability of event 2 occurring within $(t_2, t_2 + dt_2)$, i.e., by $\tau_2 = t_2 - t_1$ later than event 1, is

$$u(\tau_2)dt_2 = \exp(-n_1 \tau_2) n_1 dt_2, \quad \text{where} \quad t_1 \le t_2 < \infty.$$

This may be repeated up to the $(n+1)$th event occurring in the interval $(t_{n+1}, t_{n+1} + dt_{n+1})$:

$$u(\tau_{n+1})dt_{n+1} = \exp(-n_1 \tau_{n+1}) n_1 dt_{n+1} \quad , \quad \text{with}$$

$$\tau_{n+1} = t_{n+1} - t_n \quad \text{and} \quad t_n \le t_{n+1} < \infty.$$

As the time spans τ_k between events are independent, the joint probability for the occurrence of $n+1$ events at times $t_1 \le t_2 \le \ldots \le t_{n+1}$ is

$$\begin{aligned} w_{n+1}(\tau_1, \ldots, \tau_{n+1}) dt_1 \ldots dt_{n+1} &= \prod_{k=1}^{n+1} u(\tau_k) dt_k \\ &= \exp[-n_1(\tau_1 + \ldots + \tau_{n+1})] n_1^{n+1} dt_1 \ldots dt_{n+1} \\ &= \exp(-n_1 t_{n+1}) n_1^{n+1} dt_1 \ldots dt_{n+1} \quad . \end{aligned}$$

We now look for the probability $P(n)$ of the occurrence of n events in the interval $(0, T)$ with the nth event still occuring inside $(0, T)$, but the $(n+1)$th one beyond the boundaries of $(0, T)$. Therefore, $P(n)$ is equal to the integral of w_{n+1}

over the $(n+1)$-dimensional area bounded by $0 \le t_1 \le t_2 \le \ldots t_n \le T \le t_{n+1} < \infty$:

$$P(n) = \int_T^\infty dt_{n+1} \int_0^T dt_n \int_0^{t_n} dt_{n-1} \ldots \int_0^{t_2} w_{n+1} dt_1$$

$$= \int_T^\infty \exp(-n_1 t_{n+1}) n_1 dt_{n+1} \cdot n_1^n \int_0^T dt_n \int_0^{t_n} dt_{n-1} \ldots \int_0^{t_2} dt_1 .$$

The first integral (with respect to t_{n+1}) is $\exp(-n_1 T)$ and constitutes the probability that the $(n+1)$th event will occur after the instant T. The other integrals yield $(n_1 T)^n / n!$, which is the probability that the n preceding events occurred within $(0,T)$. As a result we get

$$P(n) = \frac{\bar{n}^n}{n!} e^{-\bar{n}} \quad , \quad \text{where} \quad \bar{n} = n_1 T,$$

i.e., the Poisson law (2.11).

3.7.14. Show that the Poisson law with $\bar{n} = \int_0^T n_1(t) dt$ holds for a flow of independent events with a nonuniform distribution $n_1(t) dt$ of the probability of an event to occur within the interval $(t, t+dt)$.

Solution. The probability of the empty interval $(0,t)$ is now $\exp[-\int_0^t n_1(t) dt]$, so that the probability of the occurrence of the nth event taking place in the interval $(t_n, t_n + dt_n)$ after the $(n-1)$th event at t_{n-1} will be

$$u(t_n, t_{n-1}) dt_n = \exp\left\{ - \int_{t_{n-1}}^{t_n} n_1(t_n) dt_n \right\}, \quad \text{where} \quad t_{n-1} \le t_n < \infty .$$

The joint probability of $n+1$ events at times $t_1, t_2, \ldots, t_{n+1}$ is by virtue of their independence $(t_0 = 0)$:

$$w_{n+1}(t_1, \ldots, t_{n+1}) dt_1 \ldots dt_{n+1} = \prod_{k=1}^{n+1} u(t_k, t_{k-1}) dt_k$$

$$= \exp\left\{ - \sum_{k=1}^{n+1} \int_{t_{k-1}}^{t_k} n_1(t) dt \right\} \cdot \prod_{k=1}^{n+1} n_1(t_k) dt_k$$

$$= \exp\left\{ - \int_0^{t_{n+1}} n_1(t) dt \right\} \cdot \prod_{k=1}^{n+1} n_1(t_k) dt_k .$$

Following the same reasoning as in the preceding exercise, we arrive at the expression for the probability of n events in the interval $(0,T)$:

$$P(n) = \int_T^\infty \exp\left\{-\int_0^{t_{n+1}} n_1(t)dt\right\} n_1(t_{n+1})dt_{n+1} \int_0^T n_1(t_n)dt_n$$

$$\times \int_0^{t_n} n_1(t_{n-1})dt_{n-1} \ldots \int_0^{t_2} n_1(t_1)dt_1 = \frac{\overline{n}}{n!}e^{-\overline{n}} \quad ,$$

where

$$\overline{n} = \int_0^T n_1(t)dt \quad .$$

3.7.15. Assuming that the logarithm of the characteristic function of a univariate random variable x defined in the interval $(-\infty, \infty)$ can be expanded into an infinite series

$$\ln \varphi(u) = \sum_{m=1}^{\infty} \frac{\lambda_m}{m!}(iu)^m,$$

show that for the probability density $w(x)$ the following relation is valid:

$$\frac{\partial w(x)}{\partial \lambda_k} = \frac{(-1)^k}{k!} \frac{\partial^k w(x)}{\partial x^k}. \tag{3.75}$$

Solution. We write the characteristic function as

$$\varphi(u) = \exp\{\ln \varphi(u)\} = \exp\left\{\sum_{m=1}^{\infty} \frac{\lambda_m}{m!}(iu)^m\right\} \quad .$$

If the cumulant λ_k is non-zero, this yields the relation

$$\frac{\partial \varphi(u)}{\partial \lambda_k} = \frac{(iu)^k}{k!}\varphi(u).$$

Differentiating (3.17) with respect to λ_k gives

$$\frac{\partial w(x)}{\partial \lambda_k} = \frac{1}{2\pi}\int_{-\infty}^{+\infty} e^{-iux}\frac{\partial \varphi(u)}{\partial \lambda_k}du = \frac{1}{2\pi}\int_{-\infty}^{+\infty} \frac{(iu)^k}{k!} e^{-iux}\varphi(u)du.$$

On the other hand, according to (3.17), we have

$$\frac{\partial^k w(x)}{\partial x^k} = \frac{1}{2\pi}\int_{-\infty}^{+\infty} (-iu)^k e^{-iux}\varphi(u)du.$$

A comparison of the last two equations leads to (3.75).

3.7.16. Use equation (3.75) of the preceding exercise to show that for any function $f(x)$ the following equation holds [3.11]:

$$\frac{\partial^n \langle f(x)\rangle}{\partial \lambda_k^n} = \frac{1}{(k!)^n}\left\langle \frac{d^{nk} f(x)}{dx^{nk}} \right\rangle. \tag{3.76}$$

Solution. Differentiating the equality

$$\langle f(x)\rangle = \int_{-\infty}^{+\infty} f(x) w(x) dx$$

with respect to λ_k and using (3.75) gives

$$\frac{\partial \langle f(x)\rangle}{\partial \lambda_k} = \int_{-\infty}^{+\infty} f(x)\frac{\partial w(x)}{\partial \lambda_k} dx = \frac{(-1)^k}{k!}\int_{-\infty}^{+\infty} f(x)\frac{\partial^k w(x)}{\partial x^k} dx.$$

Integrating by parts k times and bearing in mind that $w(x)$ together with its derivatives becomes zero as $|x| \to \infty$, we arrive at

$$\frac{\partial \langle f(x)\rangle}{\partial \lambda_k} = \frac{1}{k!}\left\langle \frac{d^k f(x)}{dx^k} \right\rangle. \tag{3.77}$$

Applying this formula n times in succession gives (3.76). Differentiating now (3.76) m times with respect to *another* cumulant λ_l, and using (3.76) once more yields

$$\frac{\partial^{n+m} \langle f(x)\rangle}{\partial \lambda_k^n \partial \lambda_l^m} = \frac{1}{(k!)^n (l!)^m}\left\langle \frac{d^{nk+ml} f(x)}{dx^{nk+ml}} \right\rangle. \tag{3.78}$$

One application of the cumulant equations (3.75–78), is the fast computation of the coefficients in expansions of the moments $\alpha_k = \langle x^k \rangle$ of a random variable x in the cumulants λ_m of this variable [3.12]. For example, in such an expansion of α_4 the coefficient at $\lambda_1\lambda_3$, from (3.78), is

$$\frac{\partial^2 \langle x^4\rangle}{\partial \lambda_1 \partial \lambda_3} = \frac{1}{1!3!}\left\langle \frac{d^4 x^4}{dx^4} \right\rangle = \frac{4!}{3!} = 4,$$

and at $\lambda_1^2\lambda_2$

$$\frac{1}{2}\frac{\partial^3 \langle x^4\rangle}{\partial \lambda_1^2 \partial \lambda_2} = \frac{1}{2(1!)^2 2!}\left\langle \frac{d^4 x^4}{dx^4} \right\rangle = \frac{4!}{2\cdot 2!} = 6.$$

The derivation of cumulant equations given in the two preceding exercises for the one-dimensional random variable can be easily generalized to multidimensional variables. Using the characteristic function of the two-dimensional variable $\xi = \{x, y\}$ [see (3.14)]

$$\varphi(u,v) = \int_{-\infty}^{+\infty} e^{i(ux+uy)} w(x,y)dx\,dy$$

and its logarithm expressed as a series

$$\ln \varphi(u,v) = \sum_{m+n\geq 1}^{\infty} \frac{(iu)^m (iv)^n}{m!n!} \lambda_{m+n} \quad ,$$

where m relates to x, and n to y ($m+n$ is the order of the cumulant), it can readily be shown that for the joint probability density $w(x,y)$ the following relationship holds

$$\frac{\partial w(x,y)}{\partial \lambda_{p+q}} = \frac{(-1)^{p+q}}{p!q!} \left\langle \frac{\partial^{p+q} w(x,y)}{\partial x^p \partial y^q} \right\rangle . \tag{3.79}$$

Furthermore, an arbitrary function $f(x,y)$ can be shown to obey

$$\frac{\partial \langle f(x,y) \rangle}{\partial \lambda_{p+q}} = \frac{1}{p!q!} \left\langle \frac{\partial^{p+q} f(x,y)}{\partial x^p \partial y^q} \right\rangle . \tag{3.80}$$

Applying this formula m times gives

$$\frac{\partial^m \langle f(x,y) \rangle}{\partial \lambda_{p+q}^m} = \frac{1}{(p!q!)^m} \left\langle \frac{\partial^{m(p+q)} f(x,y)}{\partial x^{mp} \partial y^{mq}} \right\rangle \tag{3.81}$$

and differentiating (3.81) with respect to *another* cumulant $\lambda_{p_1+q_1}$ yields

$$\frac{\partial^{m+n} \langle f(x,y) \rangle}{\partial \lambda_{p+q}^m \partial \lambda_{p_1+q_1}^n} = \frac{1}{(p!q!)^m (p_1!q_1!)^n} \left\langle \frac{\partial^{m(p+q)+n(p_1+q_1)} f(x)}{\partial x^{mp+np_1} \partial y^{mq+nq_1}} \right\rangle . \tag{3.82}$$

Cumulant equations such as (3.81) and (3.82) are very useful in dealing with nonlinear transformations of random variables [3.12]. An example of their application is given elsewhere (see [3.5], Chap. 3, Exercise 3).

4. Random Functions

In this chapter we introduce the consistent exposition of the theory of random functions. In accordance with the key objective of the *Principles* this is not a general and mathematically rigorous treatment, but rather a description of the theory at the level of rigor adhered to in physics. We consider first the *one-component* function $\xi(t)$, but not an ensemble of these functions such as $\xi(t)$, i.e., $\xi_i(t)$, $i = 1, 2, \ldots, N$. Moreover we treat the argument t from the very beginning as time (that is, we deal here with *processes*) and therefore take t to be a real variable.

In Sect. 4.1 we explain what we mean by a complete definition of a random *process* and what distinguishes it from the definition of a random *variable*. We introduce n-variate (n-point) distributions and consider the requirements of symmetry and consistency to which they are subjected. Sections 4.2,3 give definitions of the two classes of random processes that are most important from the point of view of application: *Markov and stationary* processes.

Since the distributions of random processes are time-dependent, it becomes necessary to consider anew the properties of the *statistical moments* of the processes; above all, the dependence of the moments of different orders on time (Sect. 4.4).

In Sect. 4.5 we give an idea of the widely used *correlation theory* which does not rely on the complete definition of a random process but on a description involving only moments of first and second order. A detailed discussion of the correlation theory is given in [4.1].

Random processes necessarily call for a revision of the concept of convergence to a limit, and hence also of the concepts of continuity, differentiability, and integrability. The appropriate treatment is provided in Sect. 4.6, where various definitions of probabilistic convergence are given. Section 4.7 is concerned with the important question of the relationship between ensemble averaging and time averaging. The answer to this question depends to a large extent on whether or not the random process is *ergodic*.

4.1 General Definitions

Illustrations of random processes come from various branches of physics and engineering: Oscillograms of the shot current of a vacuum tube or of thermal fluctuations of voltage across a resistance, the Brownian motion of a particle, earthquake records, variations of pressure or temperature in the atmosphere are all examples of random processes. A recording of an observable, i.e., the time development of the values actually assumed by it in a given experiment is one possible *realization* of such a process.

Imagine a set of identical replicas of a system subject to the same conditions (an *ensemble*). (The physical meaning of *identical* will be addressed to in Sect. 4.7). Provided the technique of process registration is the same, the observed realizations will in general be different. This is a manifestation of the random nature of the process. The degree of randomness is different for different realizations. If there is no randomness at all, then we have the special case of a fully deterministic process.

In a wider, somewhat loose, sense a random process $\xi(t)$ is any process in time that is governed by statistical laws. More specifically, the random function $\xi(t)$ is a function whose value at any possible instant of time t is a random variable.

The mathematical theory does not require that the argument t be interpreted in a specific way. The real or complex variable t may take on continuous or discrete values. But applications deal mostly with random functions of *time* and/or *space coordinates*. Therefore, we will be interested only in those cases where the possible discrete or continuous values of t are real. For example, we may measure the coordinate ξ of a Brownian particle only at certain times $t_1, t_2, \ldots$, thus obtaining the sequence of random variables $\xi(t_1), \xi(t_2), \ldots$. In what follows, t generally represents the time. The possible values of the function $\xi(t)$ at any fixed time $t = t_k$ are denoted by x_k.

The function $\xi(t)$ is said to be a *random* (or *stochastic* or *probabilistic*) *process*, if t is a *continuous* variable. It is called a *random sequence* (process with a *discrete* parameter, or *discrete* time) if t takes on a *countable* set of values. The term *random function* incorporates both cases, hence, it is used when t varies arbitrarily.

The set of possible values of x, of the random function $\xi(t)$ may also be either continuous or discrete. We thus have four types of random functions in accordance with the four combinations of discrete or continuous variations in x and t. Sometimes these four versions are termed random processes referring to a discrete or continuous process (depending on the character of x) with discrete or continuous time t, respectively. Clearly, mixed, i.e., discrete-continuous, processes are also possible, but it is more convenient to treat each of these forms separately.

If the argument t assumes a *finite* set of values $t_1, t_2, \ldots, t_n$, then the random sequence reduces to the collection of n random variables $\xi(t_1), \xi(t_2), \ldots, \xi(t_n)$. Thus it may be thought of as an n-dimensional random variable, specified by its n-variate probability distribution $w_n\ (x_1, x_2, \ldots, x_n)$. However, the treatment of an infinite (countable or continuous) set of values of t is beyond the scope of the classical theory of probability. It necessitates a special definition of what we mean when we say that the random function $\xi(t)$ is given.

Proceeding from the set of all possible realizations of a random function, we can obtain its comprehensive characteristic by specifying the probability distributions of these realizations. An equivalent treatment, due to Slutsky, relies on the fact that, at any fixed value of t, $\xi(t)$ is a random variable. The *variable* is defined completely if we know its distribution

$$w_1(t,x)dx = P\{x<\xi(t)\leq x+dx\},$$

i.e., the instantaneous probability density $w_1(t,x)$ that is generally time dependent. Of course, instead of $w_1(t,x)$ the corresponding characteristic function may be specified:

$$\varphi_1(t,u) = \int_{-\infty}^{+\infty} e^{iux} w_1(t,x)dx.$$

It is easily seen, however, that a knowledge of the distribution of $\xi(t)$ for any t by no means provides the complete characteristic of the random *function*. There may be functions $\xi(t)$ that depict the same distributions $w_1(t,x)$, but different statistical relationships between the values $\xi(t_1)$ and $\xi(t_2)$ at different instants of time, $t_1 \neq t_2$. In other words, with the same distributions $w_1(t,x)$ these functions may have different bivariate distributions

$$\begin{aligned} &w_2(t_1,x_1;t_2,x_2)dx_1\,dx_2 \\ &\quad = P\{x_1<\xi(t_1)\leq x_1+dx_1; x_2<\xi(t_2)\leq x_2+dx_2\} \end{aligned}$$

and, in particular, different covariances

$$\psi(t_1,t_2) = \overline{\xi(t_1)\xi(t_2)} - \overline{\xi(t_1)}\cdot\overline{\xi_j(t_2)}.$$

Of course,

$$w_1(t,x) = \int_{-\infty}^{+\infty} w_2(t,x;t',x')dx' = \int_{-\infty}^{+\infty} w_2(t',x';t,x)dx'$$

i.e., if we know the bivariate distribution, we also know the univariate one, but *not vice versa*. Hence, a coincidence of univariate distributions for random functions gives no clue as to their bivariate distributions.

On the other hand, if we have a bivariate distribution, i.e., the probability density $w_2(t_1,x_1;t_2,x_2)$ or the characteristic function

$$\varphi_2(t_1,u_1;t_2,u_2) = \int_{-\infty}^{+\infty} \exp\left[i(u_1x_1+u_2x_2)\right] w_2(t_1,x_1;t_2,x_2)dx_1\,dx_2$$

which are generally dependent on the two parameters t_1 and t_2, this characterizes $\xi(t)$ in a more comprehensive though not complete way. The bivariate distribution gives no indication of the statistical relationships between the values of $\xi(t)$ taken at any *three* times t_1, t_2, and t_3. These relations may be different though bivariate (and hence univariate, too) distributions are the same. Evidently this resoning may be taken further, to show then that for any fixed number of times t_i $(i = 1, 2, \ldots, n)$, the n-variate distribution

$$w_n(t_1, x_1; \ldots; t_n, x_n) dx_1 \ldots dx_n$$
$$= P\{x_i < \xi(t_i) \leq x_i + dx_i; \quad i = 1, 2, \ldots, n\}$$

gives us complete information about $\xi(t)$ extending also to smaller n. But it tells us nothing about the situation for larger n. This accounts for the principal distinction between a random function, with the infinite set of values taken on by its argument, and the special case with a set of finite (n) values giving rise to an n-dimensional random variable.

For practical purposes we might confine ourselves to the n-variate distribution with fairly large n, but such an approach would be just as unsatisfactory as, for example, the use of discrete variables in differential calculus. It may however be assumed that we know *finite-variate* (n-variate) distributions, but for an *arbitrarily large value of* n. Thus, we come to the following complete way of specifying a random function: *a random function is specified if its finite-variate distribution* $w_n(t_1, x_1; t_2, x_2; \ldots, t_n, x_n)$ *is known for any number* n *of arbitrarily selected values* $t_1, t_2, \ldots, t_n$. As the event in question (the joint occurrence of n inequalities $x_i < \xi(t) \leq x_i + dx_i$) is independent of the order in which these inequalities are listed, the $w_n(t_1, x_1; \ldots; t_n, x_n)$ should be *symmetric* under permutations of the pairs of arguments (t_i, x_i). For example,

$$p\{x_i < \xi(t_i) \leq x_i + dx_i; \quad i = 1, 2\}$$
$$= w_2(t_1, x_1; t_2, x_2) dx_1 \, dx_2 = w_2(t_2, x_2; t_1, x_1) dx_1 \, dx_2.$$

Assuming this condition to be satisfied, we shall often suppress the arguments of the functions w_n.

In addition to the symmetry condition, all finite-variate probability densities must be hierarchically interrelated, so that each k-variate distribution must be deduced from any n-variate distribution with $n > k$

$$w_k(t_1, x_1; \ldots; t_k, x_k)$$
$$= \int_{-\infty}^{+\infty} w_n(t_1, x_1; \ldots; t_k x_k; \ldots; t_n, x_n) dx_{k+1} \ldots dx_n.$$

In the theory of random *functions*, where w_n depend on the parameters t_i $(i = 1, \ldots, n)$, this condition is by no means trivial, as it limits the class of permissible w_n. In fact, if we integrate w_n with respect to redundant superfluous variables $x_{k+1}, \ldots, x_n$, we will automatically arrive at a result independent of the parameters $t_{k+1}, \ldots, t_n$ appearing in the integrand.

The symmetry and hierarchy conditions are to be met for *any* random function. If we subject w_n to some additional requirement, we will end up with special classes or types of these functions. There are relatively few classes of practical value for which a specific advanced theory is available. Among these we should first of all mention *random processes of the Markov type* (or those without a probabilistic memory effect) and *stationary processes*[1]. To begin with, we shall give definitions of both classes of random processes and then go on to consider them in more detail. It should be emphasized from the very beginning that these classes are distinguished by *different* criteria so that they do not necessarily coincide. A Markov random process may be stationary or nonstationary but a stationary process does not have to be Markovian.

4.2 Markov Processes

We take n successive instants of time $t_1<t_2<\ldots t_n$. According to the definition of the conditional probability we may write

$$\begin{aligned} &w_n(t_1,x_1;\ldots;t_n,x_n)dx_1\ldots dx_n \\ &\quad = w_{n-1}(t_1,x_1;\ldots;t_{n-1},x_{n-1})dx_1\ldots dx_{n-1} \\ &\qquad \times v_n(t_n,x_n|t_1,x_1;\ldots;t_{n-1},x_{n-1})dx_n. \end{aligned} \tag{4.1}$$

If the random function $\xi(t)$ assumes at times t_i the values x_i ($i = 1,2,\ldots, n-1$) then the multiplier $v_n dx_n$ on the r.-h.-s. of (4.1) is the conditional probability for finding at t_n the value of $\xi(t)$ within the interval (x_n, x_n+dx_n). The conditional probability v_n of the *state* x_n at time t_n thus depends on the prehistory of the random function, i.e., on the set of preceding states from which, depending on the number of measurements n, we select $n-1$ states $x_1,\ldots,x_{n-1}$.

Let us denote by $\xi_r(\theta)$ the value of $\xi(t)$ *actually taken on* at times $\theta\leq t_0$. For any $t>t_0$ the distribution will in general depend on the history of the process $\xi_r(\theta)$. This situation is illustrated in Fig. 4.1, which shows the variation of v with x and t. Above the curve of actual values before time t_0, as represented by $\xi_r(\theta)$ lying in the (x,t) plane, we see an infinitely high fence of delta-functions. The behavior of v as a function of x for any $t>t_0$ depends on the overall configuration of the delta fence, i.e., being a function of x and t, v is a functional of $\xi_r(\theta)$. If for the same value of x_0 at time t_0 we take some other form of $\xi_r(\theta)$, i.e., if we bend the delta-fence in some other way, then the surface $v(x,t)$ will take a different shape for $t>t_0$. It is then said that $\xi(t)$ is subjected to a *probabilistic memory-effect*.

[1] In the remainder of the book we will deal mainly with processes, i.e., random functions of a continuous argument. The general theory of random processes of the Markov type was developed by Kolmogorov in the early 1930s.

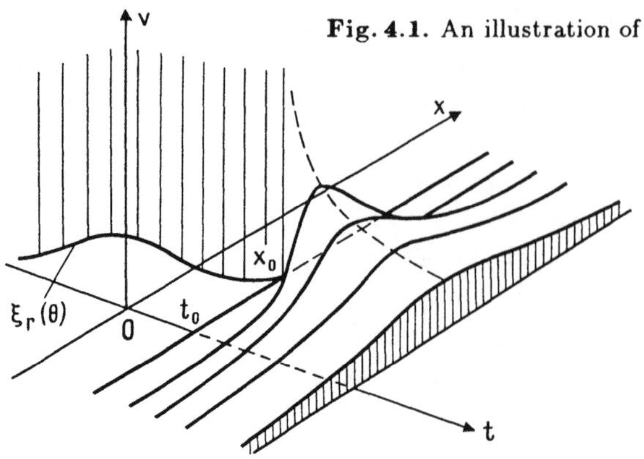

Fig. 4.1. An illustration of the definition of the Markov process

Different parts of the curve $\xi_r(\theta)$ may have a different influence on the subsequent probability distribution. In particular the process may have a damped memory effect: the influence of the values of $\xi_r(\theta)$ is the weaker the earlier the respective events occurred. This loss of memory can be gradual or sharp. For example, the influence of all $\xi_r(\theta)$ at $\theta<t_0-\tau$ may be absent, implying that only those values that occurred during the time period τ up to t_0 are of significance. In a more special, yet, physically rather important case, the conditional probability has for all $t>t_0$ the form $v_2(t,x|t_0,x_0)dx$ and is uniquely determined by the value of x_0 at time t_0. It is absolutely independent of the time history. This is essential a *Markov process* or a *process without a memory effect.*

It should be noted that such a process may be regarded as a direct generalization of a dynamic law described by a first-order differential equation of motion. The solution to this equation satisfying the condition $x=x_0$ at $t=t_0$ is

$$x=f(t,x_0,t_0), \quad \text{where} \quad x_0\equiv f(t_0,x_0,t_0).$$

The deterministic function f determines uniquely the value of x at time t for a specified value of x_0 at time t_0. Therefore the density of the conditional probability of the value x at time t can in this particular case be represented by the delta-function

$$v_2(t,x|t_0,x_0)=\delta[x-f(t,x_0,t_0)] \quad .$$

An example of such a dynamic process is the motion of a particle in a viscous liquid, the reaction being represented as the Stokes friction, i.e., the force $F=-hu$, where u is the velocity of the particle and h the friction coefficient. We have then for u (which corresponds to the variable x in this example) the equation of motion $m\dot{u}+hu=0$ with the solution

$$u=u_0\exp[-(t-t_0)h/m]\equiv f(t,u_0,t_0).$$

It is known, however, that the expression $F = -hu$ for the friction is just a first approximation and only suitable for small accelerations. In the general case, perturbations (viscous waves) caused in the liquid by the motion of a particle, affect its motion for a long period of time. Consequently, the force F depends at any instant of time t not only on the velocity u of the particle at this moment, but on all its *preceding* motion (viscous memory effect). Thus, a more rigorous statement of the problem yields an integro-differential equation for the motion of a particle in a viscous liquid. As a result the state u at time t is not determined uniquely by the state u_0 at t_0 (Sect. 6.3).

We can easily visualize an appropriate transition to a statistical phenomenon, for example, the Brownian motion of a particle in a liquid. With a Stokes-type force, this is a Markov process, but if we include the viscous memory effect, we obtain a probabilistic memory effect: the *probability distribution* u at a time t will depend on the time history of the particle motion.

We return now to the general formula (4.1). It has been stated above that for a Markov process, the conditional probability density for the state (x_n, t_n) with respect to a chain of preceding states, $(x_{n-1}, t_{n-1})\ldots, (x_1, t_1)$, coincides with the conditional probability density relative to the last of the preceding states (x_{n-1}, t_{n-1}). It is independent of the earlier history. In other words, the statistical behavior of the process at a time t_n is conditioned by the state at the time t_{n-1} and has no dependence whatsoever on any information about the course of the process up to time t_{n-1},

$$v_n(t_n, x_n|t_{n-1}, x_{n-1};\ldots;t_1, x_1) = v_2(t_n, x_n|(t_{n-1}, x_{n-1})$$

for any $t_1 < t_2 < \ldots < t_n$ [2]. Hence w_n is related to w_{n-1} by

$$w_n(t_1, x_1;\ldots;t_n x_n) \\ = w_{n-1}(t_1, x_1;\ldots;t_{n-1}, x_{n-1}) v_2(t_n, x_n|t_{n-1}, x_{n-1}).$$

Applying this formula repeatedly to w_{n-1}, w_{n-2} up to w_2,

$$w_2(t_1, x_1; t_2, x_2) = w_1(t_1, x_1) v_2(t_2, x_2|t_1, x_1), \tag{4.2}$$

we obtain

$$w_n(t_1, x_1;\ldots;t_n, x_n) \\ = w_1(t_1, x_1) v_2(t_2, x_2|t_1, x_1) v_2(t_3, x_3|t_2, x_2) \\ \ldots v_2(t_n, x_n|t_{n-1}, x_{n-1}). \tag{4.3}$$

[2] This case is sometimes called a *first-order* Markov process. The name kth order Markov process is reserved for a process in which the conditional probability of a state depends on a fixed number k of preceding states. For example, for the second-order process it has the form $v_3(t_3, x_3|t_2, x_2; t_1, x_1)$. We are not going to deal with these more general cases.

Thus, in order to be able to write the n-variate probability density of a Markov process we need only *two* functions: the univariate *probability density* w_1 and the conditional probability density v (the subscript 2 at v has been suppressed).

Clearly (4.3) can also be put at the foundation of the definition of the Markov process. If it is valid, v is said to be the *transition probability* (from one state to the next state) which is justified by the absence of a memory effect.

Expression (4.2) for the *bivariate* distribution may be written for any process, simply by the definition of the conditional probability. But already at $n = 3$ the availability of the transition probabilities is a severe constraint on the form of the possible distribution functions[3].

Thus a random function of the Markov type is completely defined by specifying the univariate distribution function and the transition probability between any two t_1 and t_2. This is in contrast to the general case in which all the finite-variate distributions have to be specified.

A random function of the *Markov* type is said to be homogeneous in t, if the transition probability $v(t_2, x_2|t_1, x_1)$ depends solely on $t_2 - t_1$. Then we have

$$w_2(t_1, x_1; t_2, x_2) = w_1(t_1, x_1)v(x_2|t_2 - t_1, x_1). \tag{4.4}$$

Generally speaking, the univariate distribution depends on t_1. Hence, the bivariate distribution depends on t_1 and t_2 separately. However, there are cases in which a *stationary* univariate distribution $w(x)$ exists which is the same for all t. Then we have

$$w_2(t_1, x_1; t_2, x_2) = w(x_1)v(x_2|t_2 - t_1, x_1) = w_2(x_1, t_2 - t_1, x_2), \tag{4.5}$$

i.e., the bivariate distribution depends now only on $t_2 - t_1$. Since the distribution w is valid for any t, it is necessary that

$$\int w_2(x_1, t_2 - t_1, x_2)dx_1 = w_1(t_2 - t_1, x_2) = w(x_2)$$

holds, i.e., $w(x)$ should obey the condition

$$\int w(x_1)v(x_2|t_2 - t_1, x_1)dx_1 = w(x_2).$$

[3] This limitation may be reduced to a certain integral relationship for w_n. As an illustration we consider the case of $n = 3$. If we take into account that

$$v(x_2|x_1) = \frac{w_2(x_2, x_2)}{w_1(x_1)}, \quad v(x_3|x_2) = \frac{w_2(x_2, x_3)}{w_1(x_2)},$$

and express w_2 and w_1 in terms of the integrals of $w_3(x_1, x_2, x_3)$, with respect to "redundant" variables, we obtain from the equality $w_3 = w_1(x_1)v(x_2|x_1)v(x_3|x_2)$ the *condition*

$$w_3 \int w_3 dx_1\, dx_3 = \int w_3 dx_3 \int w_3 dx_1.$$

In this case the homogeneous random function of the Markov type is referred to as *stationary*. The appropriate definition of stationarity will be given in the next section.

4.3 Stationary Processes

An idea of a stationary process may be obtained from the observation of some fluctuations under unchanged macroscopic conditions. Electrical noise in a resistance at constant temperature, turbulence in a steady-state flow, shot effect at a constant anode current, etc., are all stationary random processes. For practical purposes, it suffices to maintain constant conditions for a finite period of time, which is specified in such a way that all possible transition processes in the system have already terminated before this time and did not start until this period had ended.

All the stochastic characteristics of a stationary random function should not therefore change as the time origin varies. This means that

$$w_n(t_1, x_1; \ldots; t_n, x_n) \equiv w_n(t_1 + \tau, x_1; \ldots; t_n + \tau, x_n),$$

i.e., w_n (and hence the corresponding characteristic functions) may vary only with the *differences* of values $t_1, \ldots, t_n$, but not with these values themselves. A random function satisfying this condition is said to be *homogeneous in its argument* t. This generic term is commonly used when t refers to a space coordinate. But if the parameter t is time, then the temporal homogeneity is more often than not referred to as *stationarity*.

The definition of the stationary random function suggests that its univariate probability density w_1 is totally independent of t, the bivariate density w_2 may depend on $t_2 - t_1$ only, the trivariate on $t_2 - t_1$ and $t_3 - t_1$, and so forth. Consequently, the conditional probability of x_2 with given x_1 for a stationary function also depends solely on $t_2 - t_1$,

$$v(t_2, x_2 | t_1, x_1) = \frac{w_2(x_1, t_2 - t_1, x_2)}{w_1(x_1)} = v(x_2 | t_2 - t_1, x_1).$$

In many cases the conditional probability for large $\tau = t_2 - t_1$ has a stationary limit

$$\lim_{\tau \to \infty} v(x_2 | \tau, x_1) = \frac{1}{w_1(x_1)} \lim_{\tau \to \infty} w_2(x_1, \tau, x_2) = w_1(x_2).$$

This implies that the dependence of the distribution at a time $t_1 + \tau$ on its value at time t_1 becomes weaker as τ increases and vanishes in the limit $\tau \to \infty$. The conditional probability transforms into the stationary univariate probability of a state. This ensures the so-called *ergodicity* to be discussed later (Sect. 4.7).

It is worth emphasizing once more that Markov and stationary processes do not constitute two nonoverlapping classes. For a process without memory effect the n-variate probability density has the form (4.3), implying that it is in general not stationary. As we stated above, stationarity will occur when (i) the Markov process is homogeneous (i.e., the probability v for a transition from x_1 to x_2 only depends on $t_2 - t_1$) and (ii) a stationary univariate distribution function $w_1(x)$ exists. On the other hand, the stationary process is described by finite-variate probability densities w_n that depend only on the difference $t_i - t_k$. In general, those w_n cannot be represented by the transition probabilities v between two successive states.

By analogy with multidimensional random variables all the definitions and properties described above can readily be extended to include *multivariate random functions*, i.e., collections of an arbitrary number N of univariate random functions $\boldsymbol{\varrho}(t) = \{\xi^{(1)}(t), \ldots, \xi^{(N)}(t)\}$. Designating possible values $\boldsymbol{\varrho}(t)$ at a time t_i by $\boldsymbol{r}_i$, we may repeat what was said above regarding the N_n-variate distribution:

$$w_n(t_1, \boldsymbol{r}_1; \ldots; t_n, \boldsymbol{r}_n) d\boldsymbol{r}_1 \ldots d\boldsymbol{r}_n.$$

For a Markov multivariate process, w_n is expressed in terms of the state probability $w_1(t_1, \boldsymbol{r}_1)$ and transition probability $v(t_2, \boldsymbol{r}_2 | t_1, \boldsymbol{r}_1)$. For a stationary function, probability densities w_n depend only on the differences between the t_i. For discrete and continuous t, respectively, we distinguish between multivariate random sequences and processes.

4.4 Moments of Random Functions

Moments of a random function $\xi(t)$ are introduced in exactly the same manner and with exactly the same purpose as in the case of a system of random variables. These are means of the form

$$\overline{\xi(t_1) \ldots \xi(t_n)},$$

such that arbitrarily selected $t_1, \ldots, t_n$ may coincide with one another either partially (m various values of t_i, $m \leq n$) or completely ($t_1 = \ldots = t_n$, i.e., $m = 1$). The number of terms n is called the order of the moment, so that $\overline{\xi(t)}$ is the first-order moment, $\overline{\xi(t_1)\xi(t_2)}$, $\overline{\xi^2(t_1)}$, $\overline{\xi^2(t_2)}$ are the (mixed and two mean squares) second-order moments, and so forth. A fundamentally new feature here is the fact that moments of random variables are simply numbers, whereas moments of random functions are functions of selected values of the parameter t, as the distribution functions depend on these values.

Clearly, to calculate the moment of any order n it is sufficient to know the m-variate distribution, where m is still the number of different values t_i; $1 \leq m \leq n$. For example, the nth moment $\overline{\xi^n(t_1)}$ is determined using the univariate distribution

$$\overline{\xi^n(t_1)} = \int x_1^n w_1(t_1, x_1) dx_1 = M_n(t_1). \tag{4.6}$$

The mixed moment of the same order n that contains t_1 and t_2 is determined with the help of the bivariate distribution

$$\overline{\xi^{\alpha}(t_1)\xi^{\beta}(t_2)} = \int x_1^{\alpha}x_2^{\beta}w_2(t_1,x_1;t_2,x_2)dx_1\,dx_2$$
$$= M_{\alpha+\beta}(t_1,t_2), \quad \text{where} \quad \alpha+\beta=n, \tag{4.7}$$

and so forth, up to

$$\overline{\xi(t_1)\xi(t_2)\ldots\xi(t_n)} = \int x_1x_2\ldots x_nw_ndx_1\ldots dx_n = M_n(t_1,\ldots,t_n). \tag{4.8}$$

This follows naturally from the hierarchical dependence of all the finite-variate distributions: when averaging n factors, of which only $m \leq n$ are different, we may make use of any distribution function w_N with $N \geq m$; but the integration over redundant variables will immediately reduce it to w_n. For example,

$$\int x_1^{\alpha}x_2^{\beta}w_Ndx_1\ldots dx_N = \int x_1^{\alpha}x_2^{\beta}dx_1\,dx_2\int w_Ndx_3\ldots dx_N$$
$$= \int x_1^{\alpha}x_2^{\beta}w_2dx_1\,dx_2.$$

If we assume some of the values t_i in (4.8) to be equal, we can obviously obtain from (4.8) the complete collection of all the nth moments. In the remainder of the book we shall often use the following short-hand notation of the moments:

$$\overline{x_1^n},\ \overline{x_1^{\alpha}x_2^{\beta}},\ \overline{x_1x_2\ldots x_n},\quad \text{etc.}$$

The *central* moments are defined as the means of *pulsations* or *fluctuations*, $\tilde{\xi}(t_i) = \xi(t_i) - \overline{\xi(t_i)}$. The expression for the nth central moment, analogous to (4.8), has the form

$$\overline{\tilde{\xi}(t_1)\ldots\tilde{\xi}(t_n)} = \overline{[\xi(t_1)-\overline{\xi(t_1)}]\ldots[\xi(t_n-\overline{\xi(t_n)}]}$$
$$\equiv \overline{(x_1-\overline{x}_1)(x_2-\overline{x}_2)\ldots(x_n-\overline{x}_n)},$$

which may conveniently be written as the algebraic sum of products of simple moments of orders from 1 to n.

As far as the role of moments of a random function is concerned, the situation is the same as in the case of random variables. They clearly furnish less complete information about the random function than do its distribution functions, but in a number of cases of practical importance this characteristic appears to be sufficient. Moreover, the derivation of moments is more often than not simpler than the calculation of finite-variate distribution functions. Finally, the lower-order moments, and especially first and second ones, are predominant as far as practical purposes and a number of theoretical considerations are concerned.

The first moment, or the *mean*, of the function is

$$\overline{\xi(t)} = \int xw_1(t,x)dx \equiv \overline{\xi}(t).$$

The second moments, for which it is useful to introduce special notations, are the *mixed moment*,

$$\overline{\xi(t_1)\xi(t_2)} = \int x_1 x_2 w_2(t_1, x_1; t_2, x_2) dx_1\, dx_2 \equiv B(t_1, t_2), \tag{4.10}$$

that for coinciding arguments $t_1 = t_2 = t$ is equal to the *mean square* of the function at time t,

$$\overline{\xi^2(t)} = B(t,t) \equiv \overline{\xi^2}(t), \tag{4.11}$$

and the central second moment, called *covariance* (sometimes autocorrelation function)[4],

$$\begin{aligned}&\overline{[\xi(t_1) - \overline{\xi}(t_1)][\xi(t_2) - \overline{\xi}(t_2)]}\\ &\quad = B(t_1, t_2) - \overline{\xi}(t_1)\overline{\xi}(t_2) \equiv \psi(t_1, t_2),\end{aligned} \tag{4.12}$$

that for $t_1 = t_2 = t$ is equal to the variance of $\xi(t)$ at a time t

$$\overline{\xi^2}(t) - \overline{\xi}^2(t) = \psi(t,t) \equiv D[\xi(t)]. \tag{4.13}$$

The mean determines the center of gravity of a distribution but, unlike a random variable, at each instant of time t. The variance is the simplest measure of dispersion, or spread, about the center of gravity (again at each instant of time t). Lastly, the covariance characterizes the *statistical relation* between the values of the random function at any two times t_1 and t_2. The *correlation coefficient* defined as

$$K(t_1, t_2) = \frac{\psi(t_1, t_2)}{\sqrt{\psi(t_1, t_1)\psi(t_2, t_2)}} = \frac{\overline{\xi(t_1)\xi(t_2)} - \overline{\xi}(t_1)\overline{\xi}(t_2)}{\sqrt{D[\xi(t_1)]D[\xi(t_2)]}}, \tag{4.14}$$

serves the same purpose. It is proportional to the covariance and normalized to unity at $t_1 = t_2$. If $\xi(t_1)$ and $\xi(t_2)$ are statistically independent, i.e., the bivariate distribution is the product of two univariate distributions,

$$w_2(t_1, x_1; t_2, x_2) = w_1(t_1, x_1) w_1(t_2, x_2),$$

then

$$B(t_1, t_2) = \overline{\xi(t_1)\xi(t_2)} = \overline{\xi}(t_1)\overline{\xi}(t_2)$$

holds, with the result that both covariance and correlation coefficient vanish. The *statistical independence results in the absence of correlations*, but just as in the case of random variables, the reverse is not in general true.

[4] Unfortunately, no common terminology is as yet available. In the English-language literature $\psi(t_1, t_2)$ is usually called the covariance (in contrast to the variance) and the correlation function refers to the moment $B(t_1, t_2)$ and the correlation coefficient $K(t_1, t_2)$; see (4.14). Lately, owing to the increasing use in optics of the correlation theory for coherent oscillations and waves, the mixed moment $B(t_1, t_2)$ is often referred to as the *coherence function* (second order) and is denoted by $\Gamma^{(2)}(t_1, t_2)$.

Note that using (4.2), $B(t_1,t_2)$ can always be represented in the form

$$\begin{aligned} B(t_1,t_2) &= \int x_1 x_2 w_2 dx_1\, dx_2 \\ &= \int x_1 w_1(t_1,x_1)dx_1 \int x_2 v(t_2,x_2|t_1,x_1)dx_2. \end{aligned}$$

The second integral is the *conditional mean* of $\xi(t_2)$, i.e., the mean at time t_2, given that at time t_1 the function is x_1. Denoting this conditional mean which depends on t_1, t_2 and x_1, by

$$\bar{x}_2(t_2|t_1,x_1),$$

we thus obtain

$$B(t_1,t_2) = \int x_1 \bar{x}_2(t_2|t_1,x_1) w_1(t_1,x_1)dx_1. \tag{4.15}$$

For a multivariate random function $\varrho(t) = \{\xi^{(1)}(t)\ldots, \xi^{(N)}(t)\}$, we must consider moments containing products of different compounds $\xi^{(k)}(t)$. If we confine ourselves to second-order moments, then we have for a multivariate function with the means $\overline{\xi^{(k)}(t)}$ to deal not only with moments of the form $B_{kk}(t_1,t_2) = \overline{\xi^{(k)}(t_1)\xi^{(k)}(t_2)}$, and hence covariances $\psi_{kk}(t_1,t_2)$, but also with *mutual* second-order moments containing products of two different functions $\xi^{(k)}(t)$. We thus arrive at the *matrix* of second-order moments with the elements

$$B_{ik}(t_1,t_2) = \overline{\xi^{(i)}(t_1)\xi^{(k)}(t_2)}, \quad i,k = 1,2,\ldots,N, \tag{4.16}$$

and the *covariance matrix* with the elements

$$\begin{aligned} \psi_{ik}(t_1,t_2) &= \overline{[\xi^{(i)}(t_1) - \bar{\xi}^{(i)}(t_1)][\xi^{(k)}(t_2) - \bar{\xi}^{(k)}(t_2)]} \\ &= B_{ik}(t_1,t_2) - \bar{\xi}^{(i)}(t_1)\bar{\xi}^{(k)}(t_2). \end{aligned} \tag{4.17}$$

4.5 Correlation Theory

The theory of random processes concerned only with univariate and bivariate distributions is called the *correlation theory* of such functions. The distributions of higher dimensions are beyond its scope. Consequently, correlation theory enables us to find moments of various orders containing no more than two values of the parameter t, including the mean of the random function and its covariance. Generally speaking, we can calculate the mean of any deterministic function of the form $F[\xi(t_1), \xi(t_2)]$, if, of course, this mean exists

$$\overline{F[\xi(t_1), \xi(t_2)]} = \int F(x_1,x_2) w_2(t_1,x_1;t_2,x_2)dx_1\, dx_2.$$

What can be said about the first and second order moments for the case of Markov and stationary processes?

The Markov nature of the process seems to introduce nothing specific here, as (4.2) enables us always to express a bivariate distribution through the conditional probability $v(t_2, x_2|t_1, x_1)$. Only for n-variate distributions with $n \geq 3$ it is of significance that for a Markov process v is the transition probability. However, these distributions are not considered by correlation theory.

On the contrary, the stationarity of a process markedly affects the first and second order moments that are of interest to us. In fact, since the univariate probability density $w_1(x)$ is in these cases independent of t, the moments of any order derived for *one* value of t will also be independent of t, i.e., they will be numerical constants.

$$\overline{\xi^n(t)} = \int x^n w_1(x)dx = \overline{\xi^n} = \text{const.}$$

In particular, the mean $\bar{\xi}$, the mean square $\overline{\xi^2}$ and the variance $D[\xi] = \overline{\xi^2} - \bar{\xi}^2$, will be constants. Furthermore, as the bivariate probability density of a stationary process depends only through the difference $\tau = t_2 - t_1$ on t_1 and t_2, the multivariate moment of the second order and hence the covariance will be functions of the same variable, τ:

$$\begin{aligned} &\overline{\xi(t_1)\xi(t_2)} = B(\tau), \quad \overline{\xi^2} = B(0), \\ &B(\tau) - \bar{\xi}^2 = \psi(\tau), \quad D[\xi] = \psi(0), \\ &K(\tau) = \frac{\psi(\tau)}{\psi(0)} = \frac{\psi(\tau)}{D[\xi]}. \end{aligned} \tag{4.18}$$

We note that $B(\tau)$, and correspondingly $\psi(\tau)$ and $K(\tau)$, are always even functions of τ. This follows from the invariance under translation of the origin of time t by τ:

$$B(\tau) = \overline{\xi(t)\xi(t+\tau)} = \overline{\xi(t-\tau)\xi(t)} = B(-\tau).$$

As a constant mean is normally a readily measurable quantity, it suffices in many cases to consider only the fluctuation $\tilde{\xi} = \xi - \bar{\xi}$, i.e., without any loss of generality we can set $\bar{\xi}$ equal to zero. Then $\psi(\tau)$ coincides with $B(\tau)$

$$\psi(\tau) = B(\tau) = \overline{\xi(t)\xi(t+\tau)}, \tag{4.19}$$

and the correlation coefficient will be

$$K(\tau) = \frac{\psi(\tau)}{\psi(0)} = \frac{\overline{\xi(t)\xi(t+\tau)}}{\overline{\xi^2}}.$$

For stationary random functions the special role of the covariance is also associated with the fact that it determines the *harmonic spectrum* of the process, but this point is considered in [4.1].

Correlation theory allows a wider definition of stationarity. The term *wide-sense stationary* (or simply *stationary* according to *Khintchine* [4.2]) refers to those random functions $\xi(t)$ whose *mean is constant*, $\overline{\xi(t)} = \overline{\xi} =$ const, *and the second order moment B which depends only on* $\tau = t_2 - t_1$ *and is finite for* $\tau = 0$, i.e., the mean square $\overline{\xi^2} = B(0)$ is finite. As $D[\xi] = \overline{\xi^2} - \overline{\xi}^2 \geq 0$ holds, it follows that the constant mean $\overline{\xi}$ is also finite. The earlier definition which assumed the invariance of all finite-variate distributions under translation of time t (*strict-sense stationarity*) did not require finiteness for the moments of the first two orders and was in this respect less severe. But the wide-sense stationarity does not mean that all w_n at $n>2$ are invariant under translation of time t, and in this respect this definition is indeed a wider one.

We now consider the conditions under which the function $\xi(t) = A\cos(\omega t + \theta)$, where in the general case both A and θ are random variables, may be wide-sense stationary.

If only the amplitude A is random, and the phase θ has a fixed value, then

$$\overline{\xi(t)} = \overline{A}\, \cos(\omega t + \theta)$$

and the mean is only constant for $\overline{A} = 0$ for which $\overline{\xi(t)}$ also vanishes. The mixed moment is

$$\begin{aligned}\overline{\xi(t)\xi(t+\tau)} &= \overline{A^2}\, \cos(\omega t + \theta)\, \cos\left[\omega(t+\tau) + \theta\right] \\ &= \frac{\overline{A^2}}{2}\left[\cos(2\omega t + \omega\tau + 2\theta) + \cos\omega\tau\right],\end{aligned}$$

i.e., $B = B(t, \tau)$, and thus, even at $\overline{A} = 0$ we have no stationarity.

Now, let the phase θ be random as well, and independent of A. Then

$$\overline{\xi(t)} = \overline{A}\,\overline{\cos(\omega t + \theta)} = \overline{A}(\cos\omega t\,\overline{\cos\theta} - \sin\omega t\,\overline{\sin\theta})$$

is valid. Evidently, the condition for $\overline{\xi(t)}$ to be independent of t, i.e., $\overline{\xi(t)} = 0$, can now be met not only at $\overline{A} = 0$, but also at $\overline{\cos\theta} = \overline{\sin\theta} = 0$. The latter holds if the probability density for the phase $w_\theta(\theta)$ is on the interval $(0, 2\pi)$ orthogonal to $\cos\theta$ and $\sin\theta$ so that it may be represented by the Fourier series

$$w_\theta(\theta) = \frac{1}{2\pi} + \sum_{n=2}^{\infty}(a_n \cos n\theta + b_n \sin n\theta) \geq 0.$$

To get a mixed moment

$$\overline{\xi(t)\xi(t+\tau)} = \frac{\overline{A^2}}{2}\left[\overline{\cos(2\omega t + \omega\tau + 2\theta)} + \cos\omega\tau\right]$$

that depends only on τ, we have to demand that there are no terms with $n = 2$ in the Fourier series for $w_\theta(\theta)$, i.e., $\overline{\cos 2\theta} = \overline{\sin 2\theta} = 0$. Then

$$\overline{\xi(t)\xi(t+\tau)} = \frac{\overline{A^2}}{2}\cos\omega\tau = B(\tau).$$

Thus random and independent A and θ ensure that the random function $\xi(t) = A\cos(\omega t + \theta)$ will be wide-sense stationary, provided that $\overline{A^2}$ is finite and that the distribution of θ has the form

$$w_\theta(\theta) = \frac{1}{2\pi} + \sum_{n=3}^{\infty}(a_n \cos n\theta + b_n \sin n\theta).$$

It is easily seen that if this Fourier series included only terms with $n>N$, then all the mixed moments $\xi(t)$ up to order N inclusive will only depend on differences between the appropriate times. In particular, if the distribution is uniform on the interval $(0, 2\pi)$:

$$w_\theta(\theta) = 1/2\pi,$$

then all the above refers to moments of any order. This amounts to stating that $\xi(t)$ is stationary in the strict sense.

As was pointed out above, it is for all practical purposes often sufficient to know the distributions w_1 and w_2, yet, in the general case these are no substitute for a complete specification of the random function. There is, however, a very important exception that makes correlation theory substantially more valuable than it might otherwise be.

Frequently the random functions under consideration belong to the class of *normal* (or Gaussian) processes, i.e., all w_n are essentially n-variate Gaussian distributions. This occurs because the conditions of the central limit theorem are satisfied, as is the case with shot noise (Sect. 3.3), the distribution of microparticle velocities (Sect. 2.7), and so on. But in the n-variate Gaussian distribution the parameters are only the means and second order moments (both of them always being finite); and to obtain these moments it is sufficient to know only w_2 (and hence w_1). The normal random function $\xi(t)$ is thus completely specified by the probability density w_2. Knowing w_2, we may write any n-variate distribution function and, consequently, answer any question about any deterministic function of the variables $\xi(t_1), \ldots, \xi(t_n)$.

It is also clear that, for a normal random function,the wide-sense stationarity coincides with strict-sense stationarity, as the latter follows from the former: if w_1 is independent of t, and w_2 is only dependent on $\tau = t_2 - t_1$, then all w_n will be invariant under translation of time t. Finally, for normal functions the absence of correlation implies statistical independence, which does not correspond to the general case.

It is a straightforward matter to extend any of the definitions of stationarity to cover multivariate random functions. In correlation theory it is still reasonable to use the wide-sense stationarity i.e., to refer the stationarity of the N-variate function $\boldsymbol{\varrho}(t) = \{\xi^{(1)}(t), \ldots, \xi^{(N)}(t)\}$ to the finiteness and constancy of all $\overline{\xi^{(k)}(t)}$ and $\overline{[\xi^{(k)}(t)]^2}$ and the dependence of terms of the matrix B_{ik} (or covariance matrix ψ_{ik}) on $\tau = t_2 - t_1$ alone.

4.6 Probabilistic Convergence

In the remainder of the book we shall make extensive use of derivatives and integrals of random processes. Both operations, differentiation and integration, are known to presuppose that a sequence of variables converges to a limit. But for random variables that are not specified in a deterministic way, but through their probability distributions, the concept of convergence to a limit (and also of continuity, differentiability, integrability for random functions) can not have the same meaning as in calculus. For a sequence of random variables $\varsigma_N (N = 1, 2, \ldots)$ only the *probabilistic* definition of convergence to a limit is possible, which *inter alia* offers more possibilities in the selection of the definition itself. The probabilistic convergence is also significant for the discussion of the so-called *ergodicity* of random functions, which will be taken up in the next section.

For simplicity, we consider first various types of convergence of a sequence of random variables ς_N to a (nonrandom) number a.

One of the types of probabilistic convergence is the *convergence in the mean square* (m.s.), by which we mean that the mean square deviation ς_N from the number a tends to zero as $N \to \infty$

$$\lim_{N\to\infty} \langle(\varsigma_N - a)^2\rangle = 0, \tag{4.20}$$

which is written as

$$\lim_{N\to\infty} \varsigma_N = a\,(\text{m.s.}) \quad , \quad \text{or} \quad \varsigma_N \xrightarrow{\text{m.s.}} a \quad \text{as} \quad N\to\infty, \quad \text{or}$$

$$\underset{N\to\infty}{\text{l.i.m.}}\ \varsigma_N = a,$$

where l.i.m. stands for "limit in the mean square". This type of convergence is especially suitable in cases where quadratic combinations of random variables are involved (in particular, when representing the energy).

The equality (4.20) clearly requires that $\overline{\varsigma_N^2}$ be finite, thereby ensuring that the mean $\bar{\xi}_N$ is also finite as $D[\varsigma_N] = \overline{\varsigma_N^2} - \bar{\varsigma}_N^2 \geq 0$.

Substracting and adding $\bar{\varsigma}_N$ on the left-hand side of (4.20) gives

$$\lim_{N\to\infty} \langle(\varsigma_N - \bar{\varsigma}_N + \bar{\varsigma}_N - a)^2\rangle = \lim_{N\to\infty} \{D[\varsigma_N] + (\overline{\varsigma_N} - a)^2\} = 0.$$

But the limit of the sum of two nonnegative variables may be zero if, and only if, the limits of both terms are zero, i.e.,

$$\lim_{N\to\infty} \bar{\varsigma}_N = a, \qquad \lim_{N\to\infty} D[\varsigma_N] = 0.$$

Thus, a is the limit of the sequence of means ς_N, and the limit of the variance ς_N is zero.

Another type of probabilistic convergence of ς_N to a is the *convergence in probability* (in pr.). It is defined as follows:

$$\lim_{N\to\infty} P\{|\xi_N - a|<\varepsilon\} = 1, \tag{4.21}$$

where, as usual, ε is any arbitrarily small positive number. This is written in the form

$$\lim_{N\to\infty} \varsigma_N = a \text{ (in pr.)} \quad \text{or} \quad \varsigma_N \overset{\text{in pr.}}{\longrightarrow} a \quad \text{as } N\to\infty.$$

The equality (4.21) means that the probability of ς_N being found anywhere outside an arbitrarily small interval $(a-\varepsilon, a+\varepsilon)$ is zero in the limit $N\to\infty$. As ε is arbitrarily small, this in turn means that the probability density $w_N(x)$ of the random quantity ς_N tends for $N\to\infty$, to $\delta(x-a)$. But this suggests by no means that a is the limit of the sequence $\bar{\varsigma}_N$ and that $D[\varsigma_N]$ tends to zero. Moreover, $\bar{\varsigma}_N$ and $\overline{\varsigma_N^2}$ may grow infinitely with increasing N, or even be infinite at any N. For instance, let ς_N be nonnegative and subject to the Cauchy distribution

$$w_n(x) = \frac{2N}{\pi(1+N^2x^2)} \quad (x\geq 0, \quad N = 1,2,\ldots).$$

For any $x>0$ the limit of $w_N(x)$ is zero as $N\to\infty$, whereas $w_N(0) = 2N/\pi$. At the same time, the normalization condition is always met

$$\int_0^\infty w_N(x)dx = \frac{2}{\pi}\operatorname{arc\,tg}(Nx)\Big|_0^\infty = 1,$$

so that $w_N(x)$ tends to $\delta(x)$ as $N\to\infty$. However, it is readily seen that, $\bar{\varsigma}_N$ and $\overline{\varsigma_N^2}$ are infinite for any N.

The convergence in probability is often referred to as the *convergence by the law of large numbers*. Random variables ς_N are said to be constant in the limit, if a sequence of constants a_N exists, such that

$$\lim_{N\to\infty} P\{|\xi_N - a_N|<\varepsilon\} = 1.$$

If all a_N are the same (equal to a), then this equality reduces to (4.21), i.e., ς_N converges in probability to a, or the difference $\varsigma_N - a$ converges in probability to zero.

The convergence in probability should be clearly distinguished from the conventional convergence

$$\lim_{N\to\infty} |\xi_N - a| = 0 \quad \text{or} \quad \xi_N\to a.$$

In fact, nothing can be proved mathematically concerning the behavior of *empirical numbers* – the values of ς_N. We can only prove statements referring to *theoretical concepts*, including the concept of probability as defined in the initial axioms. The convergence in probability tells us not that $\varsigma_N \to a$ as $N \to \infty$, but that the *probability of the event* $|\varsigma_N - a| < \varepsilon$ tends to unity. This concept is related to experiment by the measurement axiom which states that the probability is *measured* in terms of the relative frequency of occurrence of an event in a sufficiently extensive series of tests, in a sufficiently large ensemble of systems, and so forth.

To clarify this fundamental point, we would like to consider certain limit theorems of probability theory known collectively as the *law of large numbers*, viz., theorems applicable to the case where ς_N in (4.21) is an arithmetic mean of N random variables ξ_k

$$\varsigma_N = \frac{1}{N} \sum_{k=1}^{N} \xi_k. \tag{4.22}$$

We perform a *series* of N tests and use their results $\xi_1, \xi_2, \ldots, \xi_N$ to calculate the mean (4.22). Let us denote as B_N the event characterized by

$$|\varsigma_N - a| < \varepsilon.$$

To measure the probability of the event B_N we have to perform a *very large number M of series of measurements with N tests in each*, i.e., we need a *collection of such series.* The law of large numbers (4.21) states that the larger these series are (the larger N), the closer is $P\{B_N\}$ to unity. According to the measurement axiom, this implies that more of these series will yield B_N (in the limit, practically all of them):

$$\frac{m(B_N)}{M} \to 1 \quad \text{as} \quad N \to \infty.$$

This statement is therefore meaningful, but this comes about only due to a clear connection between the mathematical definition of probability and the empirical concept of relative frequency. In other words, the law of large numbers is no more than a theorem that follows from a system of axioms for the variable P defined as a completely additive, nonnegative function of subsets that is normalised to unity.

This point, which has already been touched upon in Sect. 2.1, is not infrequently presented in textbooks in a somewhat inconsistent manner, without a clear statement of the fact that the measurement axiom relating the notions of probability theory with reality, i.e., experiment and practice, is *not contained* in the mathematical theory as such. One may encounter statements that the law of large numbers provides the foundation for successful application of the concept of probability to various problems in the natural sciences and engineering. If this were really so, then this would mean that extraction of *practical* results is a *corollary* of some abstract axioms and that these mathematical axioms as such determine the behavior of the empirical variables.

It is in principle possible to proceed from other axioms and develop another theory of probability whose conclusions, though different from those of the present theory, would be just as logically faultless and unrelated to real events. The situation here is the same as with various possible geometries. But once a mathematical theory is *supplemented* by definite ways of measuring the quantities concerned so that it becomes a physical theory, the situation changes. The correctness or incorrectness of the theory ceases then to be a question of its selfconsistency alone, and becomes a question of its correspondence to real things and phenomena. It is now possible to pose the question in respect to the truth of the axioms themselves since they may now be verified by experimental tests.

Before such a test is undertaken, however, an internal harmony is required between both parts of the physical theory: the techniques for measuring the variables to be obtained must be consistent with the equations obeyed by these variables in the mathematical part of the theory. For example, the Newtonian equations of motion assume that force is a vector; therefore they are incompatible with a method of measuring the force that will determine only its magnitude. It may be that, in actual fact, force is not a vector but, say, a tensor; however, this is *another* question, i.e., one of how adequately the given physical theory as a whole describes reality. Here we are doing no more than stating that a disagreement between the mathematical and measuring parts of the physical theory renders it inadequate even before any test of its implications by experiment.

As will be seen later, the law of large numbers differs from this point of view from other – logically equivalent – theorems of probability theory only in that it shows particularly clearly and explicitly the *compatibility* of the mathematical definition of probability with the frequency approach to measuring it. It shows that the frequentistic "measurement axiom" does not contradict the mathematical theory, yet the latter does not, and can not replace this axiom.

Proofs of various theorems concerned with the law of large numbers normally rely on Chebyshev's inequality, contained in his thesis (1846). Let a random variable ς have a finite variance $D[\varsigma]$; Chebyshev's inequality states

$$P\{|\varsigma - a|\geq\varepsilon\}\leq\frac{\overline{(\varsigma - a)^2}}{\varepsilon^2}. \tag{4.23}$$

If, in particular, $a = \bar{\varsigma}$, then the inequality (4.23) takes the form[5]

$$P\{|\varsigma - \bar{\varsigma}|\geq\varepsilon\}\leq\frac{D[\varsigma]}{\varepsilon^2}. \tag{4.24}$$

Although inequalities (4.23) and (4.24) provide a rather crude estimate of P (if the distribution law for ς is known then a more accurate one may be obtained), they are very convenient for theoretical considerations.

[5] The most general form of Chebyshev's inequality is $P\{f(\varsigma)\geq K\}\leq\overline{f(\varsigma)}/K$, where $f(\varsigma)$ is a nonnegative deterministic function of the random variable ς, and K is any positive number. For $f(\varsigma) = (\varsigma - a)^2$ and $K = \varepsilon^2$ this reduces to (4.23).

In the case when the ς in Chebyshev's inequality is an arithmetic mean (4.22) for N random variables ξ_i, the inequality (4.24) allows us to prove *Chebyshev's theorem*, that is, a fairly general form of the law of large numbers. It states that if ξ_i is a sequence of mutually independent random variables having uniformly limited variances ($D[\xi_i] \leq C$), then

$$\lim_{N\to\infty} P\{|\varsigma_N - \bar{\varsigma}_N| < \varepsilon\} = 1. \tag{4.25}$$

Really,

$$D[\varsigma_N] = D\left[\frac{1}{N}\sum_1^N \xi_i\right] = \frac{1}{N^2}\sum_1^N D[\xi_i] \leq \frac{C}{N}.$$

By Chebyshev's inequality

$$P\{|\varsigma_N - \bar{\varsigma}_N| \geq \varepsilon\} \leq \frac{C}{N\varepsilon^2},$$

which gives for the probability of the opposite event the theorem (4.24), i.e., the convergence in probability of ς_N to $\bar{\varsigma}_N$.

A special case of Chebyshev's theorem is Poisson's theorem. Let $\xi_i = \varepsilon_i$ be random variables resulting from the ith test ($\varepsilon_i = 1$ or 0, according to whether or not an event A occurs in the ith test for which $P\{A\} = p_i$). Then

$$\varsigma_N = \frac{1}{N}\sum_1^N \varepsilon_i = \frac{n}{N}, \quad \bar{\varepsilon}_i = p_i, \quad D[\varepsilon_i] = p_i q_i \leq \tfrac{1}{4} \equiv C$$

holds and whence

$$\bar{\varsigma}_N = \frac{1}{N}\sum_1^N p_i$$

is obtained. Chebyshev's theorem now gives

$$\lim_{N\to\infty} P\left\{\left|\frac{n}{N} - \frac{1}{N}\sum_1^N p_i\right| < \varepsilon\right\} = 1. \tag{4.26}$$

This relation is Poisson's theorem. We obtain one more special case if $p_i = p$. This leads us to one of the earliest formulations of the law of large numbers, namely Bernoulli's theorem:

$$\lim_{N\to\infty} P\left\{\left|\frac{n}{N} - p\right| < \varepsilon\right\} = 1. \tag{4.27}$$

Let us consider this simplest form of the law in some detail. Theorem (4.27) states that as the number of tests N increases, the relative frequency of the event A, represented by the empirical variable $n(A)/N$, converges *in probability* to p, i.e., to the probability of the event A. If this were not so, it would make no sense to measure probabilities in terms of the relative frequency. But this being the case, the frequentistic way of measuring the probabilities p (in terms of the relative frequency $n(A)/N$ of the occurrence of the event A in a series of N tests) and P (in terms of the relative frequency $m(B_N)/M$ for the occurrence of the event $B_N = \{|n/N - p|<\varepsilon\}$ in a collection of M series with N tests in each) can be accepted as a supplement to the mathematical theory, as it is consistent with the latter. Now we can do both pose questions and test by experiment whether the resulting physical theory reflects real statistical laws.

It is curious that for the theorem (4.27) to be valid for any values of p, i.e., for the convergence in probability

$$\lim_{N\to\infty} \frac{n}{N} = p \text{ (in pr.)},$$

it is sufficient to require that this convergence occur only for $p \ll 1$ (the relative frequency of the events having low probability must be small).

We now write Chebyshev's theorem for the case where all $\bar{\xi}_i = a$. Then

$$\bar{\zeta}_N = \frac{1}{N} \sum_1^N \bar{\xi}_i = a$$

and the theorem takes the form

$$\lim_{N\to\infty} P\left\{\left|\frac{1}{N}\sum_1^N \xi_i - a\right|<\varepsilon\right\} = 1 \quad \text{i.e.,} \quad \lim_{N\to\infty} \frac{1}{N}\sum_1^N \xi_i = a \quad G\text{(in pr.)},$$

which is the basis for the use of the arithmetic mean in the evaluation of measurements. The individual ξ_i may substantially deviate from a, but as $N\to\infty$ we have $\zeta_N\to a$ with the probability $P\to 1$. This happens because in calculating the mean, the random deviations of individual terms are compensated for and in the majority of cases the deviation $|\zeta_N - a|$ appears to be very small.

Deviations of ξ_i from a may be *random* errors of measurement. But if the *accuracy of reading* in measuring is no less than δ, i.e., if there is a *systematic* error associated with scale division, then the accuracy $\zeta_{N'}$ too, is no less than δ for any N. Hence it is useless to appeal to the law of large numbers in an attempt to obtain a for $N\to\infty$ with an error less than δ. There is a fairly widespread misconception that the arithmetic mean enables the bounded-from-below accuracy of measurement to be surpassed and the reading of current within microamperes to be obtained using, say, a switchboard ammeter.

Another situation is also possible: the variable measured may itself be random (noise current, etc.). Then we see that (for $P\to 1$) $\xi_N\to a$ as $N\to\infty$ i.e., the arithmetic mean tends to the mathematical expectation of the random variable.

Generally speaking, the condition of mutual independence of measurements ξ_i of a random variable requires that the latter be measured at sufficiently large time intervals. However, for the law of large numbers to be true, this condition of independency is not necessary, because Chebyshev's inequality only demands that $D[\varsigma_N]\to 0$ as $N\to\infty$. We shall not consider more general theorems nor necessary and sufficient conditions for the law of large numbers to be valid for the arithmetic mean ς_N, (4.22), as these conditions refer to the very variable ς_N and are consequently of less practical interest than the narrower conditions pertaining to individual terms ξ_i.

In 1909 Borel (and later – in a more general form – Cantelli, and then Kolmogorov) proved a stronger statement than the law of large numbers. By Bernoulli's theorem we have

$$\lim_{N\to\infty} P\left\{\left|\frac{n}{N}-p\right|<\varepsilon\right\}=1.$$

According to Borel (the *strong* law of large numbers)

$$P\left\{\lim_{N\to\infty}\left|\frac{n}{N}-p\right|=0\right\}=1, \tag{4.28}$$

i.e., the limit of the relative frequency is with certainty (or, as is commonly said, almost surely) the probability p. This is an even more sound reason for the probabiliy to be measured in terms of the relative frequency.

Based on (4.28) we may introduce one more kind of probabilistic convergence – the convergence in terms of the *strong* law of large numbers, also known as the convergence with probability $P=1$, or *almost sure convergence* (a.s.)

$$P\left\{\lim_{N\to\infty}\varsigma_N=a\right\}=1. \tag{4.29}$$

This is often denoted by

$$\lim_{N\to\infty}\varsigma_N=a \text{ (a.s.)} \quad \text{or} \quad \varsigma_N\xrightarrow{\text{a.s.}}a \quad \text{as } N\to\infty.$$

The definition (4.29) sometimes gives rise to confusion since it contains a *conventional* limit of a sequence of *random* variables. It would seem that here we abandon the statement made above that convergence of random variables can only have a probabilistic sense. But this is exactly the case of such a convergence. Among the various realizations of the sequence ς_N those which converge to a conventional limit a are also possible. It can be shown that the majority of such realizations has the definite probability P [4.3]. The almost sure convergence means that this probability, i.e., the probability of the *random event* $\lim_{N\to\infty}\varsigma_N=a$, is unity. In other words, the realizations ς_N converging to a conventional limit a almost exhaust the set of all possible realizations of the sequence ς_N. Thus in (4.29) we still have the probabilistic definition of convergence, although here we deal not with the limit of probability (as in the convergence in probability) but with the probability of a limit.

We will give two of the conditions for the convergence of ς_N to be almost sure. One of them (*necessary and sufficient*) is

$$\lim_{M\to\infty} P\{|\varsigma_N-a|>\varepsilon \text{ at least for one } N\geq M\}=0.$$

This condition, however, can never be tested in practice. The other, more stringent and *sufficient*, condition is that at any $h>0$ the following series should converge:

$$\sum_{n=1}^{\infty} \overline{(\varsigma_N - a)^h}.$$

Other sufficient conditions and rigorous mathematical treatments in general of issues pertaining to probabilistic convergence are to be found in [Ref. 4.3, Chap. 3 and 4.4, Chap. 1].

The convergence in the mean square leads, by Chebyshev's inequality, to the convergence in probability. If all ς_N are almost surely uniformly bounded in modulus, then the converse holds too, i.e., the convergence in probability suggests the convergence in the mean square. The almost sure convergence also implies the convergence in probability, but not the convergence in the mean square. At the same time the convergence in the mean square does not result in the almost sure convergence.

The above reasoning regarding the convergence of a sequence of random variables ς_N to a *number a* might be generalized in a natural way to the convergence to a random variable ς, which is defined on the same set of possible values as the variables ς_N. So, for example, the convergence of ς_N to ς in the mean square is defined as follows

$$\lim_{N\to\infty} \langle(\varsigma_N - \varsigma)^2\rangle = 0, \tag{4.30}$$

and the convergence in probability as

$$\lim_{N\to\infty} P\{|\varsigma_N - \varsigma|<\varepsilon\} = 1. \tag{4.31}$$

They differ from (4.20,21) in that averaging in (4.30) and computing the probability P in (4.31) now require a knowledge of the *joint* distribution of the variables ς_N and ς. If the probability density for this distribution is $w_N(x,y)$, then (4.30) implies

$$\lim_{N\to\infty} \int\int (x-y)^2 w_N(x,y)dx\,dy = 0,$$

and (4.31)

$$\lim_{N\to\infty} \int dy \int_{y-\varepsilon}^{y+\varepsilon} w_N(x,y)dx = 1.$$

As ε is arbitrarily small, the last equation suggests that $w_N(x,y)$ transforms into $w_\varsigma(y)\delta(x-y)$ as $N\to\infty$.

We now turn to random *processes*, i.e., random functions $\xi(t)$ with a continuous argument t.

According to the above three definitions of probabilistic convergence, we may also define the continuity of the random *function* $\xi(t)$ at a point t. This necessitates the convergence of $\xi(t + \Delta t)$ to $\xi(t)$ as $\Delta t \to 0$. Consequently, the continuity in the mean square implies

$$\lim_{\Delta t \to 0} \overline{\{\xi(t + \Delta t) - \xi(t)\}^2} = 0, \qquad \text{or}$$

$$\text{l.i.m.}\,\xi(t + \Delta t) = \xi(t).$$

It should be noted that the continuity of a random function in probability by no means supposes the continuity of its possible values. The latter may also be discrete, e.g., the charge coming to the anode in the shot noise problem. Possible values of the charge $q(t)$ are multiples of the electron charge e, i.e., $q(t)$ varies stepwise. Nonetheless, the function $q(t)$ is taken to be continuous in probability, i.e., as $\Delta t \to 0$

$$P\{|q(t + \Delta t) - q(t)| > \varepsilon\} < \eta$$

is to hold. In other words, the probability that in the time Δt the charge $\Delta q = q(t + \Delta t) - q(t)$, equal to one or several times e, will come to the anode is assumed to tend to zero as $\Delta t \to 0$.

Further, we define the differentiability of the random function $\varsigma(t)$ analogously to the probabilistic continuity, viz., as the existence of a limit $\dot{\xi}(t)$ (in any of the above senses) of the expression

$$\frac{\xi(t + \Delta t) - \xi(t)}{\Delta t}$$

for any limiting process $\Delta t \to 0$. For example, the differentiability in the mean square means that

$$\lim_{\Delta t \to 0} \overline{\left\{\frac{\xi(t + \Delta t) - \xi(t)}{\Delta t} - \dot{\xi}(t)\right\}^2} = 0, \quad \text{or}$$

$$\underset{\Delta t \to 0}{\text{l.i.m.}} \frac{\xi(t + \Delta t) - \xi(t)}{\Delta t} = \dot{\xi}(t),$$

and the almost sure differentiability means that

$$P\left\{\lim_{\Delta t \to 0} \frac{\xi(t + \Delta t) - \xi(t)}{\Delta t} = \dot{\xi}(t)\right\} = 1.$$

A word of caution is in order here, however. One should not jump to the conclusion that the transition from random sequences to random processes is a trivial step. As far as processes are concerned, questions which are perfectly justified, but not at all easy to answer, may be posed. For example, the continuity (or differentiability) of the process $\xi(t)$ on an *interval* of t (i.e., on an *uncountable* set of points t) is generally not a random event whose probability is uniquely determined by finite-variate distributions $\xi(t)$. However, this is beyond the scope of the book and the reader is again referred to mathematical texts [4.3,4].

In terms of the same three convergences, we can define the limit of the sum

$$R_N = \sum_{0}^{N-1} \xi(t_i)(t_{i+1} - t_i),$$

where $t_0 = a$, $t_N = b$, i.e., the Riemann integral of the random function $\xi(t)$.

The function $\xi(t)$ is then said to be R-integrable (in probability, almost surely, or in the mean square according to the type of convergence of R_N to R involved). In an analogous way, we may define the Stieltjes integrals

$$S = \int_a^b \xi(t) df(t), \quad S = \int_a^b f(t) d\xi(t),$$

where $f(t)$ is a deterministic function, as limits (in any of the three senses) of the integral sums

$$S_N = \sum_{i=0}^{N-1} \xi(t_i)[f(t_{i+1}) - f(t_i)],$$

$$S_N = \sum_{i=0}^{N-1} f(t_i)[\xi(t_{i+1}) - \xi(t_i)].$$

Further, for two independent random functions $\xi(t)$ and $\eta(t)$ the Stieltjes integral

$$S = \int \xi(t) d\eta(t)$$

will be the limit of the integral sum

$$S_N = \sum_{i=0}^{N-1} \xi(t_i)[\eta(t_{i+s}) - \eta(t_i)].$$

4.7 Ergodicity of Random Processes

The concept of ergodicity discussed in this section is important because it gives rise to remarkable relationships *between the distribution function and the time that the random function $\xi(t)$ has a value belonging to a given interval, and between statistical means and time averages.*

By the measurement axiom the distribution functions are inherently *stochastic*. They are *relative frequencies in an ensemble of equal systems*, i.e., systems such that each features the same conditions under which a given random process proceeds, and the same ways of registering or observing it. With fluctuations for example, the *macroscopic* characteristics of all systems constituting the ensemble must be the same. If we have an ensemble of systems, we thereby have a large collection of *realizations* of the random function $\xi(t)$ at hand. Using appropriate probabilities, i.e., distributions of systems belonging to the ensemble over the possible values of $\xi(t)$, we can then find $\overline{\xi(t)}$, $\psi(t_1, t_2)$, and so on.

Theoreticians like to deal with ensembles, but experimentalists normally have only one laboratory and one installation not 10^6 or 10^9 ones[6]. Over a given period of time $(0, T)$ an experimentalist can only obtain *one* realization of a random process of interest, and he prefers therefore to *average over time* using *one* realization $\xi(t)$, e.g., one oscillogram or the like. The question arises as to a connection between time averaging and ensemble averaging.

To look ahead a little, we would like to indicate at this point that, for *stationary* processes exhibiting *ergodicity*, both ways of averaging coincide for sufficiently large T. This is due to the fact that the stationary probability of a state is in this case equal to the relative time the system stays in a given state. Accordingly, the statistical mean is equal to the mean over a sufficiently large time interval[7]. By using the words "equal" and "coincide" we are, of course, being a little inaccurate, as we are only dealing here with stochastic convergence – in probability, in the mean square, or almost surely.

We now proceed to formulate these statements more accurately and to prove them [4.4].

Let $f(\xi)$ be a deterministic function of a random function $\xi(t)$. We denote its time average over a period $(0, T)$ by subscript T together with a wavy line.

$$\widetilde{f_T[\xi(t)]} = \frac{1}{T} \int_0^T f[\xi(t)] dt. \tag{4.32}$$

As stated above, this is an empirical quantity derived from the processing of an oscillogram $f(\xi)$ taken in the interval $(0, T)$. At the same time this is a random variable that is different for various realizations of $\xi(t)$ in the interval $(0, T)$. Let $w_1(x, t)$ and $w_2(t, x; t', x')$ be univariate and bivariate probability densities of $\xi(t)$ and calculate the mean and variance of the random variable $f_T(\xi)$. We have

[6] There are, of course, exceptions to this rule, where an ensemble is realized in one installation. Recall, for instance, the diffusion of a drop of paint in water. It is true, though, that members of the ensemble here are not complex macroscopic systems, but small particles.

[7] It is for this reason that stationary probability has the same meaning as probability for a state in thermodynamic equilibrium, and the statistical mean implies a thermodynamical equilibrium value.

$$\begin{aligned}\overline{\widetilde{f_T(\xi)}} &= \frac{1}{T}\int\limits_0^T \overline{f(\xi)}\, dt = \frac{1}{T}\int\limits_0^T dt \int\limits_{-\infty}^{+\infty} f(x) w_1(t,x) dx \\ &= \int\limits_{-\infty}^{+\infty} f(x) dx \frac{1}{T}\int\limits_0^T w_1(t,x) dt = \int\limits_{-\infty}^{+\infty} f(x) p_T(x) dx, \end{aligned} \tag{4.33}$$

where we have introduced the effective probability density

$$p_T(x) = \frac{1}{T}\int\limits_0^T w_1(t,x) dt. \tag{4.34}$$

Further, the relations

$$\begin{aligned} D[\widetilde{f_T(\xi)}] &= \langle \{\widetilde{f_T(\xi)} - \overline{\widetilde{f_T(\xi)}}\}^2 \rangle \\ &= \frac{1}{T^2}\int\limits_0^T\int\limits_0^T \{\overline{f(\xi) f(\xi')} - \overline{f(\xi)} \cdot \overline{f(\xi')}\} dt\, dt' \\ &= \frac{1}{T^2}\int\limits_0^T\int\limits_0^T \psi_f(t,t') dt\, dt', \end{aligned} \tag{4.35}$$

hold where $\psi_f(t,t')$ is the covariance of $f(\xi)$

$$\begin{aligned} \psi_f(t,t') = \iint\limits_{-\infty}^{+\infty} & f(x) f(x') \{w_2(t,x;t',x') \\ & - w_1(t,x) w_1(t',x')\} dx\, dx'. \end{aligned} \tag{4.36}$$

Thus for all $f(\xi)$ with finite $\langle f^2(\xi)\rangle$ the requirement that

$$\lim_{T\to\infty} \frac{1}{T^2}\int\limits_0^T\int\limits_0^T \psi_f(t,t') dt\, dt' = 0 \tag{4.37}$$

is the necessary and sufficient condition for the convergence in the mean square

$$\underset{T\to\infty}{\text{l.i.m.}} \{\widetilde{f_T(\xi)} - \overline{\widetilde{f_T(\xi)}}\} = 0. \tag{4.38}$$

The necessity follows directly from the definition of this type of convergence. By Chebyshev's inequality, for the random variable $\widetilde{f_T(\xi)}$ we have

$$P\{|\widetilde{f_T(\xi)} - \overline{\widetilde{f_T(\xi)}}| \geq \varepsilon\} \leq \frac{D[\widetilde{f_T(\xi)}]}{\varepsilon^2},$$

and the condition (4.37) is also sufficient for the convergence in probability

$$\lim_{T\to\infty} |\widetilde{f_T(\xi)} - \overline{f_T(\xi)}| = 0 \quad \text{(in pr.)},$$

or, in expanded form,

$$\lim_{T\to\infty} \left| \frac{1}{T} \int_0^T f[\xi(t)]dt - \int_{-\infty}^{+\infty} f(x)p_T(x)dx \right| = 0 \quad \text{(in pr.)},$$

where $p_T(x)$ is given by (4.34). This could be called the law of large numbers for continuous observation.

Now we proceed to prove that under condition (4.37) the *relative time for* $\xi(t)$ *to fall into the interval* $(x, x+dx)$ *converges in probability to* $p_\infty(x)dx$ *as* $T\to\infty$, where

$$p_\infty(x) = \lim_{T\to\infty} p_T(x),$$

if, of course, this limit exists. The relative time is the ratio of the total time $T_{(x,x+dx)}$ during which $\xi(t)$ is found in $(x, x+dx)$ to the total duration of the interval T, i.e., the ratio of the sum of all the sections of the abscissa marked by the bold lines in Fig. 4.2 to the whole interval from 0 to T.

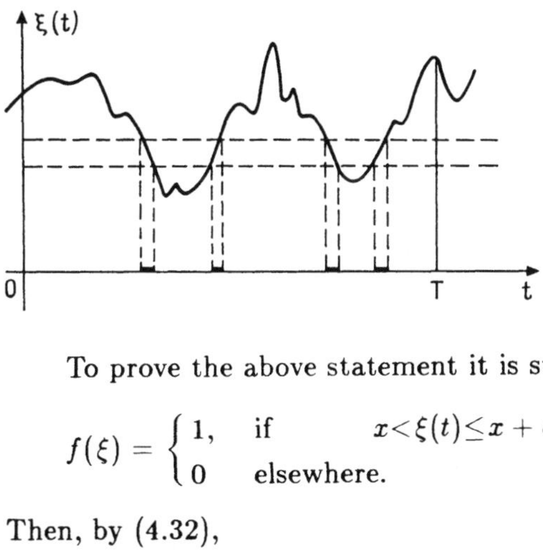

Fig. 4.2. An illustration of the definition of the relative residence time

To prove the above statement it is sufficient to take $f(\xi)$ to be given by

$$f(\xi) = \begin{cases} 1, & \text{if} \quad x < \xi(t) \le x + dx, \\ 0 & \text{elsewhere.} \end{cases}$$

Then, by (4.32),

$$\widetilde{f_T(\xi)} = \frac{T_{(x,x+dx)}}{T},$$

and by (4.33),

$$\overline{f_T(\xi)} = p_T(x)dx,$$

i.e., the distribution $p_T(x)dx$ which can be derived from the univariate probability density $w_1(t,x)$ by averaging over the parameter t, see (4.34), implies the ensemble average over the occupation time (or residence-time)

$$\frac{\overline{T_{(x,x+dx)}}}{T} = p_T(x)dx = \frac{dx}{T}\int_0^T w_1(t,x)dt. \tag{4.39}$$

Now theorem (4.38) takes the form

$$\underset{T\to\infty}{\text{l.i.m.}}\left\{\frac{T_{(x,x+dx)}}{T} - p_T(x)dx\right\} = 0. \tag{4.40}$$

All the above holds for the random process $f[\xi(t)]$, meeting the condition (4.37), i.e., the so-called ***ergodicity condition*** for this process. The necessary and sufficient condition (4.37) can be replaced by more stringent requirements – the *sufficient* conditions of ergodicity, which are fulfilled in applications of the theory in most cases. For example, the condition (4.37) will be met if the covariance $\psi_f(t,t')$ is bounded everywhere and decreases for any t with increasing $|t'-t|$, as $|t'-t|^{-\alpha}$ where $\alpha>0$. According to (4.36), this will be satisfied, if for any t, $|t'-t|$ increases sufficiently fast so as to lead to the statistical independence between $\xi(t)$ and $\xi(t')$,

$$w_2(t,x;t',x')\to w_1(t,x)w_1(t',x') \quad \text{as} \quad |t'-t|\to\infty. \tag{4.41}$$

The last requirement indicates in particular that the system in question forgets its history sufficiently quickly. By virtue of (4.2) it may also be formulated in terms of the conditional probability:

$$v(t',x'|t,x)\to w_1(t',x') \quad \text{as} \quad t'-t\to\infty. \tag{4.42}$$

For a Markov process this signifies that the transition probability into the state x' at time t' is already independent of its earlier states, and the probability simply becomes the univariate probability of the state. Rather often the ergodicity conditions are thought to be simply the requirements (4.41) or (4.42), although if these appear not to be satisfied, the necessary condition (4.37) may still hold.

Consider some examples of the distribution $p_T(x)dx$.

Let $\xi(t) = F(t)\eta(t)$, where $F(t)$ is a deterministic function of time, and $\eta(t)$ a *stationary* random process with the univariate distribution function $w(y)dy$. If we introduce as well a "distribution function" for $F(t)$, viz., $w_F(z)dz = \delta[z-F(t)]dz$, then we may use the well-known relationship for the distribution of the product of two random variables to derive the univariate probability density for $\xi(t)$. We have

$$\begin{aligned} w_1(t,x) &= \int_{-\infty}^{+\infty} w\left(\frac{x}{y}\right) w_F(y)\frac{dy}{|y|} \\ &= \int_{-\infty}^{+\infty} w\left(\frac{x}{y}\right)\delta[y-F(t)]\frac{dy}{|y|} = w\left[\frac{x}{F(t)}\right]\frac{1}{|F(t)|}. \end{aligned} \tag{4.43}$$

Substituting this into (4.34) gives for the mean relative time of residence

$$p_T(x)dx = \frac{dx}{T}\int_0^T w\left[\frac{x}{F(t)}\right]\frac{dt}{|F(t)|}. \tag{4.44}$$

In radio engineering an important role is played by the so-called *periodically-nonstationary processes* whose univariate probability density, and hence the moments $\overline{\xi^n}$, are periodic functions of t. When calculating $p_T(x)$ it is in this case apparently advisable to think of T as the period of $w_1(x,t)$. This renders $p_T(x)$ independent of T and equal to $p_\infty(x)$.

Let $w_1(t,x)$ be a normal law with the harmonically varying standard deviation $\sigma = \sigma_0(1+k\ \cos\omega t)$

$$w_1(t,x) = \frac{1}{\sqrt{2\pi}\sigma_0(1+k\ \cos\omega t)} \times \exp\left\{-\frac{x^2}{2\sigma_0^2(1+k\ \cos\omega t)^2}\right\} \quad \text{and} \quad k\leq 1.$$

Introducing $T = 2\pi/\omega$ and the integration variable $\varphi = \omega t$ gives

$$p_T(x) = \frac{1}{(2\pi)^{3/2}\sigma_0}\int_0^{2\pi} \exp\left\{-\frac{x^2}{2\sigma_0^2(1+k\ \cos\varphi)^2}\right\}\frac{d\varphi}{1+k\ \cos\varphi}.$$

The substitution

$$\frac{1-k^2}{1+k\ \cos\varphi} = 1-k\ \cos\theta$$

gives

$$p_T(x) = \frac{2}{(2\pi)^{3/2}\sigma_0\sqrt{1-k^2}}\int_0^{\pi} e^{-\beta(1-k\ \cos\theta)^2}d\theta,$$

$$\beta = \frac{x^2}{2\sigma_0^2(1-k^2)^2}.$$

If $k\to 0$, then $w_1(t,x)$ becomes the standard Gaussian distribution $w_1(x)$ with the variance σ_0^2 and then $p_T(x)$ coincides with $w_1(x)$. If $k\to 1$, i.e., $k = 1-\varepsilon$ with $\varepsilon\to 0$, the integral is dominated by small values of θ. We can now use the approximate relation

$$1-k\ \cos\theta = \varepsilon + \theta^2/2 = \varepsilon(1+s^2) \quad \text{with} \quad s = \theta/\sqrt{2\varepsilon}$$

and extend the integration limits with respect to s from 0 to ∞ to obtain

$$p_T(x) \approx \frac{2}{(2\pi)^{3/2}\sigma_0} \int_0^\infty \exp\left\{-\frac{x^2}{8\sigma_0^2}(1+s^2)^2\right\} ds.$$

This expression is valid for sufficiently small ε (i.e., $k \approx 1$). For $x \gg \sigma_0$ we can discard s^4 in the exponent to obtain

$$p_T(x) \approx \frac{1}{\sqrt{2\pi}|x|} \exp\left\{-\frac{x^2}{8\sigma_0^2}\right\} \quad (x \gg \sigma_0).$$

For $x \ll \sigma_0$ it is only the s^4-term in the exponent that is of significance. Hence, we get

$$p_T(x) \approx \frac{2}{(2\pi)^{3/2}\sigma_0} \int_0^\infty \exp\left\{-\frac{x^2 s^4}{8\sigma_0^2}\right\} ds = \frac{\sqrt[4]{2}\,\Gamma(1/4)}{\sqrt[4]{\pi^3 \sigma_0 |x|}} \quad (x \ll \sigma_0).$$

It is seen that if the standard deviation becomes approximately zero even for a short time, this markedly increases the relative occupation time of $\xi(t)$ in the range of small values of x.

How will the earlier results be affected if the *stationary* process $\xi(t)$ is ergodic? In this case w_1 is independent of t, and w_2 is only dependent on the difference $t'-t$, whereby the covariance for $f(\xi)$ is also dependent only on $t'-t$. The necessary and sufficient ergodicity condition (4.37) takes now the form

$$\lim_{T\to\infty} \frac{1}{T^2} \int_0^T \int_0^T \psi_f(t'-t)\,dt\,dt' = 0. \tag{4.45}$$

Hence another form of the condition is readily obtained that contains the single integral of $\psi_f(\tau)$ alone. Considering that $\psi_f(t'-t)$ is an even function of $\tau = t'-t$, we have

$$\begin{aligned}\frac{1}{T^2} \iint_0^T \psi_f(t'-t)\,dt\,dt' &= \frac{2}{T^2} \int_0^T dt \int_0^t \psi_f(t'-t)\,dt' \\ &= \frac{2}{T^2} \int_0^T dt \int_0^t \psi_f(\tau)\,d\tau. \end{aligned} \tag{4.46}$$

But for the covariance of the wide-sense stationary process the following inequality is valid $[\sigma^2 = \psi_f(0)]$:

$$\frac{2\sigma^2}{T^2} \int_0^{2T} dt \int_0^t \psi_f(\tau)\,d\tau \geq \left[\frac{2}{T} \int_0^T \psi_f(\tau)\,d\tau\right]^2 \tag{4.47}$$

(see [4.6] and [4.1], Chap. 1, Exercise 3). By (4.46,47), if the condition (4.45) is met, the condition

$$\lim_{T\to\infty} \frac{1}{T} \int_0^T \psi_f(\tau)d\tau = 0, \tag{4.48}$$

is also valid, thus becoming a necessary condition of ergodicity for a stationary random process. But it is also sufficient, since (4.46) can be written as

$$\frac{1}{T^2} \iint_0^T \psi_f(t'-t)dt\,dt' = \frac{2}{T^2} \int_0^T \frac{t}{T} dt \Big(\frac{1}{t} \int_0^t \psi_f(\tau)d\tau \Big)$$

whence it follows that, if the condition (4.48) is satisfied, so is (4.45).

Relation (4.48) – that is, the necessary and sufficient condition for ergodicity of a wide-sense stationary random process – is known as the Slutsky condition (see [4.7]). It assumes that $\psi_f(\tau)$ does not tend to zero as $\tau\to\infty$, but contains terms such as $a \cos \omega\tau$ [8]. In this case

$$\frac{1}{T} \int_0^T \psi_f(\tau)d\tau = \frac{a \sin \omega T}{\omega T},$$

i.e., (4.48) is met.

If the integral of $\psi_f(\tau)$ exists between the limits 0 and ∞, then we can introduce the effective correlation time τ_0

$$\tau_0 = \frac{1}{\sigma_f^2} \int_0^\infty \psi_f(\tau)d\tau \quad \text{with} \quad \sigma_f^2 = \psi_f(0)$$

(in fact, $\psi_f(\tau)$ may have several characteristic time scales). In this case the left-hand side of the condition (4.47) is for sufficiently large T (i.e., for $T \gg \tau_0$) approximately $\sigma_f^2\tau_0/T$.

The *sufficient* conditions of ergodicity of the stationary process may still be formulated either as the requirement that

$$\lim_{\tau\to\infty} \psi_f(\tau) = 0, \tag{4.49}$$

or as in (4.41) or (4.42)

$$\lim_{\tau\to\infty} w_2(x,\tau,x') = w_1(x)w_1(x'),$$

$$\lim_{\tau\to\infty} v(x'|\tau,x) = w_1(x'). \tag{4.50}$$

[8] It has already been mentioned (Sect. 4.5) that a covariance of this kind is inherent to harmonic oscillations with random and independent amplitude and phase, in particular, if the phase is uniformly distributed over the interval $(0, 2\pi)$.

But perhaps the most remarkable thing is that for the stationary process, the distribution $p_T(x)$ coincides with the univariate distribution function $w_1(x)$, as the latter is independent of t [see (4.34)]. Thus, the theorems (4.38–40) state that by *the ergodicity condition* (4.48) we have,

$$\underset{T\to\infty}{\text{l.i.m.}}\, \widetilde{f_T(\xi)} = \overline{f(\xi)}, \tag{4.51}$$

i.e., the time average of $f(\xi)$ converges in the mean square (and in probability) to the statistical mean, and in particular

$$\underset{T\to\infty}{\text{l.i.m.}}\, \widetilde{\xi_T} = \underset{T\to\infty}{\text{l.i.m.}}\, \frac{1}{T}\int_0^T \xi(t)dt = \overline{\xi}. \tag{4.52}$$

Further,

$$\underset{T\to\infty}{\text{l.i.m.}}\, \frac{T_{(x,x+dx)}}{T} = w_1(x)dx, \tag{4.53}$$

i.e., the *relative occupation time converges in the mean square (and in probability) to a univariate distribution function.* The latter is simply the statistical mean of the relative occupation time

$$\frac{\overline{T_{x,x+dx}}}{T} = w_1(x)dx. \tag{4.54}$$

This gives the experimentalist dealing with a stationary ergodic process a direct method for measuring the univariate distribution function: the probability of $\xi(t)$ belonging to an interval (a, b) is measured in terms of the relative occupation time of $\xi(t)$ in this interval over a sufficiently long period of time T.

So far we have been concerned with ergodicity conditions as applied to the time average of the random function $f[\xi(t)]$, i.e., a function dependent on the value of $\xi(t)$ at some moment of time (sometimes called the ergodicity of the first order). What can be said about the ergodicity of the second order, i.e., time averages for the functions $f[\xi(t), \xi(t')]$? Here we confined ourselves to the special case of the stationary process $\xi(t)$ and the function f equal to the product $\xi(t)\xi(t')$. What is important here, is the conditions under which the time average

$$\widetilde{[\xi(t)\xi(t+\tau)]}_T = \frac{1}{T}\int_0^T \xi(t)\xi(t+\tau)dt$$

converges to its statistical mean (as $T\to\infty$), which is readily seen to be equal to the covariance of the process $\xi(t)$[9],

[9] For simplicity we consider that $\overline{\xi} = 0$.

$$\widetilde{[\xi(t)\xi(t+\tau)]}_T = \frac{1}{T}\int_0^T \overline{\xi(t)\xi(t+\tau)}dt = \frac{1}{T}\int_0^T \psi(\tau)dt = \psi(\tau).$$

Evidently, if such a convergence does exist, i.e.,

$$\underset{T\to\infty}{\text{l.i.m.}} \frac{1}{T}\int_0^T \xi(t)\xi(t+\tau)dt = \psi(\tau), \tag{4.55}$$

this means that the covariance can be measured by averaging the product $\xi(t)\xi(t+\tau)$ over a sufficiently long period of time T.

The equality (4.55) means that the variance of the random variable $\widetilde{[\xi(t)\xi(t+\tau)]}_T$ tends to zero as $T\to\infty$. If the covariance of the random variable $\eta_\tau(t) = \xi(t)\xi(t+\tau)$ is denoted by $\psi_\eta(\sigma)$:

$$\begin{aligned}\psi_\eta(\sigma) &= \langle\eta_\tau(t)\eta_\tau(t+\sigma)\rangle - \langle\eta_\tau(t)\rangle\langle\eta_\tau(t+\sigma)\rangle \\ &= \langle\xi(t)\xi(t+\tau)\xi(t+\sigma)\xi(t+\sigma+\tau)\rangle - \psi^2(\tau),\end{aligned}$$

then, as has been shown by E.E. Slutsky (1938), the above requirement reduces to

$$\lim_{T\to\infty} \frac{1}{T}\int_0^T \psi_\eta(\sigma)d\sigma = 0. \tag{4.56}$$

If $\xi(t)$ is treated as stationary in the wide sense (i.e., it is not held that the distribution functions w_n for $\xi(t)$ are invariant under a shift of the time origin at $n\geq 3$), then the statement that ψ_η is independent of t is an additional independent assumption. If $\xi(t)$ is taken to be stationary in the strict sense, then the fourth moment will only depend on differences of time, i.e., on τ and σ. It normally follows from physical considerations that the dependence between $\eta(t)$ and $\eta(t+\sigma)$ vanishes as $\sigma\to\infty$ so that $\psi_\eta(\sigma)\to 0$ and then the condition (4.56) is met. Thus the theorem (4.55) is also valid.

As ψ_η includes the fourth moment, it does not in general enable us to express (4.56), i.e. the condition under which the theorem (4.55) for the second moment is valid, in terms of the covariance $\psi(\tau)$ of the process $\xi(t)$ itself. However, in the special and important case where $\xi(t)$ is a *normal* function this is possible, and hence *all* of its moments, including the fourth, may be expressed in terms of $\bar{\xi}$ and $\psi(\tau)$. So, if $\xi(t)$ is a normal function (and for simplicity $\bar{\xi} = 0$), then the necessary and sufficient condition for (4.55) to be satisfied will be

$$\lim_{T\to\infty} \frac{1}{T}\int_0^T |\psi(\tau)|^2 d\tau = 0. \tag{4.57}$$

If the covariance $\psi(\tau)$ contains periodic terms, then condition (4.56) is satisfied, but (4.57) is no longer valid. Note that (4.57) allows the ergodic theorem to be satisfied not only for the second moments, but also for those of any order, i.e., all the time averages of the products $\xi(t+\tau_1)\xi(t+\tau_2)\ldots\xi(t+\tau_n)$ as performed over the period $(0,T)$ in probability converge for $T\to\infty$ to the corresponding moments. Moreover, for *any function* of *normal* variables $\xi(t+\tau)$

$$f[\xi(t+\tau_1),\ \xi(t+\tau_2),\ldots,\xi(t+\tau_n)]$$

the theorem is also correct (if the mean of its modulus, $\overline{|f|}$, is finite): as $T\to\infty$ we have

$$\operatorname*{l.i.m.}_{T\to\infty} f_T = \operatorname*{l.i.m.}_{T\to\infty}\frac{1}{T}\int_0^T f[\xi(t+\tau_1),\ \xi(t+\tau_2),$$
$$\ldots,\xi(t+\tau_n)]dt = \overline{f(\xi_1,\xi_2,\ldots,\xi_n)}.$$

4.8 Exercises

4.8.1. Use the answer to Exercise 5 of Chap. 3 to write the characteristic function for the n-variate distribution of a normal random process $\xi(t)$ with the mean $\overline{\xi(t)}=0$.

Solution. The elements of the covariance matrix for n random variables $\xi(t_1),\ldots,\xi(t_n)$ are

$$b_{ik}=\overline{\xi(t_i)\xi(t_k)}=\psi(t_i,t_k),\quad \text{thus}$$

$$\varphi_n(u_1,t_1;u_2,t_2;\ldots;u_n,t_n)=\exp\left\{-\frac{1}{2}\sum_{i,k=1}^{n}\psi(t_i,t_k)u_iu_k\right\}.$$

If the process $\xi(t)$ is wide-sense stationary, then

$$\varphi_n(u_1,t_1;u_2,t_2;\ldots;u_n,t_n)=\exp\left\{-\frac{1}{2}\sum_{i,k=1}^{n}\psi(t_1-t_k)u_iu_k\right\},$$

which means that for any n the n-variate probability density depends also only on differences between n times. Thus, for a normal process the stationarity in the wide sense entails the stationarity in the strict sense.

4.8.2. From a bivariate distribution w_2 of a random process $\xi(t)$ obtain the joint distribution function $\tilde{w}$ for $\xi(t)$ and $\dot{\xi}(t)$. Find $\tilde{w}$ for the stationary normal process.

Solution. We have $w_2(x,t;x_\tau,t+\tau)$, where x_τ are possible values of $\xi(t+\tau)$. Instead of x_τ we introduce a new variable, $y=(x_\tau-x)/\tau$, so that we have $y\to\dot{x}$

as $\tau \to 0$ (for example, in the mean square). To the first order in τ we may write $x_\tau = x + y\tau$, and hence

$$w_2(x,t;x_\tau,t+\tau)dx\,dx_\tau \approx w_2(x,t;x+y\tau,t+\tau)\tau\,dx\,dy.$$

In the limit as $\tau \to 0$, we have

$$\tilde{w}(x,y,t)dx\,dy = dx\,dy \lim_{\tau\to 0} \tau w_2(x,t;x+y\tau,t). \tag{4.58}$$

For the stationary normal process (for simplicity we put $\overline{x_1^2} = \overline{x_2^2} = \sigma^2$) we get

$$\varphi_2(u_1,u_2,\tau) = \exp\left\{-\frac{\sigma^2}{2}(u_1^2+u_2^2+2K(\tau)u_1u_2)\right\},$$

where $K(\tau)$ is the correlation coefficient. As $K(\tau)$ has a maximum (equal to unity) at $\tau = 0$, its power expansion is

$$K(\tau) = 1 - \tfrac{1}{2}|K''|_0 \cdot \tau^2 + \dots$$

Hence

$$\tilde{w}(x,y) = \lim_{\tau\to 0}\frac{\tau}{4\pi^2}\int\!\!\int_{-\infty}^{+\infty} \exp\left\{-\frac{\sigma^2}{2}\left[u_1^2+u_2^2+2\left(1-\frac{1}{2}|K''|_0\tau^2\right)u_1u_2\right]\right.$$
$$\left. - i[u_1x + u_2(x+y\tau)]\right\}du_1\,du_2.$$

If we introduce the integration variables $\alpha = u_1 + u_2$ and $\beta = \tau(u_2 - u_1)/2$, then we obtain

$$\tilde{w}(x,y) = \lim_{\tau\to 0}\left\{\frac{1}{2\pi}\int\!\!\int_{-\infty}^{+\infty}\exp\left[-\frac{\sigma^2\alpha^2}{2}\left(1-|K''|_0\frac{\tau^2}{4}\right)+i\alpha\left(x+\frac{y\tau}{2}\right)\right]d\alpha\right\}$$
$$\times\frac{1}{2\pi}\int_{-\infty}^{+\infty}\exp\left\{-\frac{\sigma^2\beta^2}{2}|K''|_0 + i\beta y\right\}d\beta$$
$$= \frac{1}{\sqrt{2\pi}\sigma}\exp\left\{-\frac{x^2}{2\sigma^2}\right\}\frac{1}{\sqrt{2\pi}\sigma\sqrt{|K''|_0}}\exp\left\{-\frac{y^2}{2\sigma^2}|K''|_0\right\} . \tag{4.59}$$

Thus, ξ and $\dot{\xi}$ are normal and independent (at the same instant of time), and

$$D[\xi] = \sigma^2, \quad D[\dot{\xi}] = \sigma^2|K''(0)|. \tag{4.60}$$

4.8.3. Show that for a random process $\xi(t)$ stationary in the strict sense, the derivatives $\xi^{(n)}(t)$ and $\xi^{(n+1)}(t)$, which are adjacent in order, are uncorrelated.

Solution. To calculate the nth derivative, requires the $(n+1)$-variate distribution of the process $\xi(t)$. Consequently, the stationarity and the stationary connection of *all* the derivatives require that $\xi(t)$ be stationary in the strict sense.

For adjacent derivatives at one and the same instant of time t we have

$$\langle \xi^{(n)}(t)\xi^{(n+1)}(t)\rangle = \left\langle \frac{d}{dt}\left(\frac{1}{2}[\xi^{(n)}(t)]^2\right)\right\rangle = \frac{1}{2}\frac{d}{dt}\langle [\xi^{(n)}(t)]^2\rangle = 0,$$

as the mean square of any derivative is constant. In particular,

$$\overline{\xi\dot{\xi}} = 0, \quad \overline{\dot{\xi}\ddot{\xi}} = 0,$$

but generally $\overline{\dot{\xi}\ddot{\xi}}$ is nonzero.

5. Markov Processes

For reasons discussed in Sect. 5.1, Markov processes have a wide variety of applications. In Sect. 5.2 we derive the basic equation for the theory of Markov processes, the Smoluchowski integral equation (often referred to as the Smoluchowski-Chapman equation) and draw attention to the main types of these processes, including Markov chains.

The simplest process is one with *discrete* possible states, which is described by the Kolmogorov equations (Sect. 5.3). Before turning to the general theory of processes with *continuous* possible states, it is instructive to derive some prominent relations from the Kolmogorov equations (Sect. 5.4). This is a straightforward exercise not only for a random walk in one dimension (along the x-axis) but also in two dimensions (in the xy-plane). The latter case leads to the Rayleigh distribution, which is so important for radiophysics that the whole of Sect. 5.5 is devoted to this topic and some of its generalizations; in spite of the fact that such an arrangement detracts from a smooth and consecutive presentation of Markov processes.

Markov processes may be continuous and discrete, not only in terms of the possible values attained but also in terms of their variation with time (continuous or stepwise). For processes continuous in time, the Smoluchowski equation leads to the diffusion or Einstein-Fokker-Planck equation (Sect. 5.6). Its application to specific physical problems often involves *multicomponent* random processes. Appropriate generalizations of the theory are given in Sect. 5.7. A discussion of a number of physical problems follows: fluctuations in the Thomson vacuum-tube oscillator (Sects. 5.8, 9), rotational Brownian motion, and random refraction of a ray (Sect. 5.10).

The distributions of Markov processes with continuous sets of possible states – but stepwise (instantaneous) transitions from one state to another – obey the Kolmogorov-Feller equation (Sect. 5.11).

The chapter concludes with a discussion of the first passage problem with given boundary conditions (Sect. 5.12). This problem can also be stated for non-Markov processes, but it is included in this chapter because its complete solution is only possible for Markov processes.

5.1 Preliminary Remarks

Though Markov processes (i.e., processes without stochastic memory-effects) constitute a rather peculiar class of random processes, they are of great significance because their inherent conditions appear to be met in a wide variety of applications. Random processes of a general nature can often be reduced to the scheme appropriate for a process without memory effect. For that purpose a more detailed description of the process under consideration has to be adopted, i.e., more variables are required to describe the state of the system. To clarify this motion, we return to the case of a deterministic process, allowing a dynamic system to be described by a differential equation not of the first, (as assumed in Sect. 4.2) but of the second order. Then the solution with the initial conditions

$$x = x_0, \quad \dot{x} = u_0 \quad \text{at} \quad t = t_0$$

will be

$$x = f(t, x_0, u_0, t_0).$$

If we now agree that the state of the system may be the coordinate x, the conditional probability density for x at time t may be written as

$$v(t, x|t_0, x_0, u_0) = \delta[x - f(t, x_0, u_0, t_0)]. \tag{5.1}$$

But knowing u_0 is equivalent to knowing *two* values of x at fairly close instants of time [say x_0 at $t = t_0$ and $x_0 - h$ at $t = t_0 - \tau$, so that $u_0 = \lim_{\tau \to 0}(h/\tau)$]. Thus, the conditional probability (5.1) depends in fact on *two* preliminary states,

$$v(t, x|t_0, x_0, u_0) \approx v(t, x|t_0, x_0; t_0 - \tau, x_0 - h),$$

and hence the statistical generalization (5.1) is not a Markov process.

Not only a statistical, but also a dynamic theory normally prefers to avoid a dependence of the state of the system on its behavior before a given initial moment. This is achieved by *extending the very notion of the state* at a time t by the introduction of new variables to characterize the state. In this example the procedure is used to introduce, in addition to the coordinate x, the velocity $\dot{x} = u$. If the state is specified by given x and u at a time t, we may write the conditional probability of this state for the dynamic process under consideration in the form

$$\delta[x - f(t, x_0, u_0, t_0)]\delta[u - \dot{f}(t, x_0, u_0, t_0)].$$

This is a special case of the transition probability $v(t, x, u|t_0, x_0, u_0)$. In the corresponding statistical scheme we thus arrive at a *Markov* process; but this time for *two* random functions x and u, i.e., for a bivariate random function.

In an analogous way, a k-variate random process that is not Markovian can be made Markovian by introducing a rather extended assembly of coordinates, but only for a larger number of dimensions k'. Generally speaking, this can be achieved by treating the state of the system as the combination of the value of the process at the last moment of observation t and a number of values from the history of the process of $t'<t$. However, if we require k' to be finite, then this procedure will not always be applicable. This is just the case in the previous example of a particle traveling in a liquid with a viscous memory effect (Sect. 4.5). This effect is accounted for by an integro-differential equation, i.e., it cannot be exhausted by a finite collection of derivatives of x at the time t, nor a finite collection of x values at times $t'\leq t$.

Depending on whether or not the possible values x of the random function $\xi(t)$ and of the paramter t, respectively, are discrete or continuous, the four versions referred to in Sect. 4.1 can be distinguished for Markov random functions as well.

If the time t is discrete, then instead of the values of t_n we may regard the numbers n of these values as arguments and speak about tests numbered by the integers n. The random function $\xi(t)$ is then reduced to the sequence of random variables $\xi(t_n) \equiv \xi_n$ (ξ at nth test). Although schemes with discrete t are in a certain respect more simple and pictorial, we will focus on the processes themselves, i.e., on cases of continuous t considered first by Einstein and Smoluchowski in their work on the theory of Brownian motion. As far as Markov sequences are concerned we shall confine ourselves to some remarks.

Closest to the classical theory of probability are the *discrete sequences*, i.e., *processes with discontinuous time* and with discrete possible values of the random variables $\xi_n = \xi(t_n)$. We denote these possible values by x_i ($i = 1, 2, \ldots$). The Markov nature of the process signifies that a probability exists of the transition from any of the values x_i in the lth test to any of the values x_k in nth test ($n>l$) :

$$p(n, x_k|l, x_i) = P\{\xi_n = x_k|\xi_l = x_i\}. \tag{5.2}$$

If, in particular, the number of the possible values (states) x_i is *finite* ($i = 1, 2, \ldots, N$), then the sequence is called a *simple Markov chain*.

The homogeneity of the Markov process under consideration implies that the transition probability depends on n and l through their difference only, i.e., on the number of steps $s = n - l$ made from the first test to the final one.

Evidently, Markov sequences may be thought of as a direct generalization of sequences of independent tests mentioned in Chap. 2. In terms of transition probability, it may be said that for a sequence of independent tests $p(n, x_k|l, x_i) = P\{\xi_n = x_k\}$ holds, i.e., the probability of the transition to the state x_k simply coincides with the probability of this state at the nth test, regardless of the results of other tests.

If the random variables ξ_n take on a continuous set of possible values x, then the Markov nature of this *continuous process in discrete time* suggests the existence of a transition probability that can be written both in the integral form

$$V(n, x|l, y) = P\{\xi_n \le x | \xi_l = y\},$$

and in the differential form

$$v(n, x|l, y)dx = d_x V = P\{x < \xi_n \le x + dx | \xi_l = y\}.$$

5.2 Smoluchowski Equation

The fundamental equation in the theory of random functions of the Markov type is the Smoluchowski equation [5.1], which simply states that the transition probabilities for any *three* consecutive times should be related in a certain way. To be more specific, consider a Markov *process* (with continuous t) and continuous possible values of the random function $\xi(t)$. We then have the following situation.

The transition from the state x_0 at time t_0 to the state x at time t [with the probability $v(t, x|t_0, x_0)dx$] is realized through a state y at an intermediate time θ ($t_0 < \theta < t$), i.e., it can be represented by two consecutive transitions $x_0 \to y$ and $y \to x$. The probability of the joint occurrence of the two transitions is

$$v(\theta, y|t_0, x_0)dy\, v(t, x|\theta, y)dx,$$

with the *same* conditional probability densities v (Markov processes). But the transition $x_0 \to y \to x$ is one of the mutually exclusive special cases of the transition $x_0 \to x$ (Fig. 5.1), so that the summation over all the possible states y at the time θ must give the total probability of the transition $x_0 \to x$,

$$v(t, x|t_0, x_0)dx = dx \int v(t, x|\theta, y)v(\theta, y|t_0, x_0)dy. \tag{5.3}$$

This is the Smoluchowski equation.

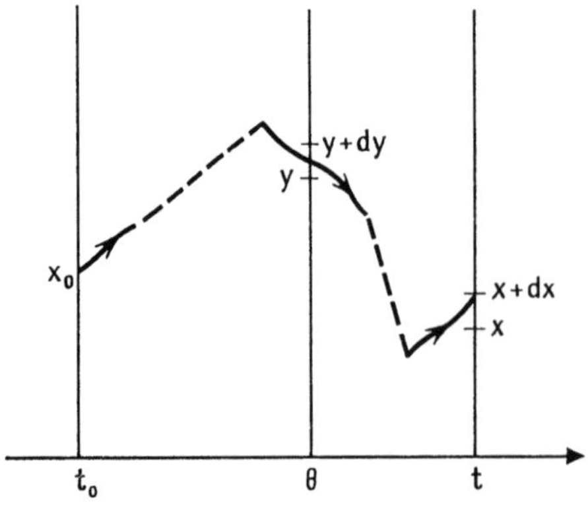

Fig. 5.1. Transition from the state x_0 at t_0 to the state $(x, x+dx)$ at t via the state $(y, y+dy)$ at some intermediate time θ

It should be noted once more that (5.3) imposes significant limitations on possible forms of the functions v : the integration with respect to y of the product of two v's should again give the function v, and should automatically exclude any dependence of the result on the intermediate time θ which enters into the integrand. In fact, this is just another manifestation of those limitations on the kind of n-variate distribution functions which characterize Markov processes. Indeed, we can arrive at the Smoluchowski equation using only the hierarchy condition for distributions of different dimensionality (Sect. 4.1) and the definition of the Markov process (4.3) (see Exercise 5.9.1).

Integrating (5.3) with respect to x from the lower boundary of the interval of possible values to x and going over from the Riemann integral with respect to y to the Stieltjes integral, gives the Smoluchowski equation for the integral transition probabilities

$$V(t,x|t_0,x_0) = \int V(t,x|\theta,y)d_yV(\theta,y|t_0,x_0) \quad (t_0<\theta<t). \tag{5.4}$$

This form holds for both continuous and discrete possible values of $\xi(t)$. In general, this distinction is secondary. Even without the use of the Stieltjes integral and dealing only with probability densities, in the discrete case we can always write these densities in terms of appropriate delta functions. Nevertheless, we will give the Smoluchowski equation for the process with discrete possible values of the random function $\xi(t)$, i.e., for the transition probabilities $p(t,x_k|t_0,x_i)$,

$$p(t,x_k|t_0,x_i) = \sum_j p(t,x_k|\theta,x_j)p(\theta,x_j|t_0,x_i) \quad (t_0<\theta<t). \tag{5.5}$$

With random Markov *sequences*, i.e., for discrete values of the parameter t, in (5.3–5) we can substitute the arguments t_n by the test numbers n. For example, instead of (5.5) we get

$$p(n,x_k|l,x_i) = \sum_j p(n,x_k|m,x_j)p(m,x_j|l,x_i) \quad (l<m<n). \tag{5.6}$$

For the case where the number of the possible discrete values is finite (Markov chains), i.e., $i,j,k = 1,2,\ldots,N$, the relation (5.6) is due to Markov and is called the *Markov equation.*

If a random function of the Markov type is ***homogeneous in t***, (5.3) becomes

$$v(x|t-t_0,x_0) = \int_{-\infty}^{+\infty} v(x|t-\theta,y)v(y|\theta-t_0,x_0)dy \quad (t_0<\theta<t), \tag{5.7}$$

and (5.6) becomes

$$p(x_k|n-l,x_i) = \sum_j p(x_k|n-m,x_j)p(x_j|m-l,x_i) \quad (l<m<n). \tag{5.8}$$

As stated earlier, we shall not concentrate on the theory of Markov sequences and chains (see [5.2]), but we shall consider the application of (5.8) to the homogeneous Markov chain as an illustrative example. We introduce a short-hand notation for the transition probability in s steps,

$$p(x_k|s, x_i) = p_{ik}(s).$$

The Markov equation (5.8) may then be written as

$$p_{ik}(s) = \sum_{j=1}^{N} p_{ij}(m)p_{jk}(s-m) \quad (0<m<s). \tag{5.9}$$

The variables $p_{ik}(s)$ form the matrix π_s of transition probabilities (the transition matrix for short) in s steps. Equation (5.9) shows that the matrix π_s is equal to the product of the matrices π_m and π_{s-m}

$$\pi_s = \pi_m \pi_{s-m} \quad (0<m<s).$$

But for $s = 2$ only $m = 1$ is possible, so that $\pi_2 = \pi_1^2$. At $s = 3$ we have $\pi_3 = \pi_1\pi_2 = \pi_2\pi_1 = \pi_1^3$, and in general

$$\pi_s = \pi_1^s. \tag{5.10}$$

This relationship expresses the transition probability in s steps in terms of the transition probability $p_{ik}(1) \equiv p_{ik}$ in one step. Note that in the transition matrix

$$\pi_1 = \begin{pmatrix} p_{11} & p_{12} \cdots p_{1N} \\ p_{21} & p_{22} \cdots p_{2N} \\ \cdots & \cdots \\ p_{N1} & p_{N2} \cdots p_{NN} \end{pmatrix}$$

all the elements are of course nonnegative, and none of the columns or rows contains only zeros (x_k only include *possible* final states), because for any initial state x_i

$$\sum_{k=1}^{N} p_{ik} = 1.$$

As an illustration, consider the one-dimensional random walk of a particle between two reflecting walls (Fig. 5.2). Everywhere, save for positions $i = 1$ and $i = N$, a step to the right has the probability p and a step to the left the probability $q = 1 - p$. At the walls the particle is bound to move one step inward. This is a typical homogeneous Markov chain with the following one-step transition probabilities $(1<i<N)$:

$$p_{ik} = \begin{cases} p & \text{if } k = i + 1, \\ q & \text{if } k = i - 1, \\ 0 & \text{elsewhere,} \end{cases}$$

Fig. 5.2. Unidimensional random walk of a particle between walls separated by N steps

$$p_{ik} = \begin{cases} 1 & \text{if } k = 2, \\ 0 & \text{elsewhere,} \end{cases}$$

$$p_{Nk} = \begin{cases} 1 & \text{if } k = N-1, \\ 0 & \text{elsewhere.} \end{cases}$$

The matrix π_1 is thus

$$\pi_1 = (p_{ik}) = \begin{pmatrix} 0 & 1 & 0 & 0\cdots 0 & 0 & 0 \\ q & 0 & p & 0\cdots 0 & 0 & 0 \\ 0 & q & 0 & p\cdots 0 & 0 & 0 \\ \cdots & \cdots & \cdots & \cdots & \cdots & \cdots \\ 0 & 0 & 0 & 0\cdots q & 0 & p \\ 0 & 0 & 0 & 0\cdots 0 & 1 & 0 \end{pmatrix}.$$

It can easily be seen that with the *inhomogeneous* Markov sequence the transition probabilities in s steps can also be expressed in terms of one-step transition probabilities. The latter are now only dependent on the test number, therefore we denote them as follows

$$p(n+1, y|n, x) = p_n(y|x).$$

According to (5.6), we have

$$p(n+2, y|n, x) = \sum_z p_{n+1}(y|z)p_n(z|x),$$

$$p(n+s, y|n, x) = \sum_{z,u,\ldots,v} p_{n+s-1}(y|z)p_{n+s-2}(z|u)\ldots p_n(v|x).$$

For the continuous function $\xi(t)$, the result can be written in terms of probability densities

$$v(n+s, y|n, x)$$
$$= \int \cdots \int v_{n+s-1}(y|z)v_{n+s-2}(z|u)\ldots v_n(w|x)dz\,du\ldots dw.$$

It follows that both relationships are simply iterated compositions of the distribution functions involved, with the results that they may be written as

$$p(n+s, y|n, x) = p_{n+s-1}(y|z) * p_{n+s-2}(z|u) * \ldots * p_n(w|x).$$

In the special case of the homogeneous Markov chain, where probabilities in a single transition do not depend on its number, this gives for the s-step transition matrix the formula (5.10).

5.3 Markov Process with Discrete States

We rewrite the Smoluchowski equation (5.5) for three successive times $t_0 < t < t + \tau$:

$$p(t+\tau, x_k|t_0, x_i) = \sum_j p(t+\tau, x_k|t, x_j) p(t, x_j|t_0, x_i). \tag{5.11}$$

We assume that for sufficiently small τ the transition probability is

$$p(t+\tau, x_k|t, x_j) = \delta_{jk} + A_{jk}\tau + o(\tau), \tag{5.12}$$

where $o(\tau)$ stands for the higher terms in τ. Equation (5.12) allows for the fact that as $\tau \to 0$, the finite state coincides with certainty with the initial one, i.e.,

$$p(t, x_k|t, x_j) = \delta_{jk}. \tag{5.13}$$

Moreover, we assume that the derivative

$$A_{jk}(t) = \left.\frac{\partial p(u, x_k|t, x_j)}{\partial u}\right|_{u=t} \tag{5.14}$$

exists. Obviously for $j \neq k$ (i.e., $\delta_{jk} = 0$) $A_{jk}(t)dt$ is the probability of the transition in time dt from the state x_j to *another* state x_k. Thus, for $j \neq k$ we have

$$A_{jk}(t) \geq 0 \quad (j \neq k). \tag{5.15}$$

As the probability of the transition to *any* of the possible states must be unity

$$\sum_k p(t+\tau, x_k|t, x_j) = 1, \tag{5.16}$$

we have

$$\sum_k A_{jk}(t) = 0, \quad A_{jj}(t) = -\sum_{k \neq j} A_{jk}(t) \leq 0. \tag{5.17}$$

We now substitute (5.12) into the right-hand side of (5.11) and pass to the limit $\tau \to 0$. This leads to the following basic equation (or rather, the system of equations) for the Markov process with discrete states[1]:

$$\frac{\partial p(t,x_k|t_0,x_i)}{\partial t} = \sum_j A_{jk}(t)p(t,x_j|t_0,x_i). \tag{5.18}$$

With initial conditions (5.13) these equations define the time dependence of the transition probabilities. If the number of possible states is finite, then it can easily be shown that for any continuous $A_{ik}(t)$ meeting the conditions (5.15) and (5.17), the equation (5.18) with the initial condition (5.13) has a unique solution which is nonnegative and satisfies (5.11) and (5.16), i.e., it defines the Markov process. For an infinite set of possible states, the question of the existence and uniqueness of a solution requires special study to be found in the mathematical literature. For practical purposes this issue is normally of no interest.

In addition to (5.18), the following differential equations with respect to the initial time t_0 are also valid [5.3]:

$$\frac{\partial p(t,x_k|t_0,x_i)}{\partial t_0} = -\sum_j A_{ij}(t_0)p(t,x_k|t_0,x_j). \tag{5.19}$$

These are readily deduced by choosing the intermediate time close to t_0.

Equations (5.18) are not only obeyed by the transition probability but also by the univariate probability of state $p_1(t,x_k)$. Indeed, if we have the initial distribution function $P(t_0,x_i)$, then

$$p_1(t,x_k) = \sum_i P(t_0,x_i)p(t,x_k|t_0,x_i). \tag{5.20}$$

Multiplying (5.18) by $P(t_0,x_i)$ and taking the summation over i gives, by (5.20),

$$\frac{\partial p_1(t,x_k)}{\partial t} = \sum_j A_{jk}(t)p_1(t,x_j). \tag{5.21}$$

These equations must be integrated with the initial conditions

$$p_1(t_0,x_k) = P(t_0,x_k). \tag{5.22}$$

If the Markov process under consideration is homogeneous, i.e., the transition probability has the form $p(x_k|t,x_i)$, (t instead of $t-t_0$), then according to (5.14), A_{jk} are constants and (5.18) becomes

$$\frac{\partial p(x_k|t,x_i)}{\partial t} = \sum_j A_{jk}p(x_j|t,x_i). \tag{5.23}$$

[1] A rigorous derivation is given in *Kolmogorov* [5.3]. Note that, by (5.13,14) the initial assumption of $t>t_0$ may be discarded, i.e., (5.18) also allows $t=t_0$.

Assuming that as $t \to \infty$ the transition probabilities "forget" the initial condition and become stationary probabilities of state (the existence of ergodicity)[2],

$$\lim_{t\to\infty} p(x_k|t, x_i) = P(x_k),$$

by (5.16) and (5.23)[3], we obtain for $P(x_k)$,

$$\sum_k P(x_k) = 1, \quad \sum_j A_{jk} P(x_j) = 0. \tag{5.24}$$

Consider two examples referring to *homogeneous* processes.

1. *Bilateral reaction.* A system may be in either of the two states 1 and 2. In particular, we can consider the disintegration of a particle (ionization, dissociation, chemical decomposition, radioactive decay), with state 1 being the undecomposed particle and state 2 the decomposed particle.

In the general case we can have both the disintegration process (the transition $1 \to 2$ with probability $\alpha\, dt$ in time dt), and the recovery process (the transition $2 \to 1$ with probability $\beta\, dt$ in time dt). Thus, $A_{12} = \alpha$, $A_{21} = \beta$ and, by (5.17), $A_{11} = -\alpha$, $A_{22} = -\beta$, so that equation (5.21) for the probabilities of the state

$$p_1(t, 1) \equiv P_1(t) \quad \text{and} \quad p_1(t, 2) \equiv P_2(t)$$

takes the form

$$dP_1/dt = -\alpha P_1 + \beta P_2, \quad dP_2/dt = \alpha P_1 - \beta P_2.$$

We could just as well dispense with the second equation. Instead, we could eliminate P_2 from the first equation using the fact that $P_1 + P_2 = 1$:

$$dP_1/dt = -\lambda P_1 + \beta \quad (\lambda = \alpha + \beta) \quad .$$

Let $P_1 = 1$ (correspondingly $P_2 = 0$) hold for $t = 0$, i.e., with certainty the system is initially in state 1. The solution is then

$$P_1(t) = e^{-\lambda t} + \frac{\beta}{\lambda}(1 - e^{-\lambda t}), \quad P_2(t) = \frac{\alpha}{\lambda}(1 - e^{-\lambda t}).$$

As $t \to \infty$, we arrive at stationary values of both probabilities, which are independent of the initial conditions

$$p_1 = \beta/\lambda, \quad P_2 = \alpha/\lambda,$$

[2] This will be the case, in particular, if *all* $A_{jk} \neq 0$, i.e., all the transitions are possible for small τ.

[3] It is assumed that $\partial p/\partial t \to 0$ as $t \to \infty$, which generally speaking does not follow from the existence of constant limits for the p themselves.

i.e., the process is ergodic. If the recovery is impossible ($\beta = 0$), as is the case, for instance, in radioactive decay, then

$$P_1(t) = e^{-\lambda t}, \quad P_2(t) = 1 - e^{-\lambda t}.$$

2. *Two-step relay.* Let such a relay be subject to a random sequence of control pulses having plus and minus signs with equal probability. Let the positive pulse create or maintain state 1, and the negative pulse, state 2. Then $A_{12} = A_{21} = \alpha$ (i.e., $\alpha\, dt$ is the probability of a *change* in the state in time dt) and, by (5.17), $A_{11} = A_{22} = -\alpha$. This appears to be a special case (with $\beta = \alpha$) of the scheme considered above. But now we write (5.23) for four *transition probabilities* ($x = 1, 2$)

$$dp(1|t,x)/dt = -\alpha p(1|t,x) + \alpha p(2|t,x),$$

$$dp(2|t,x)/dt = \alpha p(1|t,x) - \alpha p(2|t,x),$$

which are to be solved with the initial conditions (5.13). The respective solutions are

$$p(2|t,1) = p(1|t,2) = \tfrac{1}{2}(1 - e^{-2\alpha t}),$$

$$p(1|t,1) = p(2|t,2) = \tfrac{1}{2}(1 + e^{-2\alpha t}).$$

We see again that the process is ergodic: the transition probabilities tend to stationary values as $t \to \infty$ (as is to be expected, since all $A_{jk} \neq 0$),

$$P(1) = P(2) = \tfrac{1}{2}.$$

In both examples the probabilities tend monotonously to their limiting values. However, there may also be an oscillatory regime, i.e., damped oscillations. If we are seeking a particular solution of equations (5.23) proportional to $\exp(\lambda t)$, then for N possible values we find λ from the following characteristic equation (of degree N in λ) :

$$\begin{vmatrix} A_{11} - \lambda & A_{21} & \cdots & A_{N1} \\ A_{12} & A_{22} - \lambda & \cdots & A_{N2} \\ \cdots & \cdots & \cdots & \cdots \\ A_{1N} & A_{2N} & \cdots & A_{NN} - \lambda \end{vmatrix} = 0.$$

Ergodicity will occur if the polynomial in powers of λ that remains on the left-hand side after dividing by a certain power of λ is a Hurwitz polynomial, i.e., if all the nonzero solutions λ have negative real parts. Obviously, damped oscillations will only take place if the solutions include complex-conjugate pairs.

In the case of the two-step relay there are two solutions, namely $\lambda = 0$ and $\lambda = -2\alpha$. *Kolmogorov* [5.3] gives the following example for $N = 3$:

$$A_{12} = A_{23} = A_{31} = \alpha,$$

$$A_{21} = A_{32} = A_{13} = 0,$$

$$A_{11} = A_{22} = A_{33} = -\alpha.$$

Here the solutions to the characteristic equation are

$$\lambda = 0, \quad \lambda = -\frac{3\alpha}{2} + i\frac{\sqrt{3}\alpha}{2}, \quad \lambda = -\frac{3\alpha}{2} - i\frac{\sqrt{3}\alpha}{2}.$$

So we have damped oscillations and the transition probabilities thus tend to the stationary values $P(1) = P(2) = P(3) = 1/3$. For other examples and generalizations of multivariate cases see [Ref. 5.4, Sect. 3.2] and [5.5].

5.4 Transition from Discrete Sequence to Processes with Continuous Sets of States. Rayleigh Distribution

Before we go on to deduce the fundamental differential equation for Markov processes with a continuous set of possible states, we will consider a simplified procedure which is sometimes used: the time t and the set of states x are broken down into fairly small intervals Δt and Δx, the equation for the Markov sequence is then worked out and a limit is found as $\Delta t \to 0$ and $\Delta x \to 0$.

We now look at the specific example of a Brownian particle that, at times $N\Delta t$ $(N = 1, 2, \ldots)$, jumps by $\Delta x = a$ to the right or left with probabilities p and $q = 1 - p$, respectively. We know already this problem as the (random walk) problem of the "drunk man" or of the addition of oscillations with random amplitudes $\pm a$ (Sect. 2.4 and 2.6). We deal here with a homogeneous Markov sequence, which is not, however, stationary (the distribution varies with the test number).

It will be recalled that the probability that, out of a total of N steps, n steps will be made to the right [the particle will move to the right by $x = (2n - N)a$, or the intensity of the sum of N oscillations will be $J = (2n - N)^2 a^2$] is given by the binomial law $P_N(n) = C_N^n p^n q^{N-n}$. But now we will look at the problem from a somewhat different angle.

Denoting the abscissa of the particle at the time $t = N\Delta t$ by $x = ma$ and assuming that at time $t = 0$ the particle left the point $x = 0$, we introduce the *transition probability* for N steps

$$p(N\Delta t, ma|0,0) \equiv p_{N,m}.$$

It is not difficult to relate $p_{N,m}$ to the binomial probability $P_N(n)$ but, as stated above, we will now seek $p_{N,m}$ without using $P_N(n)$ (see Exercise 5.13.6).

The Nth step may place the particle at point $x = ma$, if the $(N - 1)$th step brought it to $(m-1)a$ or $(m+1)a$ with probabilities p and q, respectively.

Accordingly, the relation for the total probability yields

$$p_{N,m} = pp_{N-1,m-1} + qp_{N-1,m+1}$$

which is the Markov equation for this problem. If we change the notation for the transition probability by introducing $v(t,x)a = p_{N,m}$, then the equation will read

$$v(t,x) = pv(t-\Delta t, x-a) + qv(t-\Delta t, x+a).$$

We now expand v on the right-hand side in a power series of Δt and $a = \Delta x$. Considering that $p+q=1$, we obtain

$$\begin{aligned} 0 = & -\frac{\partial v}{\partial t}\Delta t - (p-q)\frac{\partial v}{\partial x}a + \frac{\partial^2 v}{\partial x^2}\frac{a^2}{2} \\ & + (p-q)\frac{\partial^2 v}{\partial x \partial t}a\Delta t + \frac{\partial^2 v}{\partial t^2}\frac{(\Delta t)^2}{2} + \dots . \end{aligned} \tag{5.25}$$

Next we divide the equation by Δt and go over to the limits $\Delta t \to 0$, $p-q \to 0$, and $a \to 0$, assuming that the following finite limits exist:

$$\lim_{\Delta t \to 0} \frac{(p-q)a}{\Delta t} = A, \quad \lim_{\Delta t \to 0} \frac{a^2}{\Delta t} = B. \tag{5.26}$$

It is thus supposed that $a \sim (\Delta t)^{1/2}$ and $(p-q) \sim (\Delta t)^{1/2}$, whereby all the terms in (5.25), save those in the first line, tend to zero as $\Delta t \to 0$. This procedure then gives the following differential equation for the transition probability density $v(t,x)$:

$$\frac{\partial v}{\partial t} = -A\frac{\partial v}{\partial x} + \frac{B}{2}\frac{\partial^2 v}{\partial x^2}. \tag{5.27}$$

What initial conditions must be met at $t=0$? At $t=t_0$ we have in the general case with certainty $x = x_0$ implying

$$v(t_0, x|t_0, x_0) = \delta(x-x_0). \tag{5.28}$$

In this case $v(t,x) \equiv v(t,x|0,0)$, so that we require

$$v(0,x) = \delta(x)$$

to hold and, of course,

$$\int_{-\infty}^{+\infty} v(t,x)dx = 1.$$

These conditions determine the solution[4]

[4] It is apparent that the solution is simply the de Moivre-Laplace theorem for the case of continuous tests [see Sects. 2.6,7 – in particular equation (2.18)].

$$v(t,x) = \frac{1}{\sqrt{2\pi Bt}} \exp\left[-\frac{(x-At)^2}{2Bt}\right], \tag{5.29}$$

i.e., the normal distribution with $\bar{x} = At$ and variance $\overline{(x-At)^2} = Bt$. For arbitrary initial x_0 and t_0 we would obtain

$$v(t,x|t_0,x_0) = \frac{\exp\left\{-\frac{[x-x_0-A(t-t_0)]^2}{2B(t-t_0)}\right\}}{\sqrt{2\pi B(t-t_0)}}. \tag{5.30}$$

We have thus a homogeneous, though nonstationary, Markov process.

For a random walk between two *reflecting* walls separated by a distance l (there must be no systematic flow here, i.e., $A = 0$), a limiting stationary probability exists, i.e., a homogeneous distribution with density $1/l$:

$$\lim_{t\to\infty} v(t,x|t_0,x_0) = w_1(x) = 1/l.$$

In this case the process is also ergodic, as $w_1(x)$ is independent of the initial state x_0. On the contrary, for a particle traveling between *absorbing* walls (i.e., the particles stick to the walls) we have a stationary limiting distribution; but as $w_1(x)$ retains its dependence on x_0,

$$\lim_{t\to\infty} v(t,x|t_0,x_0) = w_1(x) = \left(1-\frac{x_0}{l}\right)\delta(x) + \frac{x_0}{l}\delta(x-l),$$

there is no ergodicity.

It is readily seen that the transition probability (5.30) satisfies the Smoluchowski equation (5.7).

The behavior of $v(t,x|t_0,x_0)$ with increasing t is obvious: the distribution gradually flattens out as the variance $B(t-t_0)$ increases, with the maximum (the mean) moving from x_0 at $t = t_0$ with a constant speed A. An illustration of this is the motion of a Brownian particle under the influence of gravity. The systematic velocity here is $A = -mg/h = \text{const}$, where h is the coefficient of Stokes's friction. Equation (5.30) thus describes the *diffusion in a steady flow*, which, incidentally, also follows from the initial equation (5.27). At $A = 0$ we return to the problem of the *isotropic* walk, or of the distribution of the total *amplitude* x resulting from the addition of an infinite number N of oscillations with *equiprobable* infinitesimal amplitudes $(\pm a)$. In this case we are usually not interested in x, but in the intensity $J = x^2$. Relying on (5.29) with $A = 0$, we readily find

$$\overline{J} = \overline{x^2} = Bt, \quad \overline{(\Delta J)^2} = \overline{J^2} - \overline{J}^2 = 2B^2t^2,$$

so that (cf. Sect. 2.4)

$$\frac{\sqrt{\overline{(\Delta J)^2}}}{\overline{J}} = \sqrt{2}.$$

There is no smoothening out of intensity fluctuations: no matter how many oscillations we add up, the mean spread of the intensity will be of the same order as the mean intensity.

The intensity resulting from the addition of a large number of oscillations with random amplitudes was first considered by *Rayleigh* [Ref. 5.6, Sect. 42a]. He also studied the more general case of the uniform distribution of phases of oscillations accumulated within the interval $(0, 2\pi)$, i.e., the case of the *two-dimensional* vector diagram, or the problem of a random walk in two dimensions (Fig. 5.3). This reduces to the probability of the transition from the origin 0 to the point $\boldsymbol{r}(x, y)$ in the time t.

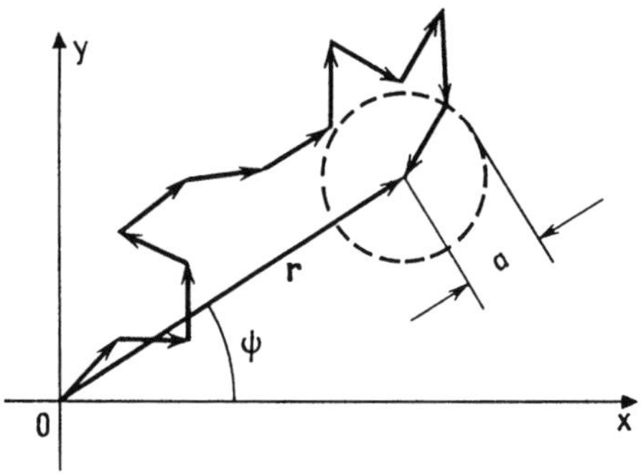

Fig. 5.3. Random walk of a particle in the plane (x, y) with each step having the same length a

If the particle at time $t - \Delta t$ is on a circle with radius a (step length) centered at (x, y) then, it may at time t come to (x, y) as a result of a transition to its center from any point on the circle. All the directions of the vector $\boldsymbol{a}$ are equiprobable (the uniform distribution of the phase φ; isotropic walk), so that

$$v(t, x, y) = \int_0^{2\pi} v(t - \Delta t, x - a\cos\varphi, y - a\sin\varphi)\frac{d\varphi}{2\pi}.$$

Expanding the integrand in powers of $a \cdot \cos\varphi$ and $a \cdot \sin\varphi$ and taking the integral gives

$$v(t, x, y) = v(t - \Delta t, x, y) + \frac{a^2}{4}\left(\frac{\partial^2 v}{\partial x^2} + \frac{\partial^2 v}{\partial y^2}\right) + o(a^3).$$

If we now assume that

$$\lim_{\Delta t \to 0} \frac{a^2}{\Delta t} = 2B,$$

then, passing to the limit $\Delta t \to 0$, we get

$$\frac{\partial v}{\partial t} = \frac{B}{2}\left(\frac{\partial^2 v}{\partial x^2} + \frac{\partial^2 v}{\partial y^2}\right), \tag{5.31}$$

i.e., the two-dimensional diffusion equation. Its normalized solution has the form

$$v(t, x, y) = \frac{1}{2\pi Bt} \exp\left[-(x^2 + y^2)/2Bt \right], \qquad (5.32)$$

i.e., it is the product of normal distributions [which reduces at $t = 0$ to $\delta(x)\delta(y)$] for the components x and y, which are thus independent. Most probable and mean values of the vector $\boldsymbol{r}$ are zero.

If we are interested in the magnitude of the vector $\boldsymbol{r}$ and its phase ψ, we introduce in the plane (x, y) the polar coordinates ($x = r \cos\psi$, $y = r \sin\psi$, $dx\,dy = r\,dr\,d\psi$), so that the transition probability may be written as

$$v(t, x, y)dx\,dy = \tilde{v}(t, r, \psi)dr\,d\psi = \frac{\exp(-r^2/2Bt)}{Bt} r\,dr \frac{d\psi}{2\pi}. \qquad (5.33)$$

It follows from (5.33) that r and ψ are independent, that the phase ψ is uniformly distributed over $(0, 2\pi)$ and that the amplitude r obeys the distribution law

$$v_r(t, r)dr = \frac{\exp(-r^2/2\sigma^2)}{\sigma^2} r\,dr, \qquad (5.34)$$

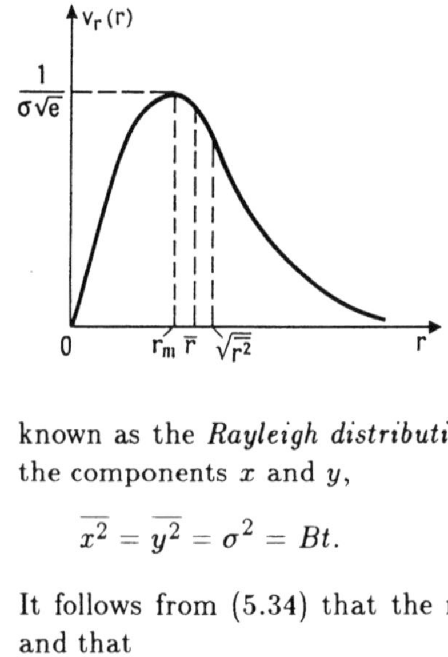

Fig. 5.4. The density of the Rayleigh distribution $v_r(r)$: $r_m = \sigma$, $\bar{r} = \sqrt{\pi/2}\sigma$ and $\sqrt{\overline{r^2}} = \sqrt{2}\sigma$

known as the *Rayleigh distribution*. Here σ^2 denotes the mean square value of the components x and y,

$$\overline{x^2} = \overline{y^2} = \sigma^2 = Bt.$$

It follows from (5.34) that the most probable value of r is $r_m = \sigma$ (Fig. 5.4) and that

$$\bar{r} = \sqrt{\pi/2}\sigma \quad \text{and} \quad \overline{r^2} = 2\sigma^2 = 2Bt. \qquad (5.35)$$

The last expression also results from the fact that $\overline{r^2} = \overline{x^2} + \overline{y^2}$.

The Rayleigh method described above is the first step towards deducing a general differential equation for the transition probability that follows from the Smoluchowski equation with the appropriate assumptions about these probabilities. This differential equation named after the scientists who were the first to derive and use it is known as the *Einstein-Fokker-Planck equation*. Kolmogorov in 1931 gave a rigoros treatment of it. The derivation and discussion of this equation will be given in Sect. 5.6, see also [Ref. 5.7, Sects. 54–56].

5.5 Some Generalizations of the Rayleigh Distribution

In this section we will digress from our treatment of Markov processes to discuss certain generalizations of the Rayleigh distribution obtained as a result of the elementary formulation of the problem of a random walk in a plane. Let us forget for a moment that in this problem the vector $\boldsymbol{r}$ is the sum of the many random steps $\boldsymbol{a}_k$. We can then summarize the results as follows. If the components x and y of $\boldsymbol{r}$ are independent and normally distributed with $\overline{x} = \overline{y} = 0$ and $\overline{x^2} = \overline{y^2} = \sigma^2$, then, by (5.34), the magnitude r and argument ψ of this vector are independent, r is distributed according to Rayleigh and ψ is distributed uniformly over the interval $(0, 2\pi)$. If we forego any of the above limitations on the distributions of x and y, this will lead to distributions for $\boldsymbol{r}$ that generalize the Rayleigh distribution. First, consider the case where it is assumed that the mean of $\boldsymbol{r}$ is nonzero all other conditions being equal. We shall treat $\boldsymbol{r}$ as a complex amplitude of oscillations in the vector diagram (r is the amplitude, ψ is the phase of oscillation).

1) Let us consider the distribution of the sum of a deterministic signal $\boldsymbol{A}$ and a normal noise $\boldsymbol{r}$. The signal is taken to have a constant amplitude A and a phase ψ_0 (harmonic signal), and the normal noise has the distribution (5.34). We are interested in the distribution of the vector $\boldsymbol{R} = \boldsymbol{A} + \boldsymbol{r}$, i.e., the joint distribution of its amplitude R and phase ψ (Fig. 5.5). Because of

$$r\ \cos\psi = R\ \cos\psi - A\ \cos\psi_0 \quad \text{and} \quad r\ \sin\psi = R\ \sin\psi - A\ \sin\psi_0,$$

we have

$$r^2 = R^2 + A^2 - 2RA\ \cos(\varphi - \psi_0) \quad \text{and} \quad r\,dr\,d\psi = R\,dR\,d\varphi.$$

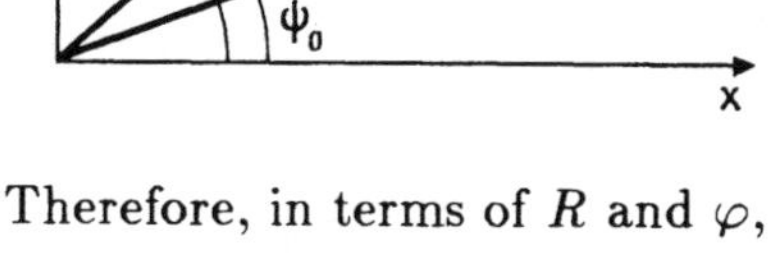

Fig. 5.5. This sketch specifies the notation for the derivation of the distribution $w_{\boldsymbol{R}}(R, \varphi)$ for the amplitude R and the phase φ of the sum $\boldsymbol{R} = \boldsymbol{A} + \boldsymbol{r}$ of the constant signal $\boldsymbol{A}$ and the normal noise $\boldsymbol{r}$

Therefore, in terms of R and φ, (5.34) gives

$$w_R(R,\varphi)dR\,d\varphi$$

$$= \exp\left\{-\frac{1}{2\sigma^2}[R^2 + A^2 - 2RA\ \cos(\varphi - \psi_0)]\right\}\frac{R\,dR\,d\varphi}{2\pi\sigma^2}, \tag{5.36}$$

which is the desired distribution. This gives us neither independent R and φ, nor a uniform distribution of φ.

Integrating (5.36) with respect to φ from 0 to 2π and using the expression for the zero-order Bessel function of the imaginary argument,

$$I_0(a) = J_0(ia) = \frac{1}{2\pi}\int_0^{2\pi} e^{a\cos\alpha}d\alpha, \tag{5.37}$$

we arrive at the distribution,

$$\begin{aligned} w_R(R)dR &= dR\int_0^{2\pi} w_R(R,\varphi)d\varphi \\ &= I_0\left(\frac{RA}{\sigma^2}\right)\exp\left[-(R^2+A^2)/2\sigma^2\right]\frac{R\,dR}{\sigma^2}, \end{aligned} \tag{5.38}$$

of the amplitude R, which is often referred to as the Rice distribution [5.8] or the generalized Rayleigh distribution. At $A = 0$, w_R goes over into w_r and, accordingly, $w_R(R)$ goes over into the Rayleigh distribution. The mean square of R is $\overline{R^2} = A^2 + 2\sigma^2$, and the moments of any (not necessarily integer) order in ν are given by

$$\overline{R^\nu} = (2\sigma^2)^{\nu/2}\Gamma\left(1+\frac{\nu}{2}\right)\cdot{}_1F_1\left(-\frac{\nu}{2},\ 1;\ -\frac{A^2}{2\sigma^2}\right),$$

where ${}_1F_1$ is the confluent hypergeometric function. For large arguments of the modified Bessel function we may use the asymptotic relation $I_0(z)\approx\exp(z)/\sqrt{2\pi z}$ for $|z|\gg 1$, to obtain from (5.38)

$$w_R(R)dR \approx \frac{1}{\sigma}\sqrt{\frac{R}{2\pi A}}\exp\left[-(R-A)^2/2\sigma^2\right]dR \quad (RA\gg\sigma^2).$$

Integrating (5.36) with respect to R from 0 to ∞, we find the distribution of the phase φ

$$w_\varphi(\varphi)d\varphi = \exp\left[-A^2/2\sigma^2\right]\left[1+\sqrt{\pi}\eta e^{\eta^2}(1+\Phi(\eta))\right]\frac{d\varphi}{2\pi}, \tag{5.39}$$

where $\Phi(\eta)$ is the so-called error integral

$$\Phi(\eta) = \frac{2}{\sqrt{\pi}}\int_0^\eta e^{-t^2}dt, \quad \text{and}$$

$$\eta = \frac{\sqrt{2}A}{\sigma}\cos(\varphi-\psi_0).$$

At $A = 0$ the distribution (5.39) becomes uniform.

2) We now take normally distributed x and y, and assume $\bar{x} = \bar{y} = 0$ in order to consider the case of *correlated x and y with dissimilar variances,*

$$\overline{x^2} = \sigma_1^2, \quad \overline{y^2} = \sigma_2^2, \quad \overline{xy} = \sigma_1\sigma_2 K,$$

i.e., the normal distribution (2.28). Assuming in addition that $x = r\ \cos\psi$, $y = r\ \sin\psi$, we get

$$w_r(r,\psi)dr\,d\psi = \exp\left\{-\frac{r^2}{4(1-K^2)}\left[\frac{1}{\sigma_1^2}+\frac{1}{\sigma_2^2}+\left(\frac{1}{\sigma_1^2}-\frac{1}{\sigma_2^2}\right)\cos 2\psi \right.\right.$$
$$\left.\left. -\frac{2K}{\sigma_1\sigma_2}\sin 2\psi\right]\right\}\frac{r\,dr\,d\psi}{2\pi\sigma_1\sigma_2\sqrt{1-K^2}}. \tag{5.40}$$

Integrating with respect to ψ from 0 to 2π and using the relation

$$I_0\left(\sqrt{a^2+b^2}\right) = \frac{1}{2\pi}\int_0^{2\pi}\exp\left[a\ \cos\alpha + b\ \sin\alpha\right]d\alpha,$$

gives the distribution

$$w_r(r)dr = I_0\left[\frac{qr^2}{4(1-K^2)}\right]$$
$$\times\exp\left\{-\frac{r^2}{4(1-K^2)}\left(\frac{1}{\sigma_1^2}+\frac{1}{\sigma_2^2}\right)\right\}\frac{r\,dr}{\sigma_1\sigma_2\sqrt{(1-K^2)}}, \tag{5.41}$$

where

$$q = \sqrt{\left(\frac{1}{\sigma_1^2}+\frac{1}{\sigma_2^2}\right)^2 - \frac{4(1-K^2)}{\sigma_1^2\sigma_2^2}}.$$

If there is no correlation between x and y ($K = 0$), then

$$w_r(r)dr = I_0\left[\left(\frac{1}{\sigma_1^2}-\frac{1}{\sigma_2^2}\right)\frac{r^2}{4}\right]^2\exp\left\{-\left(\frac{1}{\sigma_1^2}+\frac{1}{\sigma_2^2}\right)\frac{r^2}{4}\right\}\frac{r\,dr}{\sigma_1\sigma_2}. \tag{5.42}$$

However, if there is a correlation, but $\sigma_1^2 = \sigma_2^2 = \sigma^2$, then

$$w_r(r)dr = I_0\left[\frac{Kr^2}{2\sigma^2(1-K^2)}\right]\exp\left\{-\frac{r^2}{2\sigma^2(1-K^2)}\right\}\frac{r\,dr}{\sigma^2\sqrt{1-K^2}}. \tag{5.43}$$

Of course, both formulas become the Rayleigh distribution if in (5.42) we set $\sigma_1 = \sigma_2$, and in (5.43), $K = 0$.

If the correlation coefficient K is so close to unity that $1 - K^2 \ll (\sigma_1/\sigma_2 + \sigma_2/\sigma_1)^2/4$ and if the argument of the Bessel function is large, i.e., $qr^2 \gg 4(1 - K^2)$, so that we can use its asymptotic expression, then the distribution (5.41) becomes the one-sided Gaussian distribution ($r \geq 0$)

$$w_r(r)dr = 2\ \exp\left\{-\frac{r^2}{2(\sigma_1^2+\sigma_2^2)}\right\}\frac{dr}{\sqrt{2\pi(\sigma_1^2+\sigma_2^2)}}.$$

3) If we do not insist on a normal distribution for x and y, this naturally reduces the possibilities of obtaining any specific results. Consider the case of independent x and y obeying the same distribution law, so that their joint distribution has the form $w(x)dx{\cdot}w(y)dy$. In terms of the polar coordinates we obtain

$$w_r(r,\psi)dr\,d\psi = w(r\ \cos\psi)w(r\ \sin\psi)r\,dr\,d\psi,$$

and hence

$$w_r(r)dr = r\,dr\int_0^{2\pi} w(r\ \cos\psi)w(r\ \sin\psi)d\psi. \tag{5.44}$$

In the case of Gaussian densities w, relation (5.44) brings us back to the Rayleigh density $w_r(r)$.

Next we introduce the characteristic function of the components x and y,

$$\varphi(u) = \langle e^{iux}\rangle = \int_{-\infty}^{+\infty} e^{iux}w(x)dx.$$

The density $w_r(r)$ can then be written as

$$\begin{aligned} w_r(r) &= \frac{r}{4\pi^2}\iint_{-\infty}^{+\infty}\varphi(u)\varphi(v)du\,dv\int_0^{2\pi}\exp[-ir(u\ \cos\psi+v\ \sin\psi)]d\psi \\ &= \frac{1}{2\pi}\iint_{-\infty}^{+\infty}\varphi(u)\varphi(v)J_0(r\sqrt{u^2+v^2})du\,dv. \end{aligned}$$

Introducing via $u=\lambda\ \cos\alpha$ and $v=\lambda\ \sin\alpha$ ($du\,dv=\lambda\,d\lambda\,d\alpha$) polar corrdinates in the (u,v) plane as well, the latter expression reads

$$w_r(r) = r\int_0^{\infty} F(\lambda)J_0(\lambda r)\lambda\,d\lambda, \quad \text{where} \tag{5.45}$$

$$F(\lambda) = \frac{1}{2\pi}\int_0^{2\pi}\varphi(\lambda\ \cos\alpha)\varphi(\lambda\ \sin\alpha)d\alpha. \tag{5.46}$$

Thus, $w_r(r)$ is the Bessel transform of the function $F(\lambda)$, multiplied by r.

4) We now return to the random walk in a plane, where the vector $\boldsymbol{r}$ is treated as the sum of random vectors $\boldsymbol{a}_k$ (elementary steps) and the statistics of $\boldsymbol{r}$ is not specified directly but is determined by the statistics of $\boldsymbol{a}_k$. Take,

for instance, the case of the *isotropic random walk with random step lengths.* Let all the elementary vectors $\boldsymbol{a}_k$ be mutually independent and distributed in a similar fashion so that their respective polar angles (phase) θ_k are independent of their amplitudes a_k and distributed uniformly over the interval $(0, 2\pi)$ (all directions of the elementary steps are equiprobable, thus ensuring the isotropic character of the walk. Thus, for any $\boldsymbol{a}_k$, we get

$$w_a(a_k, \theta_k) da_k d\theta_k = w_a(a_k) da_k \frac{d\theta_k}{2\pi}.$$

The corresponding characteristic function is

$$\begin{aligned}\varphi_{\boldsymbol{a}}(u,v) &= \int_0^\infty w_a(a_k) da_k \cdot \frac{1}{2\pi} \int_0^{2\pi} \exp\left[ia_k(u \cos\theta_k + v \sin\theta_k)\right] d\theta_k \\ &= \int_0^\infty w_a(a) J_0(\lambda a) da \equiv \varphi_{\boldsymbol{a}}(\lambda), \qquad (5.47)\end{aligned}$$

where $\lambda = \sqrt{u^2 + v^2}$. By virtue of the statistical independence of all $\boldsymbol{a}_k$, the bivariate characteristic function of the total vector

$$\boldsymbol{r} = \sum_{k=1}^{n} \boldsymbol{a}_k \quad \text{is}$$

$$\varphi_{\boldsymbol{r}}(u,v) = \varphi_{\boldsymbol{a}}^n(\lambda) \equiv \varphi_{\boldsymbol{r}}(\lambda) \qquad (5.48)$$

and the joint probability density of the Cartesian coordinates x and y of the vector $\boldsymbol{r}$ may be written as

$$\begin{aligned}w_{\boldsymbol{r}}(x,y) &= \frac{1}{4\pi^2} \int\int_{-\infty}^{+\infty} \varphi_{\boldsymbol{r}}(u,v) \exp\left[-i(ux + vy)\right] du\, dv \\ &= \frac{1}{4\pi^2} \int\int_{-\infty}^{+\infty} \varphi_{\boldsymbol{a}}^n(\lambda) \exp\left[-i(ux + vy)\right] du\, dv.\end{aligned}$$

Going over to the polar coordinates in the planes (x, y) $(x = r \cos\psi,\ y = r \sin\psi)$ and (u, v) $(u = \lambda \cos\alpha,\ v = \lambda \sin\alpha)$, we obtain

$$\begin{aligned}\tilde{w}_{\boldsymbol{r}}(r,\psi) &= r w_{\boldsymbol{r}}(r \cos\psi,\ r \sin\psi) \\ &= \frac{r}{4\pi^2} \int_0^\infty \varphi_{\boldsymbol{a}}^n(\lambda) \lambda\, d\lambda \int_0^{2\pi} \exp\left[-i\lambda r \cos(\psi - \alpha)\right] d\alpha \\ &= \frac{r}{2\pi} \int_0^\infty \varphi_{\boldsymbol{a}}^n(\lambda) J_0(\lambda r) \lambda\, d\lambda \equiv w_{\boldsymbol{r}}(r) \frac{1}{2\pi}. \qquad (5.49)\end{aligned}$$

This joint density of r and ψ does not contain ψ, i.e., the direction of the resultant vector $\boldsymbol{r}$ does not depend on its amplitude r and is uniformly distributed over $(0, 2\pi)$. In accordance with the central limit theorem, the distribution $w_r(x,y)$ tends for $n \to \infty$ and $\overline{r^2} = n\overline{a^2} = 2\sigma^2 = \text{const.}$, to the normal law for independent x and y with $\overline{x^2} = \overline{y^2} = \sigma^2$, whereby the amplitude distribution $\tilde{w}_r(r)$ tends to the Rayleigh distribution [5.9].

If we return to the case of a fixed step length as discussed in Sect. 5.4, i.e., to the probability density $w_a(a_k) = \delta(a_k - a)$, then, by (5.47) we have $\varphi_a(\lambda) = J_0(\lambda a)$. It then follows from (5.49) that

$$w_r(r) = r \int_0^\infty J_0^n(\lambda a) J_0(\lambda r) \lambda \, d\lambda .$$

Taking

$$\lambda \int_0^R J_0(\lambda r) r \, dr = R J_1(\lambda R) = -\frac{dJ_0(\lambda R)}{d\lambda}$$

into account, we can write the probability that r will not be larger than R as

$$\begin{aligned} P(r<R) &= \int_0^R w_r(r) dr = R \int_0^\infty J_0^n(\lambda a) J_1(\lambda R) d\lambda \\ &= -\int_0^\infty J_0^n(\lambda a) \frac{dJ_0(\lambda R)}{d\lambda} d\lambda . \end{aligned}$$

Hence, at $R = a$, we find

$$P(r<a) = -\frac{[J_0(\lambda a)]^{n+1}}{n+1}\bigg|_0^\infty = \frac{1}{n+1} .$$

5) Next we consider the m-distribution of *Nakagami* [5.10]. Earlier we discussed the magnitude r of the vector $\boldsymbol{r}$ by specifying the distribution of its Cartesian components x and y ($r = +\sqrt{x^2 + y^2}$), or assuming $\boldsymbol{r}$ to be the sum of vectors $\boldsymbol{a}_k$. We also specified the distribution of $\boldsymbol{a}_k$ in a particular way. Now we take a significantly more general case in which r is an arbitrary, nonnegative, deterministic function of a $2n$-dimensional random variable $\boldsymbol{a} = \{\boldsymbol{a}_1, \ldots, \boldsymbol{a}_n\}$:

$$r = \Phi(\boldsymbol{a}) .$$

Making no assumptions as to the form of the $2n$-variate distribution of $\boldsymbol{a}$ (i.e., the joint distribution of the Cartesian components of all $\boldsymbol{a}_k$ or the distribution of all magnitudes of $\boldsymbol{a}_k$ and their phases θ_k), we have

$$w_r(r) = \langle \delta[r - \Phi(a)] \rangle , \tag{5.50}$$

where the angle brackets mean, as usual, the averaging over the distribution $\boldsymbol{a}$

(see Exercise 2.8.9). For $\phi \geq 0$ the delta-function may be represented as a Hankel transform,

$$\delta(r-\Phi) = r^{\nu+1} \int_0^\infty \frac{J_\nu(\lambda\Phi)}{\Phi^\nu} J_\nu(\lambda r)\lambda\, d\lambda, \quad \mathrm{Re}\,\nu \geq -\frac{1}{2}.$$

Substituting this into (5.50) yields

$$\begin{aligned} w_r(r) &= r^{\nu+1} \int_0^\infty \left\langle \frac{J_\nu(\lambda\Phi)}{\Phi^\nu} \right\rangle J_\nu(\lambda r)\lambda\, d\lambda \\ &= \frac{r^{\nu+1}}{2^\nu \Gamma(\nu+1)} \int_0^\infty F_\nu(\lambda) J_\nu(\lambda r) \lambda^{\nu+1} d\lambda, \end{aligned} \tag{5.51}$$

where

$$F_\nu(\lambda) = 2^\nu \Gamma(\nu+1) \left\langle \frac{J_\nu(\lambda\Phi)}{(\lambda\Phi)^\nu} \right\rangle. \tag{5.52}$$

Note that (5.45) corresponds to the special case of $\nu = 0$ with $F_0(\lambda) = \langle J_0(\lambda\Phi)\rangle$.

From the known power series,

$$\frac{J_\nu(z)}{z^\nu} = \frac{1}{2^\nu} \sum_{k=0}^\infty \frac{(-1)^k (z/2)^{2k}}{k!\,\Gamma(\nu+k+1)},$$

and from (5.52), it follows that $F_\nu(\lambda)$ contains only *even* powers of λ and *even* moments of Φ

$$F_\nu(\lambda) = \Gamma(\nu+1) \sum_{k=0}^\infty \frac{(-1)^k (\lambda/2)^{2k} \overline{\Phi^{2k}}}{k!\,\Gamma(\nu+k+1)} \tag{5.53}$$

Nakagami [5.10] treats ν in (5.51) and (5.53) as a real quantity, so that the condition on the $\mathrm{Re}\,\nu$ reduces to the requirement that $\nu \geq -1/2$. But even given this limitation, the relations (5.51) and (5.53) seem to cover an exceedingly large variety of distributions of r. This comes about because we have not only a selection of values of ν, but also of distributions of the multidimensional random variable $\boldsymbol{a}$ which is the argument of the function $\Phi(\boldsymbol{a})$. However, Nakagami follows a different procedure: he simply *constructs* the function $F_\nu(\lambda)$, ignoring the details of the distribution involved. On the one hand, he imposes the condition that $F_\nu(\lambda)$ be dependent only on two parameters related to the moments $\overline{\Phi^2}$ and $\overline{\Phi^4}$, i.e.,

$$\Omega = \overline{\Phi^2}, \quad m = \frac{\overline{(\Phi^2)}^2}{D[\Phi^2]} = \frac{\Omega^2}{\overline{\Phi^4} - \Omega^2}. \tag{5.54}$$

On the other hand, he replaces the well-known Lyapunov's inequality

$$\overline{\Phi^{2k+2l}} \geq \overline{\Phi^{2k}} \cdot \overline{\Phi^{2l}}$$

by an *equality* expressing the higher-order even moments of Φ in terms of $\overline{\Phi^2}$ and $\overline{\Phi^4}$ (e.g., $\overline{\Phi^6} = \overline{\Phi^2} \cdot \overline{\Phi^4}$, $\overline{\Phi^8} = (\overline{\Phi^4})^2$, etc.) i.e., in terms of Ω and m. Equation (5.53) may then be reduced to a series that, for $\nu = 0$, adds up to the function

$$F_0(\lambda) = \exp[-\Omega\lambda^2/4m]P_{m-1}(\Omega\lambda^2/4m), \tag{5.55}$$

where P_{m-1} are the Laguerre polynomials. If ν is nonzero, then it is *chosen to be equal to* $m-1$. In this case, (5.53) reduces to a series that is to a good approximation described by the function

$$F_\nu(\lambda) \approx \exp[-\Omega\lambda^2/4(\nu+1)], \quad \nu = m-1. \tag{5.56}$$

Although it was assumed that $m \neq 1$, but is easily seen that (5.55) and (5.56) coincide at $m = 1$ ($\nu = 0$), since $P_0 = 1$.

Substituting (5.55) (at $\nu = 0$) and (5.56) (at $\nu = m-1$) into (5.51) and taking the integrals gives the same expressions for $w_r(r)$,

$$w_r(r) \equiv M(r, m, \Omega) = \frac{2m^m r^{2m-1} e^{-mr^2/\Omega}}{\Gamma(m)\Omega^m}, \tag{5.57}$$

which Nakagami calles the m-distribution density. Note that for $\nu = m-1$, it follows from the condition $\nu \geq -1/2$ that $m \geq 1/2$. By (5.54) we have $m = \Omega^2/(\overline{r^4} - \Omega^2)$, with $\Omega = \overline{r^2}$, hence $\overline{r^4} \leq 3(\overline{r^2})^2$.

Reference [5.10] discusses the many applications and generalizations of the m-distribution, but we will confine ourselves here to two remarks. It is seen from (5.57) that the m-distribution at $m = 1/2$ becomes the one-sided Gaussian distribution, and at $m = 1$ the Rayleigh distribution. Further, if in (5.57) we transform to the variable $x = mr^2/\kappa\Omega$ ($\kappa > 0$) and denote m by α, then we obtain for x the gamma-distribution (3.36). (The generalizations of the Rayleigh distribution are discussed in [5.11].)

5.6 Continuous Markov Processes. The Einstein-Fokker-Planck Equation

In the Smoluchowski equation (5.3), the intermediate time θ can be selected arbitrarily between t_0 and t. We take it to be fairly close to t, make $\theta = t - \tau$ (Fig. 5.6), and assume that the following limits exist:

$$\lim_{\tau \to 0} \frac{\overline{x-y}}{\tau} = \lim_{\tau \to 0} \frac{1}{\tau} \int_{-\infty}^{+\infty} (x-y) v(t, x | t-\tau, y) dx = A(y, t). \tag{5.58}$$

The significance of this limit is self-evident; $\overline{x-y}$ is the conditional mean of the displacement during the time τ from the fixed point y. Hence, $A(y,t)$ is the

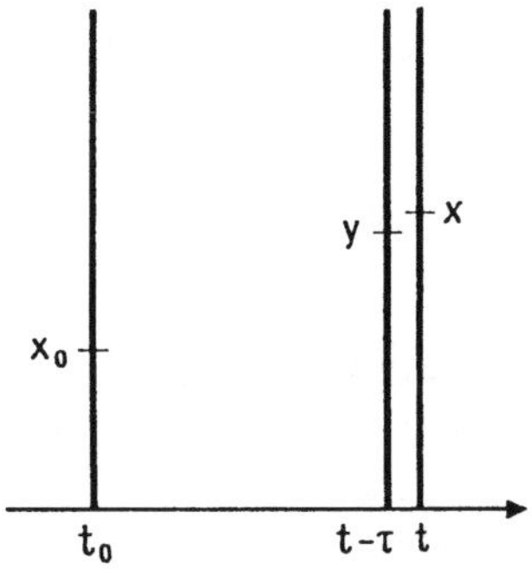

Fig. 5.6. This drawing is used in the derivation of the Einstein-Fokker-Planck equation, see the text

mean rate of state variation at time t at point y (the so-called drift coefficient).

Next, we assume that

$$\lim_{\tau\to 0}\frac{\overline{(x-y)^2}}{\tau}=\lim_{\tau\to 0}\frac{1}{\tau}\int\limits_{-\infty}^{+\infty}(x-y)^2 v(t,x|t-\tau,y)dx=B(y,t). \tag{5.59}$$

The quantity $\overline{(x-y)^2}$ is the measure of the spread of the final points x relative to the fixed initial point y. It is thus supposed that this spread increases within the time interval from $t-\tau$ to t, i.e., it varies with τ according to the diffusion law. Assumptions (5.58,59) reproduce the conditions (5.26) adopted when discussing the scheme with discrete jumps (Sect. 5.4). The coefficient $B(y,t)$, or rather $B/2$, is said to be the diffusion coefficient.

Later on we shall view the Markov process from another angle, using the notions of random impacts or *random forces* acting on the system in question. We shall see then that $B(y,t)$ characterises the *intensity of impacts*.

Finally, we assume that

$$\lim_{\tau\to 0}\frac{\overline{|x-y|^3}}{\tau}=\lim_{\tau\to 0}\frac{1}{\tau}\int\limits_{-\infty}^{+\infty}|x-y|^3 v(t,x|t-\tau,y)dx=0. \tag{5.60}$$

We thus consider that the probability of notable changes in $|x-y|$ tends to zero sufficiently rapidly as $\tau\to 0$, i.e., so rapidly that $\overline{|x-y|^3}$ falls off faster than τ. It is this requirement that allows us to regard x in a system subject to random forces as a *continuous* variable, i.e., as the mean over a time that is much longer than the gap between random impacts. Conclusions drawn from the assumption (5.60) can not therefore be applied to times τ which are short as compared with the time between impacts. For instance, the probability of collisions of gas molecules during the time τ that is small as compared with the free path time ϑ is $1-\exp(-\tau/\vartheta)\approx\tau/\vartheta$. In a collision, the molecular velocity u (corresponding here to the variable x describing the state of the molecule) changes on the average by a finite amount $\Delta u=\pm a$. Consequently, for $\tau\ll\vartheta$

$$\overline{|\Delta u|^3}\approx a^3\tau/\vartheta$$

and the condition

$$\lim_{\tau\to 0}\frac{\overline{|\Delta u|^3}}{\tau}=0$$

is not satisfied.

To accept condition (5.60) means confining ourselves to those Markov processes where we have not just a set of continuous possible values, but also a continuous evolution of the process in time, i.e., the states change continuously (in the stochastic sense) and without jumps. Such Markov processes are often referred to as *diffusion* processes. The case of stepwise variations will be discussed later (Sect. 5.11).

Putting $\theta = t-\tau$ in the Smoluchowski equation (5.3) we multiply it by an arbitrary function $q(x)$ that is zero, along with $q'(x)$, at the boundaries of the interval (here $x=\pm\infty$) and has a finite third derivative $|q'''(x)|<M$. Integrating with respect to x gives

$$\begin{aligned}
&\int_{-\infty}^{+\infty} q(x)v(t,x|t_0,x_0)dx\\
&\quad=\int_{-\infty}^{+\infty} v(t-\tau,y|t_0,x_0)dy\int_{-\infty}^{+\infty} v(t,x|t-\tau,y)q(x)dx\\
&\quad=\int_{-\infty}^{+\infty} v(t-\tau,y|t_0,x_0)dy\int_{-\infty}^{+\infty} v(t,x|t-\tau,y)\\
&\qquad\times\left\{q(y)+q'(y)(x-y)+q''(y)\frac{(x-y)^2}{2}+q'''(z)\frac{(x-y)^3}{6}\right\}dx,
\end{aligned}$$

where $y<z<x$. In the term with $q(y)$ the integration with respect to x gives unity. We transpose this term to the left-hand side, with the integration variable y changed into x, and divide the equation by τ. Going over to the limit $\tau\to 0$ gives

$$\begin{aligned}
&\int_{-\infty}^{+\infty} q(x)\frac{\partial v(t,x|t_0,x_0)}{\partial t}dx\\
&\quad=\int_{-\infty}^{+\infty} v(t,y|t_0,x_0)\left\{q'(y)A(y,t)+q''(y)\frac{B(y,t)}{2}\right.\\
&\qquad\left.+\frac{q'''(z)}{6}\lim_{\tau\to 0}\frac{\overline{|x-y|^3}}{\tau}\right\}dy.
\end{aligned}$$

The last term on the right-hand side is zero, since q''' is bounded and the condition (5.60) is met. Substituting now x for y on the right-hand side and taking the integral by parts [if we consider that $q(\pm\infty) = q'(\pm\infty) = 0$ and $q(x)$ is arbitrary], we obtain

$$\frac{\partial v(t,x|t_0,x_0)}{\partial t} = -\frac{\partial A(x,t)v}{\partial x} + \frac{1}{2}\frac{\partial^2 B(x,t)v}{\partial x^2}. \tag{5.61}$$

This parabolic (diffusion type) equation is known as the Einstein-Fokker-Planck equation (sometimes referred to as the Fokker-Planck or the second Kolmogorov equation). Equation (5.27) appears to be a special case of (5.61) for constant A and B.

The solution to (5.61) must be nonnegative, unity-normalized and meet the initial condition

$$v(t_0,x|t_0,x_0) = \delta(x - x_0). \tag{5.62}$$

Equation (5.61) may graphically be interpreted as follows. At time t_0 a large number (ensemble) of particles travelling independently of each other leaves point x_0. Their concentration (relative "frequency") at a point x at an instant of time t will be $v(t,x|t_0,x_0)$. The flux of particles S is composed of the *systematic* ("hydrodynamic") flux Av, where A is the velocity of the systematic motion at the point x at time t, and the *diffusion* flux $-(\partial Bv/\partial x)/2$, where $B/2$ is the diffusion coefficient

$$S = Av - \frac{1}{2}\frac{\partial Bv}{\partial x}. \tag{5.63}$$

Then the Einstein-Fokker-Planck equation is simply the continuity equation,

$$\frac{\partial v}{\partial t} + \frac{\partial S}{\partial x} = 0,$$

describing the conservation of the number of particles[5].

It is easily seen that $v(t,x|t_0,x_0)$, treated as a function of the initial variables x_0 and t_0, satisfies the *conjugate* equation (in the sense of the conjugation of the linear operators $\hat{L}$ and $\hat{M}$, i.e., $\int_{-\infty}^{+\infty}(f_1\hat{L}f_2 - f_2\hat{M}f_1)dx = 0$) :

$$\frac{\partial v(t,x|t_0,x_0)}{\partial t_0} = -A(x_0,t_0)\frac{\partial v}{\partial x_0} - \frac{B(x_0,t_0)}{2}\frac{\partial^2 v}{\partial x_0^2}. \tag{5.64}$$

This equation is known as the first Kolmogorov equation.

---

[5] Usually, the diffusion coefficient refers to the factor at $-\partial v/\partial x$. Therefore, $B/2$ in (5.63) cannot, strictly speaking, be called the diffusion coefficient. If we write the total flux as

$$S = \left(A - \frac{1}{2}\frac{\partial B}{\partial x}\right)v - \frac{B}{2}\frac{\partial v}{\partial x},$$

then $B/2$ will be the diffusion coefficient, but the systematic part of S will contain the gradient B.

If at the initial time t_0 we specify not the initial state x_0, but the *initial distribution* $w(t_0, x_0)$, then, since the bivariate probability density is

$$w_2(t, x; t_0, x_0) = w(t_0, x_0)v(t, x|t_0, x_0),$$

the univariate distribution function at time t is

$$w_1(t, x) = \int_{-\infty}^{+\infty} w(t_0, x_0)v(t, x|t_0, x_0)dx_0.$$

If we multiply (5.62) by $w(t_0, x_0)$ and integrate with respect to x_0, we shall see that the state probability $w_1(t, x)$ satisfies the same equation

$$\frac{\partial w_1(t,x)}{\partial t} = -\frac{\partial A(x,t)w_1}{\partial x} + \frac{1}{2}\frac{\partial^2 B(x,t)w_1}{\partial x^2}. \tag{5.65}$$

Of course, the initial condition here is

$$w_1(x, t_0) = w(x, t_0), \tag{5.66}$$

and the solution must also be nonnegative and normalized to unity. We now turn to the implications and special cases of (5.61).

We introduce the conditional moments

$$\overline{x^n(t)} = \int_{-\infty}^{+\infty} x^n v(t, x|t_0, x_0)dx,$$

i.e., mean values of x^n at time t, given that $\overline{x}(t_0) = x_0$. Next we multiply (5.61) by x^n and integrate with respect to x from $-\infty$ to $+\infty$. Integrating the terms with A and B by parts and assuming that the integrated expressions vanish on the boundaries (here these are $x = \pm\infty$), gives

$$\frac{d\overline{x^n}}{dt} = n\overline{Ax^{n-1}} + \frac{n(n-1)}{2}\overline{Bx^{n-2}}, \quad n = 1, 2, \ldots . \tag{5.67}$$

In particular, for $n = 1$ and 2

$$\frac{d\overline{x}}{dt} = \overline{A(x,t)}, \tag{5.68}$$

$$\frac{d\overline{x^2}}{dt} = 2\overline{A(x,t)x} + \overline{B(x,t)}. \tag{5.69}$$

It should be emphasized that the functions $A(x,t)$ and $B(x,t)$, are by their respective definitions (5.58) and (5.59), *deterministic* functions. The bar over them in (5.67–69), denotes averaging over the conditional distribution $v(t, x|t_0, x_0)dx$ of a random *argument* x of these functions.

If A and B are series (or polynomials) in the powers of x, then (5.67) is a system of ordinary linear differential equations for the moments $\overline{x^n}$. In the general case this is an infinite system of coupled equations, i.e., the equations contain moments of different orders. This also applies to (5.68). Although it does not reflect the influence of the diffusion probability flow $(-\partial Bv/\partial x)/2$, it is by no means an equation of phenomenological kinetics or dynamics that has to be closed (i.e., to contain only one moment $\overline{x}$). The sole exception is the case where $A(x,t)$ is a linear function of x : $A(x,t) = a(t)x + b(t)$. Then equation (5.68) separates and becomes closed

$$\frac{d\overline{x}}{dt} = a(t)\overline{x} + b(t).$$

In any case, we can deduce from (5.68) that dx/dt and $A(x,t)$ may only differ in an addend with zero mean:

$$\frac{dx}{dt} = A(x,t) + F(x,t) \quad \text{with} \quad \overline{F(x,t)} = 0, \tag{5.70}$$

a fact that we shall make use of later in Sect. 6.5.

Now let us look at a *stationary* Markov process. The transition probability then depends on $t - t_0$ so that by (5.58,59) A and B do not vary with t. The univariate state probability, if it exists, does not depend on t at all (Sect. 4.3). As a result, from (5.65) we obtain for $w_1(x)$ the equation

$$\frac{d}{dx}\left(Aw_1 - \frac{1}{2}\frac{dBw_1}{dx}\right) = 0.$$

If on the boundaries of the range of x the stationary flow is

$$S = Aw_1 - \frac{1}{2}\frac{dBw_1}{dx} = 0,$$

the integration shows that it is equal to zero everywhere

$$Aw_1 - \frac{1}{2}\frac{dBw_1}{dx} = 0.$$

For an infinite range of x the conditions $w_1 = w_1' = 0$ as $|x| = \infty$ occur with positive and nonzero B and at $A>0$ or $A<0$, respectively, for $x \to -\infty$ or $x \to +\infty$. In both cases $|A|>0$ [5.19][6]. Integrating the last equation gives

$$w_1(x) = \frac{C}{B(x)} \exp\left[\int_0^x \frac{2A(s)}{B(s)} ds\right]. \tag{5.71}$$

[6] If these conditions are not met, then it may well be that there are no stationary solutions, see (5.71).

The constant C is found from the normalization condition.

An example of a stationary distribution is the Brownian motion of a particle under gravity ($A = -mg/h = \text{const.}$) above a *reflecting boundary*. On the boundary the condition must be met that the flow S becomes zero, so that (5.71) is valid. With constant A and B it gives

$$w_1(x) = C \exp\left(-\frac{2mgx}{hB}\right),$$

i.e., the barometric height formula. It is known [Ref. 5.7, Sect. 53], that

$$B = 2kT/h,$$

where k is the Boltzmann constant and T the absolute temperature. Hence

$$w_1(x) = C \exp\left(-\frac{mgx}{kT}\right).$$

The stationary probability $w_1(x)$ corresponds exactly to the state probability considered in statistical thermodynamics. For an isothermal system the distribution w_1 is canonic, and for an adiabatic isolated system, microcanonic.

Consider also the *process uniform in the abscissa* (Bachelier-type process). In that case the transition probability only depends on $x - x_0$, and accordingly, by (5.58,59), A and B can depend on t, but not on x. Equation (5.61) then takes the form

$$\frac{\partial v(t, x - x, t_0)}{\partial t} = -A(t)\frac{\partial v}{\partial x} + \frac{B(t)}{2}\frac{\partial^2 v}{\partial x^2}. \tag{5.72}$$

If we change the variables, i.e.,

$$x' = x - \int_{t_0}^{t} A(\theta)d\theta \quad \text{and} \quad t' = \int_{t_0}^{t} B(\theta)d\theta,$$

the above equation becomes

$$\frac{\partial v}{\partial t'} = \frac{1}{2}\frac{\partial^2 v}{\partial x'^2}.$$

In the initial variables the solution for any values of its argument is given by

$$v = \frac{1}{\sqrt{2\pi}\sigma(t)} \exp\left\{-\frac{1}{2\sigma^2(t)}\left[x - x_0 - \int_{t_0}^{t} A(\theta)d\theta\right]^2\right\},$$

$$\sigma^2(t) = \int_{t_0}^{t} B(\theta)d\theta.$$

Pawula [5.12] considered in some detail a much more general problem: the differential equation describing the conditional probability density $v(t,x|T,X)$ for any random process with a continuous set of states. Here (T,X) denotes the collection of conditions

$$(T,X)=(t_1,x_1;\dots;t_m,x_m;t_{m+1},y_{m+1};\dots;t_k,z_k),$$

where $x(t),y(t),\dots,z(t)$ may also be different random processes, and the set T of times $t_1,\dots,t_k$ may be unordered. What matters here is that the time t does not belong to the set T.

The initial equation is not, of course, the Smoluchowski equation, but the formula for the total density of conditional probabilities

$$v(t,x|T,X)=\int_{-\infty}^{+\infty}\tilde{v}(t,x|t-\tau,y;T,X)v(t-\tau,y|T,X)dy, \tag{5.73}$$

where $\tilde{v}$ and v, are in general different functions. In [5.13] it was shown (see Exercise 5.13.16) that the conditional probability density $v(t,x|T,X)$ obeys the equation

$$\frac{\partial v(t,x|T,X)}{\partial t}=\sum_{n=1}^{\infty}\frac{(-1)^n}{n!}\frac{\partial^n A_n v}{\partial x^n}, \tag{5.74}$$

$$A_n\equiv A_n(x,t|T,X)=\lim_{\tau\to 0}\frac{1}{\tau}\int(x-y)^n\tilde{v}(t,x|t-\tau,y;T,X)dy. \tag{5.75}$$

The right-hand and left-hand limits ($\tau\to 0\pm$) may be different, but they are assumed to be finite. It has also been found that if $A_n=0$ for any even n, then $A_n=0$ for all $n\geq 3$. In other words, if (5.74) is a finite-order equation, then it *can not be higher than of second order*. In that case we obtain the generalized Einstein-Fokker-Planck equation

$$\frac{\partial v(t,x|T,X)}{\partial t}=-\frac{\partial A_1 v}{\partial x}+\frac{1}{2}\frac{\partial^2 A_2 v}{\partial x^2}. \tag{5.76}$$

In the special case of Markov processes the collection of conditions (T,X) reduces to a single preceeding (or subsequent) state of the process $x(t)$ itself, and the function $\tilde{v}$ coincides with v. Of course, the above theorem that all A_n with $n\geq 3$ vanish is still valid, so that (5.76) reduces to the diffusion equation (5.61) for the transition probability density $v(t,x|t_0,x_0)$.

In principle, *if* we know the coefficients A_n of the form

$$A_n(x_k,t_k|t_1,x_1;\dots;t_{k-1},x_{k-1}),\quad k=1,2,\dots,$$

then (5.74) and (5.76) enable us to find all the conditional densities

$$v_k(t_k, x_k | t_1, x_1; \ldots; t_{k-1}, x_{k-1}),$$

and consequently, any of the n-variate distribution functions

$$w_n(t_1, x_1; \ldots; t, x_n) = w_1(t_1, x_1) \prod_{k=2}^{n} v_k(t_k, x_k | t_1, x_1; \ldots; t_{k-1}, x_{k-1}).$$

It is assumed that w_1 exists, and is either known or can be obtained from the conditional probability densities v_k if all, or some, of the times $t_1, \ldots, t_{k-1}$ are equal to $-\infty$ (i.e., if the strong ergodicity condition is met). But in order for this scheme for evaluating w_n to work in practice we have to overcome a fundamental difficulty inherent in all the purely stochastic schemes under consideration (including the classical procedure for Markov processes). After all, the regular procedure for finding the coefficients A_n requires that we know in advance the desired conditional probability density $\tilde{v}$ (at least for arbitrarily short transition times τ). Moreover, even if the A_n are assumed to be known one way or another, we are still faced with a good many other problems, albeit of a more technical nature. These are associated with the complexity of the equations themselves and with the need for breaking up the range of possible values of $x(t)$ into sections where v and its derivatives are continuous. In addition, the correct boundary conditions for the desired solution have to be established, and so forth.

The fact that (5.74) or (5.76) also hold for non-Markov processes has been established independently in a number of papers besides [5.13], which gives the most comprehensive treatment. The theory is easily extended to include multivariate random processes. When in the next section we discuss this generalization, we will return to Markov processes and again confine ourselves to the diffusion (Einstein-Fokker-Planck) equation.

5.7 Generalization to Multivariate Random Functions

Let the state of a system at time t be described by a collection of n random variables $x_i(t)$, $i = 1, 2, \ldots, n$, i.e., a random vector $\boldsymbol{x}(t)$ in an n-dimensional space. The components of the random function $\boldsymbol{x}(t)$ may be (generalized) coordinates of the system or a set of coordinates and velocities (or momenta). The "coordinates" x_i in an n-dimensional space are generally curvilinear. But the derivation of the Einstein-Fokker-Planck equation is independent of the choice of coordinates and, in general, also of the metric of space $\boldsymbol{x}$. This is due to the fact that the Smoluchowski equation

$$v(t, \boldsymbol{x} | t_0, \boldsymbol{x}_0) d\boldsymbol{x} = d\boldsymbol{x} \int v(t, \boldsymbol{x} | t - \tau, \boldsymbol{y}) v(t - \tau, \boldsymbol{y} | t_0, \boldsymbol{x}_0) d\boldsymbol{y}$$

serving as a starting point, is given in terms of the probabilities $dP = v\, dx$ which naturally enough are required to be scalars under any coordinate transformation. The line of reasoning remains the same as in the preceeding section, the only departure being that we now deal with transition probabilities for n-dimensional random variables.

Just as in the univariate case, we make the assumptions

$$\lim_{\tau\to 0}\frac{\overline{x_i - y_i}}{\tau} = \lim_{\tau\to 0}\frac{1}{\tau}\int (x_i - y_i)v(t, z|t-\tau, \boldsymbol{y})d\boldsymbol{x} = A_i(\boldsymbol{y}, t)$$
$$\lim_{\tau\to 0}\frac{\overline{(x_i - y_i)(x_k - y_k)}}{\tau} = B_{ik}(\boldsymbol{y}, t),$$
$$\lim_{\tau\to 0}\frac{1}{\tau}\overline{|(x_i - y_i)(x_k - y_k)(x_l - y_l)|} = 0, \tag{5.77}$$

which enable us to derive the Einstein-Fokker-Planck equation in an analogous way:

$$\frac{\partial v}{\partial t} = -\sum_i \frac{\partial A_i v}{\partial x_i} + \frac{1}{2}\sum_{i,k}\frac{\partial^2 B_{ik} v}{\partial x_i \partial x_k}. \tag{5.78}$$

The solution to this equation must be nonnegative, normalized to unity,

$$\int v(t, \boldsymbol{x}|t_0, \boldsymbol{x}_0)d\boldsymbol{x} = 1, \tag{5.79}$$

and meet the initial condition

$$\begin{aligned} v(t_0, \boldsymbol{x}|t_0, \boldsymbol{x}_0) &= \delta(\boldsymbol{x} - \boldsymbol{x}_0) \\ &= \delta(x_1 - x_{10})\delta(x_2 - x_{20})\ldots\delta(x_n - x_{n0}). \end{aligned} \tag{5.80}$$

In addition, some boundary conditions may be imposed.

As was observed above, the form of equation (5.78) is the same irrespective of the choice of coordinates $\boldsymbol{x}$, i.e., it must, of course, be conserved under any transformation $x_i = x_i(\boldsymbol{x}')$ that satisfies the conditions of mutual uniqueness and continuity. Although the equation as a whole is covariant, this does not mean that the respective variables (v, A_i, B_{ik}) are invariant. Really, the fact that the probability dP is a scalar suggests that the probability density v is not invariant, since the invariant volume element is not $d\boldsymbol{x} = dx_1\ldots dx_n$ but $\sqrt{g}d\boldsymbol{x}$, where g is a determinant composed of the metric tensor elements g_{ik}. We can define the scalar probability density $\tilde{v} = v/\sqrt{g}$, i.e.,

$$dP = v\,d\boldsymbol{x} = \tilde{v}\sqrt{g}\,d\boldsymbol{x},$$

which may conveniently be used in transforming the coordinates. In particular, if the space $\boldsymbol{x}$ is a $2n$-dimensional phase space (n-coordinates $x_1, \ldots, x_n$ and n velocities $\dot{x}_1, \ldots, \dot{x}_n$ or momenta $p_1, \ldots, p_n$), then $\tilde{v}$ constitutes the *phase* probability density.

Furthermore, using (5.77) to go over to new coordinates x'_α, the coefficients A_i and B_{ik} are readily seen to transform as follows:

$$A_i = \sum_{\alpha} A'_\alpha \frac{\partial x_i}{\partial x'_\alpha} + \frac{1}{2} \sum_{\alpha,\beta} B'_{\alpha\beta} \frac{\partial^2 x_i}{\partial x'_\alpha \partial x'_\beta},$$

$$B_{ik} = \sum_{\alpha,\beta} B'_{\alpha\beta} \frac{\partial x_i}{\partial x'_\alpha} \frac{\partial x_k}{\partial x'_\beta}. \tag{5.81}$$

Hence, B_{ik} is a tensor of rank two, but A_i does not form a vector (if we do not confine ourselves to linear transformations). As a consequence, the fact that A_i are zero in one coordinate system does not imply that they are also zero in another, i.e., *A_i do not characterize the velocity of the systematic flow of probability in an invariant way.* According to what was said above, however, with adequate transformation laws for all variables (including, of course, operations of differentiation with respect to x_i) we can also see directly, i.e., without resorting to the derivation[7], that (5.78) has covariant form.

We now write (5.78) for the following two cases which we shall consider shortly: for a particle subject to *homogeneous* and *isotropic* impacts moving (1) in a *plane* and (2) on the surface of a unit *sphere*.

In the first case (plane) we will be interested in (5.78) written in terms of the polar coordinates r and φ,

$$x_1 \equiv x = r \cos\varphi, \quad x_2 \equiv y = r \sin\varphi.$$

The isotropy of impacts suggests that the mean square displacement is the same in any direction

$$\overline{(\Delta x)^2} = \overline{(\Delta y)^2} = \overline{(\Delta r)^2} = r^2\overline{(\Delta\varphi)^2} = \varepsilon^2, \tag{5.82}$$

which is only possible if mutually orthogonal displacements are not correlated

$$\overline{\Delta x \cdot \Delta y} = \overline{\Delta r \cdot r\Delta\varphi} = 0. \tag{5.83}$$

The homogeneity of impacts means that the mean square is the same anywhere on the plane, i.e., ε^2 is independent of the coordinates x, y or r, φ.

By (5.78), we have

$$\frac{\partial v}{\partial t} = -\frac{\partial A_r v}{\partial r} - \frac{\partial A_\varphi v}{\partial \varphi} + \frac{1}{2}\left(\frac{\partial^2 B_{rr} v}{\partial r^2} + 2\frac{\partial^2 B_{r\varphi} v}{\partial r\, \partial\varphi} + \frac{\partial^2 B_{\varphi\varphi} v}{\partial \varphi^2}\right), \tag{5.84}$$

where, according to (5.77)

[7] With arbitrary coordinate transformations we should, of course, distinguish between covariant and countercovariant components. For the derivation of the Einstein-Fokker-Planck equation in terms of invariant characteristics of a Markov process see [5.14,15]. Note that the quantity $\tilde{A}^i = A^i + \Gamma^i_{rs} B^{rs}/2$, where Γ^i_{rs}, are Christoffel symbols, is a vector. See also [Ref. 5.16, Sects. 2 and 3].

$$A_r = \lim_{\Delta t \to 0} \frac{\overline{\Delta r}}{\Delta t}, \quad A_\varphi = \lim_{\Delta t \to 0} \frac{\overline{\Delta \varphi}}{\Delta t}, \quad B_{rr} = \lim_{\Delta t \to 0} \frac{\overline{(\Delta r)^2}}{\Delta t},$$

$$B_{r\varphi} = \lim_{\Delta t \to 0} \frac{\overline{\Delta r \Delta \varphi}}{\Delta t}, \quad B_{\varphi\varphi} = \lim_{\Delta t \to 0} \frac{\overline{(\Delta \varphi)^2}}{\Delta t}.$$

It follows from (5.82,83) that

$$B_{rr} = r^2 B_{\varphi\varphi} = \lim_{\Delta t \to 0} \frac{\varepsilon^2}{\Delta t} = B, \quad B_{r\varphi} = 0. \tag{5.85}$$

As for A_r and A_φ, they are to be computed more accurately by including the second-order term with respect to ε (5.82,85). From the relations

$$(r + \Delta r)^2 = (x + \Delta x)^2 + (y + \Delta y)^2, \quad \varphi + \Delta\varphi = \text{arc tan} \frac{y + \Delta y}{x + \Delta x}$$

we have

$$\begin{aligned} \Delta r &= \Delta x \cos\varphi + \Delta y \sin\varphi \\ &\quad + \frac{(\Delta x)^2 \sin^2\varphi + (\Delta y)^2 \cos^2\varphi - 2\Delta x \Delta y \sin\varphi \cos\varphi}{2r} + \dots, \\ r\Delta\varphi &= \Delta y \cos\varphi - \Delta x \sin\varphi \\ &\quad + \frac{[(\Delta x)^2 - (\Delta y)^2] \sin\varphi \cos\varphi - \Delta x \Delta y (\cos^2\varphi - \sin^2\varphi)}{r} + \dots . \end{aligned}$$

Next we average these equalities, divide them by Δt and go over to the limit $\Delta t \to 0$. Using (5.82,83,85), we obtain

$$A_r = R + \frac{B}{2r}, \quad rA_\varphi = \Phi, \tag{5.86}$$

where we have introduced the quantities

$$\begin{aligned} &R = A_x \cos\varphi + A_y \sin\varphi, \quad \Phi = A_y \cos\varphi - A_x \sin\varphi, \\ &A_x = \lim_{\Delta t \to 0} \frac{\overline{\Delta x}}{\Delta t}, \quad A_y = \lim_{\Delta t \to 0} \frac{\overline{\Delta y}}{\Delta t}. \end{aligned} \tag{5.87}$$

If A_x and A_y were the Cartesian components of the *vector* $\boldsymbol{A}$, its polar components A_r and rA_φ would coincide with R and Φ, respectively. The first equation (5.86) shows, however, that A_r does not coincide with the r-component of the vector $\boldsymbol{A}$. It is easily seen that inserting (5.85,86) into (5.84) leads once more to the equation

$$\frac{\partial v}{\partial t} = -\frac{\partial R v}{\partial r} - \frac{\partial \Phi v}{r \partial \varphi} + \frac{B}{2} \left[\frac{\partial}{\partial r} \left(r \frac{\partial}{\partial r} \left(\frac{v}{r} \right) \right) + \frac{1}{r^2} \frac{\partial^2 v}{\partial \varphi^2} \right], \tag{5.88}$$

where $v = v(t, r, \varphi | t_0, r_0, \varphi_0)$ is the conventional (noninvariant) density of transition probability, i.e., $dP = v \cdot dr\, d\varphi$. It should be noted that for the scalar density $\tilde{v} = v/r$, for which $dP = \tilde{v} \cdot r\, dr\, d\varphi$, equation (5.88) takes the form

$$\begin{aligned} \frac{\partial \tilde{v}}{\partial t} &= -\frac{1}{r}\frac{\partial R r \tilde{v}}{\partial r} - \frac{\partial \Phi \tilde{v}}{r \partial \varphi} + \frac{B}{2}\left[\frac{1}{r}\frac{\partial}{\partial r}\left(r \frac{\partial \tilde{v}}{\partial r}\right) + \frac{1}{r^2}\frac{\partial^2 \tilde{v}}{\partial \varphi^2}\right] \\ &= -\mathrm{div}(\boldsymbol{A}\tilde{v}) + \frac{B}{2}\Delta \tilde{v}, \end{aligned}$$

where $\boldsymbol{A}$ is taken to be a vector with polar components R and Φ.

We now turn to the second case, i.e., the motion of a particle on a unit sphere. The coordinates of the particle are the co-latitude θ and longitude φ, so that (5.78) reads

$$\begin{aligned} &\frac{\partial v(t, \theta, \varphi | t_0, \theta_0, \varphi_0)}{\partial t} \\ &= -\frac{\partial A_\theta v}{\partial \theta} - \frac{\partial A_\varphi v}{\partial \varphi} + \frac{1}{2}\left(\frac{\partial^2 B_{\theta\theta} v}{\partial \theta^2} + 2\frac{\partial^2 B_{\theta\varphi} v}{\partial \theta \partial \varphi} + \frac{\partial^2 B_{\varphi\varphi} v}{\partial \varphi^2}\right), \end{aligned} \tag{5.89}$$

where

$$A_\theta = \lim_{\Delta t \to 0} \frac{\overline{\Delta \theta}}{\Delta t}, \quad A_\varphi = \lim_{\Delta t \to 0} \frac{\overline{\Delta \theta}}{\Delta t}, \quad B_{\theta\theta} = \lim_{\Delta t \to 0} \frac{\overline{(\Delta \theta)^2}}{\Delta t}$$

$$B_{\theta\varphi} = \lim_{\Delta t \to 0} \frac{\overline{\Delta \theta \Delta \varphi}}{\Delta t}, \quad B_{\varphi\varphi} = \lim_{\Delta t \to 0} \frac{\overline{(\Delta \varphi)^2}}{\Delta t}.$$

This equation determines the conventional density transition of probability $v(t, \theta, \varphi | t_0, \theta_0, \varphi_0)$ i.e., $dP = v\, d\theta\, d\varphi$.

By virtue of the assumed isotropy of impacts, we have

$$\overline{(\Delta \theta)^2} = \sin^2 \theta \overline{(\Delta \varphi)^2} = \varepsilon^2, \quad \sin \theta \overline{\Delta \varphi \Delta \theta} = 0,$$

so that

$$B_{\theta\theta} = B_{\varphi\varphi} \sin^2 \theta = \lim_{\Delta t \to 0} \frac{\varepsilon^2}{\Delta t} \equiv B, \quad B_{\theta\varphi} = 0,$$

where B is constant on the sphere (homogeneous impacts).

If a particle shifts by $\Delta \alpha$ from a point (θ, φ) to a point (θ', φ'), then using a formula from spherical trigonometry we get

$$\cos \theta' = \cos \theta \cos \Delta \alpha + \sin \theta \cos \psi \sin \Delta \alpha,$$

where ψ is the angle by which the arc $\Delta \alpha$ turns, counted from the meridian (Fig. 5.7). Expanding $\cos \theta'$ in terms of $\Delta \theta = \theta' - \theta$, we obtain to the second order in $\Delta \alpha$

$$\Delta \theta = \mathrm{ctn}\, \theta \sin^2 \psi \frac{(\Delta \alpha)^2}{2} - \cos \psi \delta \alpha.$$

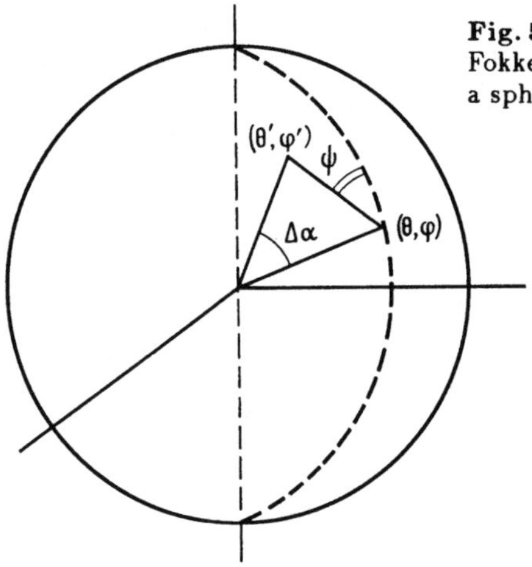

Fig. 5.7. An illustration to the use of the Einstein-Fokker-Planck equation applied to a random walk on a sphere

All the directions of ψ from 0 to 2π appear equiprobable (isotropic impacts), so that $\overline{\cos\psi} = 0$ and $\overline{\sin^2\psi} = 1/2$. On the other hand, treating the small spherical area near the point (θ, φ) as a plane, we have

$$(\Delta\alpha)^2 = (\Delta\theta)^2 + \sin^2\theta(\Delta\varphi)^2,$$

hence

$$\overline{(\Delta\alpha)^2} = \overline{(\Delta\theta)^2} + \sin^2\theta\overline{(\delta\varphi)^2} = 2\varepsilon^2.$$

This gives

$$\overline{\Delta\theta} = \frac{\varepsilon^2}{2}\operatorname{ctn}\theta \quad \text{i.e.,} \quad A_\theta = \frac{B}{2}\operatorname{ctn}\theta.$$

Arguing along the same lines we find that $A_\varphi = 0$. With

$$\begin{aligned} A_\theta &= \frac{B}{2}\operatorname{ctn}\theta, \quad A_\varphi = 0, \quad B_{\theta\theta} = B, \\ B_{\varphi\varphi} &= \frac{B}{\sin^2\theta}, \quad B_{\theta\varphi} = 0, \end{aligned} \tag{5.90}$$

(5.89) now takes the form

$$\frac{\partial v}{\partial t} = \frac{B}{2}\left[\frac{\partial}{\partial\theta}\left(\sin\theta\frac{\partial}{\partial\theta}\left(\frac{v}{\sin\theta}\right)\right) + \frac{1}{\sin^2\theta}\frac{\partial^2 v}{\partial\varphi^2}\right]. \tag{5.91}$$

For the scalar density $\tilde{v}$, i.e., the density in an element of the solid angle $\sin\theta\, d\theta\, d\varphi$ $(\tilde{v} = v/\sin\theta)$ we have

$$\sin\theta\frac{\partial\tilde{v}}{\partial t} = \frac{B}{2}\left[\frac{\partial}{\partial\theta}\left(\sin\theta\frac{\partial\tilde{v}}{\partial\theta}\right) + \frac{1}{\sin\theta}\frac{\partial^2\tilde{v}}{\partial\varphi^2}\right]. \tag{5.92}$$

If we count θ from the initial position of the particle, i.e., draw a polar axis through this position (thus $\theta_0 = 0$) then, by virtue of the symmetry, $\tilde{v}$ will be independent of φ. The equation then takes the form

$$\sin\theta \frac{\partial \tilde{v}(t,\theta|t_0,0)}{\partial t} = \frac{B}{2}\frac{\partial}{\partial \theta}\left(\sin\theta \frac{\partial \tilde{v}}{\partial \theta}\right), \tag{5.93}$$

and its solution must satisfy the initial condition

$$\tilde{v}_{t=t_0} = \frac{\delta(\theta)}{2\pi \sin\theta}.$$

The solution is expressed in terms of the Legendre polynomials (see Sect. 5.10).

5.8 Fluctuations in the Thomson Vacuum-Tube Oscillator

Now we consider how the Einstein-Fokker-Planck equation can be applied to a simple self-oscillatory system with one degree of freedom: the vacuum-tube oscillator with a Thomson oscillator in the anode network, inductive feedback, and cubic tube characteristic (softself-excitation), see Fig. 5.8. If a generator is said to be a *Thomson* system, this amounts to saying that it is close to the linear harmonic oscillator, i.e., the differential equation describing the oscillator has sufficiently small dissipative and nonlinear terms. Specific assumptions about the oscillator circuit (Fig. 5.8) are only made to simplify the dynamic model. However, the applications of the theory presented below are not confined to this system. As a matter of fact, they include the general case of the Thomson self-oscillatory system, closed (autonomous) and nonautonomous, both with one or many degrees of freedom. For one degree of freedom the dynamic equation has the form

$$\ddot{x} + x = \mu f(x, \dot{x}, t),$$

where μ is a small parameter.

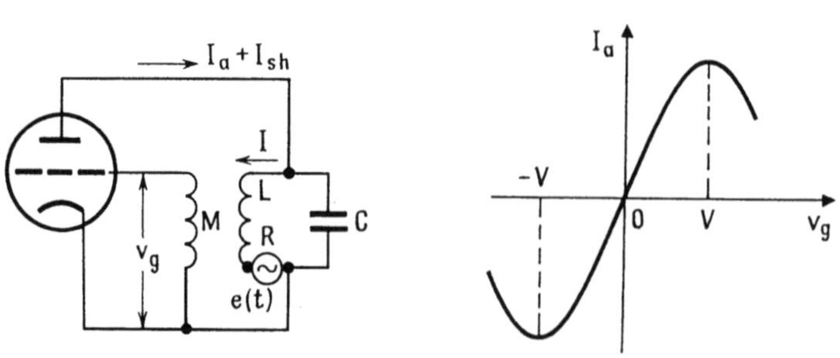

Fig. 5.8. Block diagram of the Thomson self-oscillatory system (*left*) and the current-voltage characteristic of the triode (*rihgt*)

Similarly, the origin and specific form of nonlinearity are of no significance, i.e., the theory is also applicable to other vacuum-tube circuits, semiconductor generators, and lasers (if a quasi-classical description is possible), and so forth.

We shall return to the model under consideration later [5.17], and so we include in the differential equation describing the oscillator two sources of fluctuations: (1) *shot current* flowing through the anode circuit of the tube, $I_{\text{sh}}(t)$ to be added to the instantaneous mean (i.e., the statistical average over the ensemble of identical oscillators) anode current $I_{\text{a}}(t)$; and (2) ***thermal fluctuations*** of the current and voltage in the oscillator circuit. The true source of these thermal fluctuations is the thermal motion of microcharges in the conductors of the circuit. But in [5.17], we will give a phenomenological description of thermal fluctuations in an electrical circuit as a result of the action of an equivalent random electromotive force. Therefore, the diagram in Fig. 5.8 has a generator of these fluctuational "thermal" e.m.f., $e(t)$ included in the oscillator circuit.

Using the notation of Fig. 5.8, we have

$$L\dot{I} + RI - e = \frac{1}{C}\int (I_{\text{a}} + I_{\text{sh}} - I)dt,$$

$$I_{\text{a}} = Sv_{\text{g}}\left(1 - \frac{v_{\text{g}}^2}{3V^2}\right), \quad v_{\text{g}} = M\dot{I}.$$

Having eliminated I_{a} and v_{g}, we obtain for the current I the equation

$$\ddot{I} + \frac{R}{L}\dot{I} + \frac{I}{LC} = \frac{SM}{LC}\dot{I}\left(1 - \frac{M^2\dot{I}^2}{3V^2}\right) + \frac{I_{\text{sh}}}{LC} + \frac{\dot{e}}{L}.$$

Since we treat this as a Thomson system, the nonlinear and nonconservative terms of the equation are assumed to be small, i.e., of the order of smallness of the parameter μ. We use

$$1/LC = \omega_0^2$$

and introduce the following dimensionless variables and coefficients:

$$t' = \omega_0 t, \quad x = I/I_0, \quad I_0 = 2V/\omega_0 M,$$

$$\omega_0 MS = \mu, \quad \omega_0(MS - RC) = \mu p.$$

The equation then takes the form

$$\ddot{x} + x = \mu\dot{x}\left(p - \frac{4}{3}\dot{x}^2\right) + \mu F(t'), \tag{5.94}$$

where

$$\mu F(t') = \frac{M\omega_0}{2V}\left[I_{\text{sh}}(t) + \frac{\dot{e}(t)}{L\omega_0}\right] = \frac{1}{I_0}\left[I_{\text{sh}}(t) + \frac{\dot{e}(t)}{L\omega_0}\right] \tag{5.95}$$

is the fluctuational "force" acting on the system and associated with shot current and thermal fluctuations. This force is also considered to be small, i.e., of the same order in μ as the dissipative term of equation (5.94). As already stated, we shall later make use of its concrete form and, in general, its presence in (5.94). For the moment it serves only as an illustration of how random actions may enter into the description of a dynamic system. Further development of the theory in this section, though imposing certain limitations on the statistical behavior of $F(t')$, does not explicitly use this random force i.e., it relies on the purely dynamic equation

$$\ddot{x} + x = \mu\dot{x}(p - \tfrac{4}{3}\dot{x}^2). \tag{5.96}$$

It is common knowledge that Tomson systems close to the harmonic oscillator ($\ddot{x} + x = 0$ with $\mu = 0$) can be described using the approximate method of Van der Pol, also known as the method of slow perturbation. In a first (relative to the small parameter μ) approximation, the solution to (5.96) differs from the solution for the harmonic oscillator $x = r\ \cos(t' + \varphi)$, where the amplitude r and phase φ are constants, in that r and φ appear to be *slow* functions of time; i.e., they depend on t' through the so-called "slow time" $\theta = \mu t'$

$$x = r(\theta)\ \cos\left[t' + \varphi(\theta)\right]. \tag{5.97}$$

The rates of variation of the amplitude $r(\theta)$ and phase $\varphi(\theta)$ (i.e., the addition of $\dot{\varphi}(\theta)$ to the frequency of oscillation) are thus quantities of the first order in μ

$$\dot{r} = \mu r', \quad \dot{\varphi} = \mu\varphi'$$

where the prime stands for the differentiation with respect to the slow time θ.

Recall the simplified procedure for deriving the Van der Pol (or so-called *shortened*) equations for the Thomson system, i.e., the first approximation equation for the slowly varying amplitude $r(\theta)$ and phase $\varphi(\theta)$. In this dynamic treatment we will proceed not from (5.96) for the selected model of the oscillator, but from the more general equation of the Thomson system with one degree of freedom,

$$\ddot{x} + x = \mu f(x, \dot{x}) + \mu F(t'). \tag{5.98}$$

Computing to the first order in μ the derivatives of (5.97) with respect to t', we get

$$\dot{x} \equiv \frac{dx}{dt'} = -r\ \sin(t' + \varphi) + \mu[r'\ \cos(t' + \varphi) - r\varphi'\ \sin(t' + \varphi)],$$
$$\ddot{x} + x = -2\mu[r'\ \sin(t' + \varphi) + r\varphi'\ \cos(t' + \varphi)].$$

The function $f(x, \dot{x})$ in (5.98) is already given together with the factor μ, and so a zeroth approximation for $\dot{x}$ will suffice. Dividing out by μ gives

$$-2[r' \sin(t'+\varphi) + r\varphi' \cos(t'+\varphi)]$$
$$= f[r \cos(t'+\varphi); -r \sin(t'+\varphi)] + F(t').$$

Obviously, the function which is dependent on periodic arguments in $u = t'+\varphi$, is itself periodic in u with a period of 2π, and therefore may be expanded in a Fourier series

$$f(r \cos u; -r \sin u) = \frac{a_0}{2} + \sum_{n=1}^{\infty}(a_n \cos nu + b_n \sin nu),$$

where

$$\left.\begin{matrix} a_n(r) \\ b_n(r) \end{matrix}\right\} = \frac{1}{\pi}\int_{-\pi}^{\pi} f(r \cos u; -r \sin u)\left\{\begin{matrix} \cos nu \\ \sin nu \end{matrix}\right\} du.$$

The high selectivity of the system (5.98) [nonconservative terms only enter in the order of μ and in a zeroth approximation we have the ideal (infinitely selective) oscillator $\ddot{x} + x = 0$] implies that in the entire Fourier series only those terms are of significance which oscillate in t' with the basic dimensionless frequency $n = 1$. The constant term (a_0) and higher harmonics $(n = 2, 3, \dots)$ may be discarded as being far from the resonance. The equation thus becomes

$$-2[r' \sin(t'+\varphi) + r\varphi' \cos(t'+\varphi)]$$
$$= a_1(r) \cos(t'+\varphi) + b_1(r) \sin(t'+\varphi) + F(t').$$

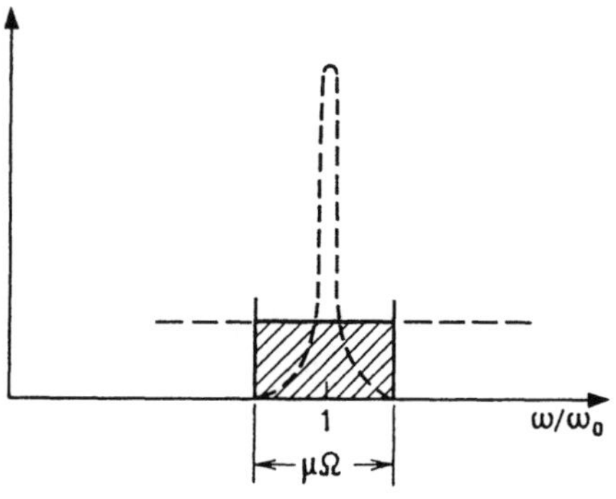

Fig. 5.9. Illustration of the definition of the random force $\tilde{F}(t')$

For the same reason, only the close neighborhood of the resonance frequency 1 is of importance in the entire spectrum of the force $F(t')$. By cuting out from the spectrum of $F(t')$ a band with a (dimensionless) width $\mu\Omega$ (which is, however, notably wider than the resonance curve for the regenerated oscillator circuit; Fig. 5.9) that straddles frequency 1, we can replace $F(t')$ by a process $\tilde{F}(t')$ whose spectrum is limited by the above narrow band. Such a narrow-band process is essentially an oscillation with a slowly varying amplitude and frequency (modulated oscillation). Therefore, it can always be written as

$$\tilde{F}(t') = F_{\|}(\theta)\ \cos\,[t' + \varphi(\theta)] + F_{\perp}(\theta)\ \sin\,[t' + \varphi(\theta)]. \tag{5.99}$$

In [5.17] we will have to discuss in more detail the behavior of the functions $F_{\|}$ and $F_{\perp}$. Here we only note that the apparent fact that these functions vanish when there is no external force, $F(t') = 0$.

Next we insert the expression (5.99) for $F(t')$ into the preceeding equation. Then we collect the terms with $\sin(t' + \varphi)$ and $\cos(t' + \varphi)$ and equate to zero the coefficients obtained. We thus find two Van der Pol equations for r and φ as functions of the slow time $\theta = \mu t'$,

$$r' = R(r) - \tfrac{1}{2}F_{\perp}(\mu),\quad r\varphi' = \Phi(r) - \tfrac{1}{2}F_{\|}(\theta),\quad \text{where} \tag{5.100}$$

$$\left.\begin{matrix} R(r) \\ \Phi(r) \end{matrix}\right\} = -\frac{1}{2}\left\{\begin{matrix} b_1(r) \\ a_1(r) \end{matrix}\right. = -\frac{1}{2\pi}\int\limits_{-\pi}^{\pi} f(r\ \cos\ u; -r\ \sin\ u)\left.\begin{matrix}\sin u \\ \cos u\end{matrix}\right\} du. \tag{5.101}$$

If the system is autonomous, i.e., $F(t') = 0$, then equations (5.100) will be

$$r' = R(r),\quad r\varphi' = \Phi(r). \tag{5.102}$$

Integrating the first of these yields for a specified initial value of $r(0)$ the variation of the amplitude $r(\theta)$. Inserting this $r(\theta)$ into the second equation gives at a specified $\varphi(0)$ the law of variation of the phase $\varphi(\theta)$. In the steady states, if these exist, the amplitude r is constant $(r' = 0)$, and the constant values of r are obtainable from the equation $R(r) = 0$. Let r_0 be any one of the solutions to this equation. The second equation (5.102) then defines the corresponding phase

$$\varphi(\theta) = \varphi(0) + \frac{\Phi(r_0)}{r_0}\theta = \varphi(0) + \mu\frac{\Phi(r_0)}{r_0}t'.$$

If $\Phi(r_0)/r_0 \neq 0$, the self-oscillations (5.97) thus occur with the frequency $1 + \mu[\Phi(r_0)/r_0]$ i.e., the natural frequency of the linear oscillator plus a perturbation.

What is the form of equations (5.102) for our simple model of an oscillator with $f(x, \dot{x}) = \dot{x}(p - 4\dot{x}^2/3)$? For $f(x, \dot{x})$ with such a simple form (polynomials in x and $\dot{x}$; in this case only in $\dot{x}$) we can even dispense with integrals (5.101), and apply elementary trigonometry. Substituting $\dot{x} = -r\ \sin\ u$ gives

$$\dot{x}(p - \tfrac{4}{3}\dot{x}^2) = -pr\ \sin\ u + \tfrac{4}{3}r^3\ \sin^3 u = -r(p - r^2)\ \sin\ u - \frac{r^3}{3}\ \sin\ 3u.$$

Discarding the third harmonic, we thus have $a_1 = 0$, $b_1 = -r(p - r^2)$ or $\Phi = 0$, $R = r(p - r^2)/2$, i.e., the equations

$$r' = \frac{r}{2}(p - r^2)\quad \text{and}\quad \varphi' = 0. \tag{5.103}$$

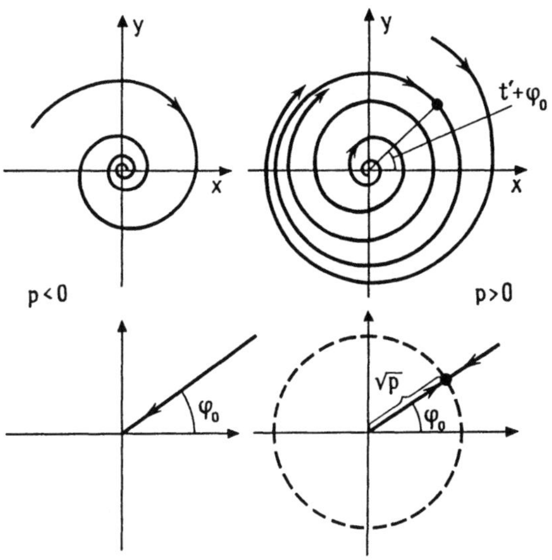

Fig. 5.10. The paths in the phase plane (*top*) and the Van der Pol plane (*bottom*) for a self-oscillatory system with ($p>0$) and without ($p<0$) excitation

The possible equilibrium states are $r = 0$ and, for $p>0$ also $r = \sqrt{p}$. Elementary examination shows that for $p<0$ the equilibrium state $r = 0$ is stable, and for $p>0$ unstable, but the limiting cycle with $r = \sqrt{p}$ is stable. The phase $\varphi = \varphi_0$ =const, i.e., it retains its initial value. Figure 5.10 shows the trajectories in the phase plane $x = r\ \cos(t' + \varphi)$, $\dot{x} \approx -r \cdot \sin(t' + \varphi)$, and in the lower part of the figure the corresponding picture is given in the Van der Pol plane, i.e., in the plane with the polar coordinates r and φ. The connection between both pictures, in the first approximation used, is very simple: if we rotate the Van der Pol plane with unit circular frequency about the origin, then slow motion of a representative point along the ray $\varphi = \varphi_0$ =const will produce a spiral, and the stable equilibrium position $r = \sqrt{p}$ will produce a circle (the limiting cycle) in the phase plane $(x, \dot{x})$. Such, in broad outline, is the solution of this dynamic problem.

How will these results change if we include random actions? What causes these random actions? Let us first tackle the second question.

The anode current of a tube fluctuates not only due to the shot effect (Sects. 2.5 and 3.3), but also because of the "flicker effect" – chaotic variations of the cathode emission. In the oscillator circuit itself, fluctuations of current and voltage are conditioned not only by the thermal motion of electrons, but also by a variety of other causes. The oscillator is subject to various external influences: mechanical (which cause in particular the vibration of tube electrodes, the microphone effect) thermal, (which may affect the parameters of the circuit), electromagnetic (electric fluctuations in power supplies, stray pick-up), and so forth. Some of these influences can in principle be removed: we may, for instance, adequately block the power supplies, we may also carefully shield the generator itself, put it in a thermostat, attach shock-absorbers, and so on. But phenomena such as shot effect and thermal fluctuations cannot be removed, as they are associated with the atomic structure of the electric charge. The flicker effect brought about by fairly involved processes of diffusion and adsorption in the surface layer of the cathode is found experimentally to be present in all cases, but, unlike shot and thermal noise, it occurs relatively slowly as compared with high-frequency oscillations of the oscillator.

Random actions can be subdivided into two kinds. Some cause slow, but marked, variations of the device parameters and corresponding frequency drifts. *Gorelik* [5.18] suggested to call these drifts and the associated nonmonochromaticity of oscillations as *technical* effects, thereby stressing that they are due to external factors that are in principle removable (though the flicker effect which contributes to slow fluctuations also falls into this category). He also proposed describing as *natural* the principally unavoidable nonmonochromaticity resulting from the extremely fast, but shallow, chaotic modulation due to shot and thermal fluctuations. In [5.17] we will return to the discussion of the influence of both types of chaotic modulation on the self-oscillation spectrum. At the moment we are interested in the time variation of these phenomena and the possibility of describing them as a *continuous* Markov process.

As has already been noted (Sect. 5.6), such a description assumes a sufficient frequency of independent random impacts: the probability of finite changes of the state during a macroscopically small time $\Delta t'$ must tend to zero faster than $\Delta t'$. At the same time the interval $\Delta t'$ must be large as compared with the time gap between random impacts, i.e., during the time $\Delta t'$ the system must be subjected to a very large number of mutually independent random actions.

What is the macroscopic scale of smallness of $\Delta t'$? Obviously, the *period* of self-oscillation (which is equal to 2π in terms of the dimensionless time t'), as other time-scales of the Thomson system (the relaxation time of the self-oscillator or of its oscillator circuit) cover very many ($\sim 1/\mu$) periods. Accordingly, $2\pi \gg \Delta t'$ which is in turn much longer than the time between impacts. But this condition is only met by shot and thermal fluctuations. Random actions resulting in technical frequency drifts are much slower than the period, or the relaxation time of the oscillator. If we stick to treating the oscillator as a system with *one* degree of freedom then it follows that these actions cannot be incorporated into the scheme of a random process without stochastic memory effects. Thus, only by treating the state as a combination of the two variables x and $\dot{x}$ (or r and φ) at a time t' and describing the random variation of their state as a continuous bivariate Markov process can we allow for the contributions of *shot* and *thermal* fluctuations.

As a result of these fluctuations, which vary fast in the course of even a small section of the limiting cycle (a small part of the period), the representation point in the phase plane (x, y) [or in the plane (r, φ) where the systematic clockwise rotation with a period of 2π is excluded] executes a sort of Brownian motion. In order to write the appropriate Einstein-Fokker-Planck equation for the transition probability $v(\theta, r, \varphi|\theta_0, r_0, \varphi_0)$, it is necessary to specify the form of the coefficients B_{ik}. We assume that the random impacts are isotropic, i.e., that the expressions (5.85) are valid,

$$B_{rr} = r^2 B_{\varphi\varphi} \equiv B, \quad B_{r\varphi} = 0.$$

Furthermore, we suppose that the impacts are *homogeneous*, i.e., B is a constant independent of the state of the oscillator.

Note in passing that the last assumption is, strictly speaking, not suitable for the vacuum-tube oscillator. It holds for thermal fluctuations in the oscillator circuit, as the intensity of these is independent of the current flowing in the conductors. But the intensity of anode current fluctuations in the tube (shot effect) is proportional to the average current, see (3.27). With self-oscillations, the average current pulsates in phase with the grid voltage oscillations. Accordingly, in terms of (r, φ) the coefficient B should vary not only with r and φ, but also with time t' (periodic nonstationarity under steady-state conditions). The assumption of constant B means that we are dealing with an oversimplified model. Nevertheless, this model still gives a certain insight into the influence of oscillations on the oscillator behavior and is fairly realistic in those cases where the fluctuations of the anode current are relatively small.

The above qualitative treatment can in no way be considered as a *proof* that it is only under conditions of "dense" shot and thermal fluctuations that a random process in the oscillator is really a continuous Markov process and obeys (with the additional assumption that random impacts are homogeneous and isotropic) the Einstein-Fokker-Planck equation (5.88). Moreover, apart from the coefficient B this equation includes the deterministic functions $R(r, \varphi)$ and $\Phi(r, \varphi)$, which are still to be specified. As yet we have no well-defined technique for finding these functions, so that in this problem, too, we will have to be guided by physical considerations.

We saw in Sect. 5.6 that for an univariate Markov process, the coefficient $A(x, t)$ in the Einstein-Fokker-Planck equation coincides with the right-hand side of the dynamic equation (5.70) up to an additive random function with zero mean. In our bivariate problem such functions could appear on the right-hand side of the dynamic equations (5.103), if in deriving them we had not discarded in the initial equation (5.94) the random force $F(t)$ (whose mean is just zero). It is only natural to assume that R and Φ in (5.88) must be the right-hand side of equations (5.103), i.e., to set

$$R = \frac{r}{2}(p - r^2), \quad \Phi = 0. \tag{5.104}$$

Then equation (5.88) for the transition probability $v(r, \varphi|\theta - \theta_0, r_0, \varphi_0)$ will be

$$\frac{\partial v}{\partial t} = -\frac{\partial Rv}{\partial r} + \frac{B}{2}\left[\frac{\partial}{\partial r}\left(r\frac{\partial}{\partial r}\left(\frac{v}{r}\right)\right) + \frac{1}{r^2}\frac{\partial^2 v}{\partial \varphi^2}\right]. \tag{5.105}$$

The process is homogeneous in time (v depends on $\theta - \theta_0$) owing to the fact that R and B are independent of θ.

All our plausible assumptions resulting in (5.105) will be justified in the next chapter which is devoted to stochastic differential equations.

Physically, it is obvious that the process is ergodic, i.e., the system has a finite memory, so that the transition probability goes over, with increasing $\theta - \theta_0$, into the univariate state probability w_1 irrespective of the initial values of r_0 and φ_0

$$v(r, \varphi | \theta - \theta_0, r_0, \varphi_0) \to w_1(r, \varphi), \quad |\theta - \theta_0| \to \infty. \tag{5.106}$$

Of course, the complete solution to equation (5.105) with the initial condition

$$v(r, \varphi | 0, r_0, \varphi_0) = \delta(r - r_0)\delta(\varphi - \varphi_0) \tag{5.107}$$

would show the validity of (5.106) without any additional assumptions whatsoever. But we will from the very outset confine ourselves to the simpler task of deriving a *stationary* solution $w_1(r, \varphi)$. It has to satisfy the equation

$$\frac{\partial R w_1}{\partial r} - \frac{B}{2}\left[\frac{\partial}{\partial r}\left(r \frac{\partial}{\partial r}\left(\frac{w_1}{r}\right)\right) + \frac{1}{r^2}\frac{\partial^2 w_1}{\partial \varphi^2}\right] = 0.$$

We will now make one point that will become completely clear in Sect. 5.9: a stationary distribution is only possible under the condition that we take for the range of the phase φ the *finite* interval $(0, 2\pi)$. This is by no means the only possible statement of the problem, but if it is accepted, then it follows immediately that all values of φ in the interval $(0, 2\pi)$ are equiprobable. This means that the distribution is uniform in φ (w_1 does not depend on φ), so that the equation for w_1 will take the form

$$\frac{d}{dr}\left(R w_1 - \frac{B}{2} r \frac{d}{dr}\left(\frac{w_1}{r}\right)\right) = 0.$$

It is readily seen that the only solution that has no singularity at the point $r = 0$ is the one for which the flow of w_1, i.e., the expression in parentheses, vanishes. Integrating a second times gives

$$w_1 = Cr \exp\left\{\frac{2}{B}\int_0^r R\, dr\right\},$$

where C is determined by the normalization condition

$$\int_0^{2\pi} d\varphi \int_0^{\infty} w_1 dr = 2\pi C \int_0^{\infty} \exp\left\{\frac{2}{B}\int_0^r R\, dr\right\} r\, dr = 1.$$

If we substitute R from (5.103) into these expressions and take the integrals, than we will get the desired stationary distribution

$$w_1(r, \varphi) dr\, d\varphi = \frac{e^{-(p-r^2)^2/4B}}{\pi\sqrt{\pi B}[1 + \mathrm{erf}(p/2\sqrt{B})]} r\, dr\, d\varphi, \tag{5.108}$$

where

$$\operatorname{erf} x = \frac{2}{\sqrt{\pi}} \int_0^x e^{-t^2} dt.$$

Figure 5.11 shows the surface $w_1(r, \varphi)$ for $p>0$ (excited oscillator), when it is crater-shaped. The most probable value of r will be

$$r_m = \sqrt{\frac{p}{2} + \sqrt{\frac{p^2}{4} + B}}, \qquad \text{i.e.,}$$

$$r_m \approx \sqrt{p}\left(1 + \frac{B}{2p^2}\right) \quad \text{for} \quad p \gg 2\sqrt{B},$$

$$r_m \approx \sqrt{\frac{B}{|p|}} \quad \text{for} \quad p \ll -2\sqrt{B}.$$

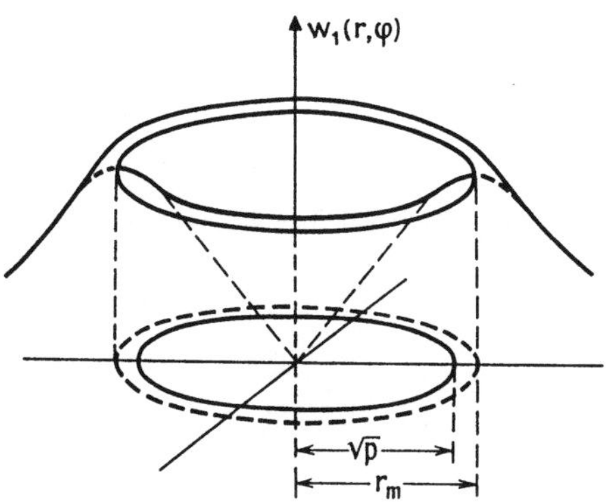

Fig. 5.11. The profile of the stationary probability density $w_1(r, \varphi)$ over the phase plane (r, φ)

In the last case, i.e., sufficiently far from the self-excitation threshold in the direction of negative p, we can ignore the fourth power of r in the exponent in (5.108) and make use of the asymptotic expansion

$$\operatorname{erf}(-x) = -1 + \frac{e^{-x^2}}{\sqrt{\pi}|x|} - \dots .$$

We obtain then

$$w_1(r, \varphi) dr\, d\varphi = \frac{|p|}{B} e^{-r^2|p|/2B} r\, dr \cdot \frac{d\varphi}{2\pi}, \tag{5.109}$$

i.e., for the amplitude r, as is to be expected in the case of the linear dissipative circuit, we have the Rayleigh distribution (5.34) where $2\sigma^2 = \overline{r^2} = 2B/|p|$.

On the contrary, if the oscillator is sufficiently excited, $(p \gg 2\sqrt{B})$, then the "roller" around the crater in Fig. 5.11 will only be notable over the ring containing the limiting cycle. If we next put $r = \sqrt{p} + \varrho$ and confine ourselves to the square of ϱ in the exponent of (5.108), then taking into accout the fact that now

$$\operatorname{erf}(p/2\sqrt{B})\approx 1,$$

we get

$$w_1(\varrho,\varphi)d\varrho\, d\varphi = \sqrt{\frac{p}{\pi B}}e^{-\varrho^2 p/B}d\varrho\frac{d\varphi}{2\pi}. \tag{5.110}$$

Thus, the radial deviation ϱ from the limiting cycle has in this case the normal distribution with the variance $\sigma^2 = \overline{\varrho^2} = B/2p$.

Figure 5.12 shows the w_1 versus r behavior for three values of p, in particular for $p = 0$ which is on the self-excitation treshold. Of course, at all values of p the maximum w_1 becomes more and more smeared out with increasing intensity of random impacts (increase in B).

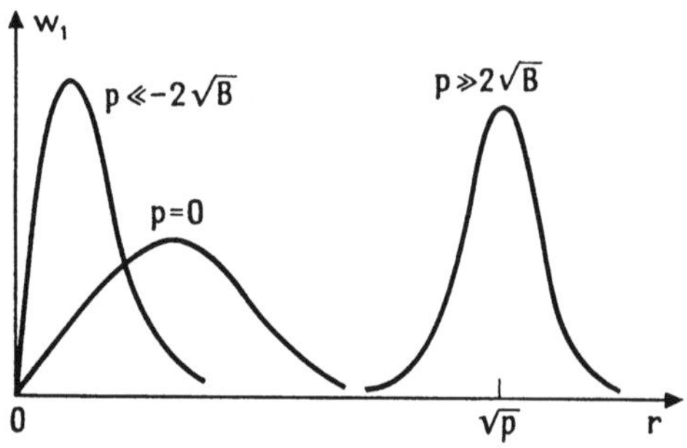

Fig. 5.12. The variation of w_1 with r is depicted for three values of the increment p

5.9 Fluctuations at Large Self-Oscillation Amplitudes

We will now take a closer look at the case of large positive values of p. The condition that $p \gg 2\sqrt{B}$ clearly suggests that the radius of the limiting cycle $\sqrt{p}$ is large as compared with the standard deviation for the amplitude fluctuation

$$\sqrt{p} \gg \sqrt{B/2p} = \sqrt{\overline{\varrho^2}}.$$

This enables the dynamic equations (5.103) to be linearized in $\varrho = r - \sqrt{p}$. These take the form

$$\varrho' = -p\varrho, \quad \varphi' = 0, \tag{5.111}$$

i.e., the deviation of the representative point from the limiting cycle behaves as if the point were connected to the cycle by a spring and moved in a viscous medium. The increment p acts as the ratio of the spring elasticity coefficient to the viscous friction coefficient and characterizes the "rigidity" of the limiting cycle – the rate of return of the representative point back to the cycle after an initial deviation from it.

According to (5.111) we can also simplify the Einstein-Fokker-Planck equation (5.105), in which we set $\partial/\partial r = \partial/\partial\varrho$, $r = \sqrt{p}$ and $R = -p\varrho$

$$\frac{\partial v}{\partial \theta} = p\frac{\partial \varrho v}{\partial \varrho} + \frac{B}{2}\left(\frac{\partial^2 v}{\partial \varrho^2} + \frac{1}{p}\frac{\partial^2 v}{\partial \varphi^2}\right). \tag{5.112}$$

The simplification is so significant that we can now obtain not only a stationary distribution, but also the solution of (5.112) for arbitrary initial conditions $\varrho = \varrho_0$, $\varphi = \varphi_0$, $\theta = \theta_0$ (in doing so, the condition $\varrho_0 \ll p$ must, of course, be fulfilled). For the transition probability $v(\varrho,\varphi|\theta - \theta_0, \varrho_0, \varphi_0)$ the above initial conditions mean that

$$v(\varrho,\varphi|0,\varrho_0,\varphi_0) = \delta(\varrho - \varrho_0)\delta(\varphi - \varphi_0). \tag{5.113}$$

If we again suppose that the possible values of φ lie in the interval $(0,2\pi)$, i.e., we look for a solution of (5.112) which is periodic in φ with a period of 2π, then the usual procedure of separating the variables will yield the following result:

$$v(\varrho,\varphi|\theta - \theta_0,\varrho_0,\varphi_0) = v_\varrho(\varrho|\theta - \theta_0,\varrho_0)v_\varphi(\varphi|\theta - \theta_0,\varphi_0), \tag{5.114}$$

i.e., at any instant of time ϱ and φ are independent, and the amplitude fluctuations ϱ are normally distributed,

$$v_\varrho(\varrho|\theta - \theta_0,\varrho_0) = \frac{1}{\sqrt{2\pi}\sigma}\exp\left[-(\varrho - \bar{\varrho})^2/2\sigma^2\right] \quad , \tag{5.115}$$

where the mean and variance, respectively, are

$$\begin{aligned} \bar{\varrho} &= \varrho_0 \exp\left[-p(\theta - \theta_0)\right], \\ \sigma^2 &= \frac{B}{2p}\{1 - \exp\left[-2p(\theta - \theta_0\right]\} \quad , \end{aligned} \tag{5.116}$$

and the distribution of the phase φ is

$$v_\varphi(\varphi|\theta - \theta_0,\varphi_0) = \frac{1}{2\pi}\left\{1 + 2\sum_{n=1}^{\infty} e^{-n^2 D(\theta-\theta_0)} \cos n(\varphi - \varphi_0)\right\}, \tag{5.117}$$

where $D = B/2p$.

At the initial instant of time $\theta = \theta_0$ we have $\bar{\varrho} = \varrho_0$, $\sigma^2 = 0$, i.e., the distribution of ϱ becomes $\delta(\varrho - \varrho_0)$, and the distribution of φ at $\theta = \theta_0$ gives $\delta(\varphi - \varphi_0)$, since

$$1 + 2\sum_{1}^{\infty} \cos n(\varphi - \varphi_0)$$

is the Fourier series for the periodic function $2\pi\delta(\varphi - \varphi_0)$. Both the amplitude and the phase are ergodic: as $\theta \rightarrow \infty$, a stationary distribution is established that is independent of the initial conditions, namely the normal distribution for ϱ with $\bar{\varrho} = 0$, the variance $\sigma^2 = B/2p$ and the uniform distribution for φ over the interval $(0,2\pi)$ as was obtained earlier, see (5.110).

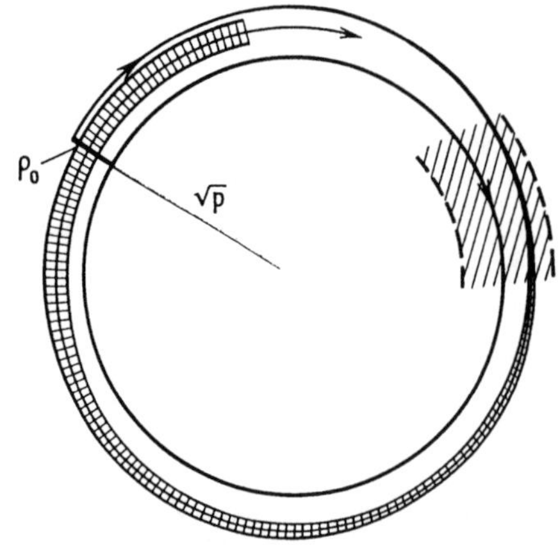

Fig. 5.13. The evolution of the amplitude distribution starting from a given initial value $\sqrt{p} + \varrho_0$: $\bar{\varrho}$ decays in an exponential manner from ϱ_0 to zero, and $D[\varrho]$ grows from zero to its stationary value

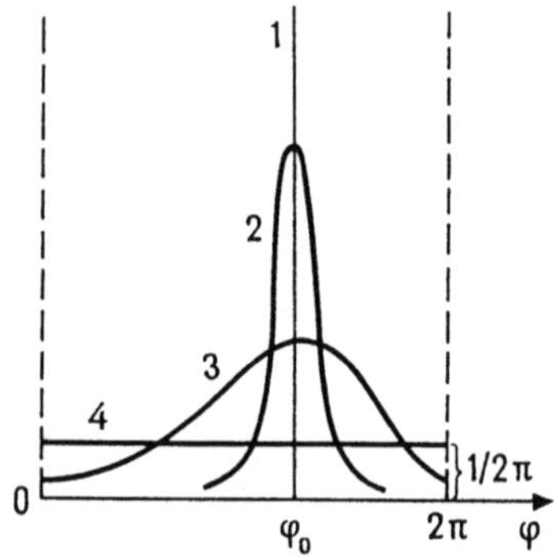

Fig. 5.14. The phase distribution diffuses from the initial distribution $\delta(\varphi - \varphi_0)$ denoted by 1, via 2 and 3 to the uniform distribution 4 within the interval $(0, 2\pi)$

Figure 5.13 shows how the amplitude distribution is reached: the mean $\bar{\varrho}$ decays in the same exponential way in which the representative point approaches the limiting cycle in a purely dynamic problem. At the same time, the availability of random impacts increases the variance $\overline{(\varrho - \bar{\varrho})^2}$ up to its established value. The phase distribution varies as shown in Fig. 5.14: the ensemble of identical oscillators, started at a time $\theta = \theta_0$ with the same phase φ_0, spreads diffusively over the limiting cycle and, after a sufficiently long time, the members of this ensemble are uniformly distributed over the whole cycle.

Note now the following. The choice of the interval $(0,2\pi)$ for φ as applied to an individual oscillator means that from its total phase progression we discard a multiple of completed cycles. But if we are to use the oscillator as a *clock*, it is not immaterial whether the phase has drifted by an angle α, or $\alpha + 2\pi$, $\alpha + 4\pi$, and so forth. In other words, we are interested here in the *total* phase progression so that its possible values are to be found within the interval from $-\infty$ to $+\infty$. In that case, the solution to (5.112) will still have the form (5.113) and for the amplitude fluctuations the relations (5.115,116) hold; but now the transition probability for the phase $v_\varphi(\varphi|\theta - \theta_0, \varphi_0)$ will not be periodic in φ, and with the initial condition $\delta(\varphi - \varphi_0)$ is easily seen to be given by

$$v_\varphi(\varphi|\theta - \theta_0, \varphi_0) = \frac{1}{\sqrt{4\pi D(\theta - \theta_0)}} \exp\left\{-\frac{(\varphi - \varphi_0)^2}{4D(\theta - \theta_0)}\right\}, \tag{5.118}$$

where $D = B/2p$.

We thus return to the case of the isotropic random walk of a particle along an infinite straight line (see Sects. 2.4, 7 and 5.4), and hence we no longer have ergodicity: the stationary distribution independent of φ_0 is not attained for any

θ, and the mean square of the fluctuation phase shift during the time $\theta - \theta_0$ grows indefinitely with $\theta - \theta_0$

$$\overline{(\varphi - \varphi_0)^2} = 2D(\theta - \theta_0). \tag{5.119}$$

The formulations of both problems are closely linked (Fig. 5.15): if the distribution (5.118), which is spread along the infinite φ-axis, is "convoluted" by transferring all the intervals $[2\pi n, 2\pi(n+1)]$ to the band $(0,2\pi)$, then in this band we get the distribution (5.117), i.e., we return to the situation shown in Fig. 5.14. Obviously, as long as the flattening of (5.118) is still not substantial, i.e., $\overline{(\varphi - \varphi_0)^2} \ll 2\pi$, the distribution laws (5.117) and (5.118) are essentially the same over the interval $(0,2\pi)$. This can also be shown in a straightforward way by replacing the sum in (5.117) by an integral.

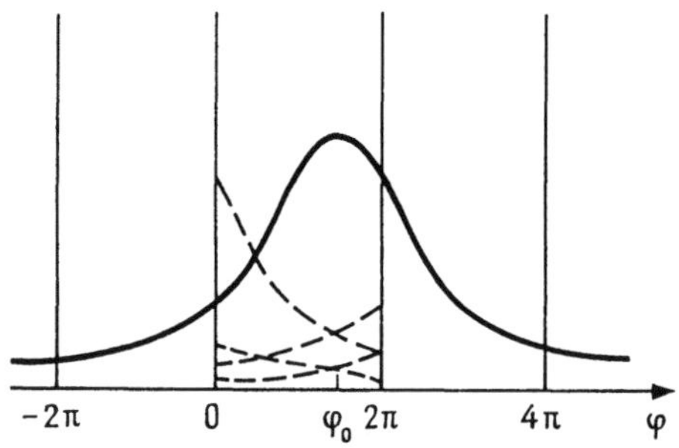

Fig. 5.15. The relation between phase distributions within the intervals $(-\infty, +\infty)$ and $(0, 2\pi)$

It follows from the assumption of homogeneous random impacts that the Markov process at hand is homogeneous in time: the transition probabilities (5.115, 117 and 118) only vary with the difference between the finite (θ) and the initial (θ_0) moments. For the initial instant of time we take $\theta = \mu t'$, and for the finite, $\theta + \tau = \mu(t' + \tau')$. The transition probabilities for ϱ and φ will depend only on $\tau = \mu\tau'$. If we denote the values of ϱ and φ at the time $\theta + \tau$ by ϱ_τ and φ_τ, we can rewrite (5.115, 117) as

$$\begin{aligned} v_\varrho(\varrho_\tau|\tau,\varrho) &= \frac{1}{\sqrt{2\pi}\sigma_\tau} \exp\left[-(\varrho_\tau - \bar{\varrho}_\tau)^2/2\sigma_\tau^2\right], \\ \bar{\varrho}_\tau &= \varrho\, e^{-p\tau}, \quad \sigma_\tau^2 = \frac{B}{2p}(1 - e^{-2p\tau}), \\ v_\varphi(\varphi_\tau|\tau,\varphi) &= \frac{1}{2\pi}\left\{1 + 2\sum_{n=1}^{\infty} e^{-n^2 D\tau} \cos n(\varphi_\tau - \varphi)\right\}. \end{aligned} \tag{5.120}$$

As $\tau \to \infty$, we obtain from these expressions the stationary univariate distributions

$$w_{1\varrho}(\varrho) = \frac{1}{\sqrt{2\pi}\sigma_\infty} \exp\left(-\varrho^2/2\sigma_\infty^2\right), \quad \sigma_\infty^2 = \frac{B}{2p}, \tag{5.121}$$

$$w_{1\varphi}(\varphi) = \frac{1}{2\pi}. \tag{5.122}$$

It will be recalled (Sect. 4.2) that transition probability and univariate probability are all that is necessary to deduce any n-variate distribution of a Markov process. In particular, the bivariate distributions ϱ and φ will be

$$\begin{aligned} w_{2\varrho}(\varrho_\tau,\tau,\varrho) &= w_{1\varrho}(\varrho)v_\varrho(\varrho_\tau|\tau,\varrho) \\ &= \frac{1}{2\pi\sigma_\infty\sigma_\tau}\exp\left\{-\frac{\varrho^2}{2\sigma_\infty^2}-\frac{(\varrho_\tau-\bar{\varrho}_\tau)^2}{2\sigma_\tau^2}\right\}, \end{aligned} \tag{5.123}$$

$$\begin{aligned} w_{2\varrho}(\varphi_\tau,\tau,\varphi) &= w_{1\varphi}(\varphi)v_\varphi(\varphi_\tau|\tau,\varphi) \\ &= \frac{1}{4\pi^2}\left\{1+2\sum_{n=1}^{\infty}e^{-n^2D\tau}\cos n(\varphi_\tau-\varphi)\right\}. \end{aligned} \tag{5.124}$$

As $w_{1\varrho}(\varrho)$ and $w_{1\varphi}(\varphi)$ are independent of θ, and the transition probabilities are uniform in θ, the Markov process in question is *stationary*.

Knowing the bivariate distributions (5.123, 124), we can compute means of any functions of ϱ, ϱ_τ and of φ, φ_τ. We will use them to calculate the covariance of the self-oscillatory process studied,

$$x(t') = r(\theta)\cos\left[t'+\varphi(\theta)\right] = (\sqrt{p}+\varrho)\cos(t'+\varphi),$$

which we shall need later when considering the oscillator spectrum. The desired covariance is

$$\psi_x = \overline{xx_{\tau'}} - \bar{x}\,\bar{x}_{\tau'}\quad ,$$

where $x \equiv x(t')$ and $x_{\tau'} \equiv x(t'+\tau') = (\sqrt{p}+\varrho_\tau)\cos(t'+\tau'+\varphi_\tau)$.

Since ϱ and φ are at any moment of time independent and the univariate phase distribution (5.121) is uniform over the interval $(0,2\pi)$, we have $\overline{\cos\varphi} = \overline{\sin\varphi} = 0$. Accordingly, for any t' we obtain

$$\bar{x} = (\sqrt{p}+\bar{\varrho})\,\overline{\cos(t'+\varphi)} = 0.$$

Considering further that, by (5.120), $\bar{\varrho} = \bar{\varrho}_\tau = 0$, we get

$$\begin{aligned} \psi_x = \overline{xx_{\tau'}} &= \overline{(\sqrt{p}+\varrho)(\sqrt{p}+\varrho_\tau)\cos(t'+\varphi)\cos(t'+\tau'+\varphi_\tau)} \\ &= \tfrac{1}{2}(p+\overline{\varrho\varrho_\tau})\left\{\overline{\cos(\tau'+\varphi_\tau-\varphi)}+\overline{\cos(2t'+\tau'+\varphi_\tau+\varphi)}\right\}. \end{aligned} \tag{5.125}$$

To find ψ_x we have thus to calculate the following means

$$\overline{\varrho\varrho_\tau},\quad \overline{\left.\begin{matrix}\sin\\ \cos\end{matrix}\right\}(\varphi_\tau-\varphi)},\quad \overline{\left.\begin{matrix}\sin\\ \cos\end{matrix}\right\}(\varphi_\tau+\varphi)}.$$

According to (5.123), we have

$$\overline{\varrho\varrho_\tau} = \int_{-\infty}^{+\infty} \varrho\varrho_\tau w_{2\varrho}(\varrho_\tau, \tau, \varrho) d\varrho\, d\varrho_\tau$$

$$= \int_{-\infty}^{+\infty} \varrho w_{1\varrho}(\varrho) d\varrho \int_{-\infty}^{+\infty} \varrho_\tau v_\varrho(\varrho_\tau|\tau, \varrho) d\varrho_\tau .$$

The internal integral is the conditional mean of $\overline{\varrho}_\tau$ that is equal to $\varrho \exp(-p\tau)$, see (5.120). Hence, by (5.121),

$$\overline{\varrho\varrho_\tau} = e^{-p\tau} \int_{-\infty}^{+\infty} \varrho^2 w_{1\varrho}(\varrho) d\varrho = \overline{\varrho^2} e^{-p\tau} = \sigma_\infty^2 e^{-p\tau} = \frac{B}{2p} e^{-p\tau} \quad .$$

Taking into account that $\overline{\varrho\varrho_{-\tau}} = \overline{\varrho\varrho_\tau}$, we can write this result for any sign of τ in the form

$$\overline{\varrho\varrho_\tau} = \frac{B}{2p} e^{-p|\tau|} \quad . \tag{5.126}$$

As $w_{2\varphi}(\varphi_\tau, \tau, \varphi)$ is an even function of $\varphi_\tau - \varphi$ [see (5.124)], it follows immediately that

$$\overline{\cos(\varphi_\tau + \varphi)} = \overline{\sin(\varphi_\tau + \varphi)} = \overline{\sin(\varphi_\tau - \varphi)} = 0. \tag{5.127}$$

we need only to calculate $\overline{\cos(\varphi_\tau - \varphi)}$, i.e., the integral [see (5.124)]

$$\overline{\cos(\varphi_\tau - \varphi)} = \int_0^{2\pi}\int_0^{2\pi} \cos(\varphi_\tau - \varphi) w_{2\varphi}(\varphi_\tau, \tau, \varphi) d\varphi\, d\varphi_\tau$$

$$= \frac{1}{4\pi^2} \int_0^{2\pi}\int_0^{2\pi} \cos(\varphi_\tau - \varphi) \left\{1 + 2\sum_{n=1}^{\infty} e^{-n^2 D\tau} \cos n(\varphi_\tau - \varphi)\right\} d\varphi\, d\varphi_\tau .$$

It is easily seen that a nonzero result is only obtained from the term of the sum with $n = 1$

$$\overline{\cos(\varphi_\tau - \varphi)} = \frac{\exp(-D\tau)}{2\pi^2} \int_0^{2\pi}\int_0^{2\pi} \cos^2(\varphi_\tau - \varphi) d\varphi\, d\varphi_\tau = \exp(-D\tau),$$

or for any sign of τ

$$\overline{\cos(\varphi_\tau - \varphi)} = \exp(-D|\tau|). \tag{5.128}$$

Inserting (5.126–128) into (5.125) gives

$$\psi_x(\tau') = \tfrac{1}{2}(p + D\, e^{-p|\tau|}) e^{-D|\tau|} \cos \tau' \tag{5.129}$$

with $\tau = \mu\tau'$ and $D = B/2p$.

The three factors in $\psi_x(\tau')$ vary at markedly different rates as functions of τ'. The function $\cos\tau'$ oscillates rather rapidly while much slower variations are exhibited by the exponential function

$$e^{-p|\tau|} = e^{-\mu p|\tau'|} \quad ,$$

that entered into the expression (5.129) through the covariance of the amplitude fluctuations (5.126). The latter decreases e times during the time $1/\mu p$. Normally the dimensionless increment μp is of the order of 10^{-2}–10^{-3}. An even slower fall off is shown by the exponential function

$$e^{-D|\tau|} = e^{-\mu D|\tau'|} \quad ,$$

where $2\mu D$ is the phase diffusion coefficient

$$\overline{(\varphi_\tau - \varphi)^2} = 2D\tau = 2\mu D\tau', \tag{5.130}$$

If the diffusion is due to natural fluctuations, the quantity $2\mu D$ is of the order of 10^{-10}–10^{-13}. This means that a mean square phase shift of about π occurs, in terms of dimensionless time, within the time interval

$$\tau' \approx \frac{\pi^2}{2\mu D},$$

and, in terms of conventional time,

$$\Delta t = \frac{\tau'}{\omega_0} \approx \frac{\pi}{4\mu D f_0} \approx (10^{10} - 10^{13}) \frac{1}{f_0} .$$

With $f_0 = 10\,\mathrm{MHz}$ this amounts to $\Delta t' \approx 10^3$–$10^6\,\mathrm{s}$.

It is common knowledge, however, that a phase progression of π for two unstabilized oscillators occurs much faster at such frequencies – in a matter of seconds. This is due to *technical* phase drifts (Sect. 5.8). It would seem that the technical drifts should completely mask the diffusion of the phase, but such a conclusion would be too facile, as the technical drifts fundamentally differ from the natural ones in that they have longer correlations. We will consider this question in [Ref. 5.17, Sect. 3.4]. Lastly, we note that the parameter B entering all the above relations and determining, in particular, the phase diffusion coefficient appeared first in the initial Einstein-Fokker-Planck equation (5.105). However, we do not yet have any relation of B to quantities describing a particular mechanism of random actions on the oscillator (shot current, thermal noise), or, more specifically, with stochastic characteristics of the random force (5.95). This is one more gap in the purely stochastic scheme presented here, which is to be filled in the next chapter.

5.10 Rotational Brownian Motion. Random Refraction of a Ray

Let a spherical particle be suspended in a liquid or gas. We assume an axis to be rigidly connected to the particle so that the direction characterizing orientation of the particle is determined by the spherical angles θ and φ. We neglect for the moment the translational and rotational motions of the particle. Then, the particle (its axis), bombarded by molecules or – which in this statement of the problem is the same – subject to fluctuations of tangential (viscous) stresses will rotate in a random way, executing rotational Brownian motion, see [5.19] and [Ref. 5.7, Sect. 57]. This implies that a point (θ,φ) undergoes translational Brownian motion over the surface of a unit sphere.

The random rotational moment seems to be isotropic (all of its directions are equivalent) and homogeneous (the intensity of pushes does not depend on the orientation of the particle). The Einstein-Fokker-Planck equation, meeting these conditions and determining the scalar density of the transition probability $\tilde{v}(\theta,\varphi|t-t_0,\theta_0,\varphi_0)$, i.e., a density such that $dP=\tilde{v}\ \sin\theta\, d\theta\, d\varphi$, has already been derived, see (5.93). It is simplified if the co-latitude θ is counted from the initial direction of the particle axis ($\theta_0=0$). Owing to the symmetry, the distribution of longitude φ then becomes uniform, and the equation for the transition probability $\tilde{v}(\theta|t-t_0,0)$ takes the form (5.93)

$$\sin\theta\frac{\partial\tilde{v}}{\partial t}=\frac{B}{2}\frac{\partial}{\partial\theta}\left(\sin\theta\frac{\partial\tilde{v}}{\partial\theta}\right),\quad \tilde{v}(\theta|0,0)=\frac{\delta(\theta)}{2\pi\ \sin\theta}. \tag{5.131}$$

The solution to equation (5.131) with this initial condition is

$$v(\theta|t-t_0,0)=\frac{1}{4\pi}\sum_{n=0}^{\infty}(2n+1)P_n(\cos\theta)\times\exp\left\{-\frac{n(n+1)}{2}B(t-t_0)\right\}, \tag{5.132}$$

where $P_n(\cos\theta)$ are the Legendre polynomials. As $t\to\infty$, the stationary distribution is reached: of the entire sum only the term with $n=0$ remains, so that

$$\lim_{t\to\infty}\tilde{v}(\theta|t-t_0,0)=\tilde{w}_1(\theta)=\frac{1}{4\pi}. \tag{5.133}$$

If we are interested in the conditional mean of any function $f(\theta)$, it can be obtained with the aid of (5.132) by expanding it in the Legendre polynomials and using their orthogonality,

$$\int_0^{\pi}P_m(\cos\theta)P_n(\cos\theta)\ \sin\theta\, d\theta=\frac{2\delta_{mn}}{2n+1}.$$

In certain cases it is more convenient not to solve (5.131), but rather to use it to derive a differential equation for the mean $\overline{f(\theta)}$ itself. Let, for instance, $f(\theta) = P_1(\theta) = \cos\theta$. We multiply (5.131) by $\cos\theta$ and integrate with respect to θ from 0 to π. With

$$\overline{\cos\theta} = \int_0^\pi \cos\theta \cdot \tilde{v} \sin\theta \, d\theta \qquad \text{we get}$$

$$\frac{d\,\overline{\cos\theta}}{dt} = -B\,\overline{\cos\theta}.$$

The solution to this equation with the initial condition $\overline{\cos\theta} = 1$ at $t = 0$ is

$$\overline{\cos\theta} = e^{-Bt} \quad . \tag{5.134}$$

As $t \to \infty$, this gives $\overline{\cos\theta} = 0$, i.e., by (5.133), all the orientations become equiprobable. For small t we have

$$\overline{\cos\theta} = 1 - \overline{\theta^2}/2 = 1 - Bt, \quad \text{i.e.,}$$

$$\overline{\theta^2} = 2BT, \tag{5.135}$$

just as for two-dimensional Brownian motion in the plane, see (5.35).

If there is some deterministic rotational moment (apart from the random one), that orients the particle axis in the planes $\varphi = \text{const}$, then the stationary distribution will no lónger be uniform. Let, for instance, the particle have a dipole moment p and be subjected to an electric field E. The rotational moment will be

$$M(\theta) = -pE \sin\theta = -\partial U/\partial\theta,$$

where $U = -pE\cos\theta$ is the energy in the external field. Neglecting the inertia of the particle, the systematic angular velcotiy is $\dot{\theta} = M/h$, where $h = 8\pi\eta a^3$ is the rotational coefficient of viscous friction. If we thus stick to the assumption that in the coefficients A_i of the Einstein-Fokker-Planck equation the terms independent of B coincide with the right-hand sides of the dynamic equations for the relevant velocities, we obtain for A_θ the expression

$$A_\theta = \frac{M}{h} + \frac{B}{2}\operatorname{ctg}\theta$$

instead of (5.90). Equation (5.93) now becomes

$$\sin\theta\frac{\partial\tilde{v}}{\partial t} = -\frac{\partial}{\partial\theta}\left(\frac{M}{h}\sin\theta\cdot\tilde{v}\right) + \frac{B}{2}\frac{\partial}{\partial\theta}\left(\sin\theta\frac{\partial\tilde{v}}{\partial\theta}\right),$$

and the equation for the stationary probability $\tilde{w}_1(\theta)$ will be

$$\frac{d}{d\theta}\left[\sin\theta\left(\frac{B}{2}\frac{d\tilde{w}_1}{d\theta} - \frac{M}{h}\tilde{w}_1\right)\right] = 0.$$

Integrating once and taking the integration constant to be zero, we will get

$$\frac{B}{2}\frac{d\tilde{w}_1}{d\theta} - \frac{M}{h}\tilde{w}_1 = \frac{B}{2}\frac{d\tilde{w}_1}{d\theta} + \frac{1}{h}\frac{dU}{d\theta}\tilde{w}_1 = 0.$$

Hence

$$\tilde{w}_1(\theta) = C\,e^{-2U/hB} \quad ,$$

where C is deduced from the normalization condition

$$2\pi\int_0^\pi \tilde{w}_1(\theta)\ \sin\theta d\theta = 1.$$

But this result must coincide with the Boltzmann distribution, whence (see Sect. 5.6) $hB/2 = kT$ or

$$B = 2kT/h.$$

It might appear that the issue of rotational Brownian motion is of no interest to the field of radiophysics. But the problems discussed earlier have shown repeatedly that one and the same stochastic scheme may cover disparate phenomena. In this case we again have an example of this kind. With certain reservations, rotational Brownian motion can be likened to the problem of the propagation of a ray in a medium with smooth random inhomogeneities. By the smoothness, or slowness, of the variation of the medium properties we mean the applicability of geometric optics (or acoustics). For sufficiently short radio waves the appropriate conditions may sometimes be fulfilled in the troposphere, ionosphere, and solar corona, so that a ray treatment of random refraction appears to be possible. But the fluctuations of the direction of ray propagation are of direct interest to radiophysics.

The behavior of a medium may be characterized by the refractive index $n(x, y, x)$ or the phase velocity of a wave $v(x, y, z)$ whose variation from point to point describes the inhomogeneities in the medium. The statement that these inhomogeneities are random means that n or v are random functions of the point, i.e., not of one parameter t, as was always the case earlier, but of the *three* parameters x, y, and z. This is an example of the so-called *random* field. In [5.20,21] the random fields will be considered in more detail, and not only in the context of refraction fluctuations.

Just as one of the principal characteristics of a random process $\xi(t)$ is its mixed moment $\overline{\xi(t)\xi(t')}$, so for the random field $\xi(x, y, z) = \xi(\boldsymbol{r})$, the moment

$\overline{\xi(\boldsymbol{r})\xi(\boldsymbol{r}')}$ is extremely important as it furnishes a measure of the statistical connection between the values of $\xi(\boldsymbol{r})$ at two points in space. Assuming that $\bar{\xi} = 0$, we will in the following identify this mixed moment with the covariance $\psi_\xi(\boldsymbol{r}, \boldsymbol{r}')$. The field is termed *homogeneous*, if the covariance is dependent only on the difference of the coordinates of these points, $\boldsymbol{R} = \boldsymbol{r} - \boldsymbol{r}'$, i.e., a shift of the origin does not affect $\psi_\xi(\boldsymbol{R}) = \overline{\xi(\boldsymbol{r})\xi(\boldsymbol{r}')}$. If, in addition, ψ_ξ depends only on the distance $R = |\boldsymbol{r} - \boldsymbol{r}'|$, but not on the direction of $\boldsymbol{R}$, then the field is referred to as statistically *isotropic*.

We suppose that the random field $v(\boldsymbol{r})$ is homogeneous and isotropic, or, in other words, that the medium has no statistically distinguished positions and directions. When referring to random refractions of the ray in a inhomogeneous medium, this assumption plays about the same role as the hypothesis of the homogeneity and isotropy of random pushes for a Brownian particle.

The spatial covariance of the homogeneous and isotropic field varies with only one argument $\psi_\xi(\boldsymbol{R}) = \psi_\xi(R)$, just as in the case of the time-dependent covariance $\psi_\xi(\tau)$ for the stationary random process. In many problems a certain characteristic time interval ϑ can be specified for $\psi_\xi(\tau)$ where the covariance is notable and decays for shifts $\tau > \vartheta$ sufficiently fast. In such cases ϑ is called the *correlation time*. In spatial problems an analogous role is played by the *correlation radius* a; as R increases beyond a, the covariance $\psi_\xi(R)$ tends to zero. If ξ is taken to mean the random refractive index of a medium n (or the random phase velocity of a wave v), then a characterizes, evidently, the mean size of the medium inhomogeneities.

The applicability of the geometric approximation presupposes that not only the condition $a \gg \lambda$ (λ is the wave length) is satisfied, but also the condition that diffraction effects be small: $z\lambda \ll a^2$ where z is the thickness of a layer traversed by the ray in the inhomogeneous medium. The last condition implies that even for those inhomogeneities that are the most distant from the observation point, the size of the Fresnel zone $\sim\sqrt{z\lambda}$ is much smaller than that of inhomogeneities a.

As far as the ray configuration is concerned, a characterizes the length of those sections of the ray where it suffers a single deviation. These random bends due to individual inhomogeneities accumulate to make up the total random refraction of interest to us. It is represented by the angular deviation θ of the ray direction from the z-axis, i.e., from the initial direction $\theta = 0$ of the ray before it entered the inhomogeneous medium.

The problem of ray refraction at random inhomogeneities was thoroughly treated in [5.22]. There it was shown that the angle θ can be regarded as a function of the Markov type, provided the following conditions are met:

1) Along all the path of the ray $\theta \ll \pi/2$ must hold (the so-called *small angle approximation*)[8]. In a first approximation this allows us not to distinguish between the thickness z of the medium layer covered and the complete ray length l in this layer. This condition is otherwise expressed as the requirement

[8] The need for this condition has been overlooked in early works on random refraction.

that the mean square of the transverse displacement of the ray from the z-axis ϱ be small as compared with z^2

$$\overline{\varrho^2} \ll z^2. \tag{5.136}$$

2) Along its path through a layer of thickness z the ray must undergo many single deviations, i.e., it must encounter many inhomogeneities

$$z \gg a.$$

The inequality must be so strong that the complete thickness z could be broken down into segments $\Delta z \ll z$, such that even within a segment the ray must meet a large number of inhomogeneities ($\Delta z \gg a$), i.e., it must be subject to many mutually uncorrelated impacts due to random gradients of the refractive index. At the same time, the mean square of the resulting ray deviation $\overline{(\Delta\theta)^2}$ over the section Δz is still small, its order of magnitude being not lower than Δz. This is quite similar to the Markov conditions for a Brownian particle: during a time Δt, the particle suffers many independent pushes, but the mean square deviation $\overline{(\Delta s)^2}$ has an order of smallness not lower than Δt. In this macroscopic sense the time Δt is infinitesimal.

Under these conditions[9] there exists a transition probability $\tilde{v}(\theta|z, 0)$ satisfying the same equation (5.131) as the probability for the axis of a spherical particle to rotate by an angle θ under the action of random homogeneous and isotropic rotational moments. It is, of course, of no significance that here the independent variable rather than the time t is the thickness z of the medium layer covered by the ray. The principal difference is that for the particle the angle of deflection θ is not limited, whereas the ray equation (5.131) is only suitable for small θ, when it is possible to substitute θ for $\sin\theta$. Consequently, for (5.134) we must substitute $\overline{\cos\theta} \simeq 1$, and for the mean square of θ we can use (5.135) alone, i.e.,

$$\overline{\theta^2} = 2Bz \ll 1.$$

With the same limitation the mean square of the ray transverse displacement is

$$\overline{\varrho^2} = \tfrac{2}{3}Bz^3 = \tfrac{1}{3}\overline{\theta^2}z^2,$$

whence (5.136) is derived.

[9] We will return to these conditions in Sect. 6.4 to see where they originate.

5.11 Stepwise Markov Processes. The Kolmogorov-Feller Equation

The Einstein-Fokker-Planck equation was derived for Markov processes $\xi(t)$ that are continuous not only in the sense of the continuity of possible values x of the random function $\xi(t)$, but also in the sense of the stochastic continuity of variations of $\xi(t)$ in time (almost all realizations of $\xi(t)$ are continuous in t in the conventional sense). If we forego the last limitation, i.e., the condition (5.60), we may treat the problem as a Markov process, in which variations occur *in instantaneous steps*. The general case here will be a *mixed* situation, where the state may vary both continuously and in steps, but we will first consider purely stepwise transitions. An example of such a process has been given earlier, when referring to the fact that the condition (5.60) is not met in collisions of gas molecules: the molecular velocity changes stepwise in collisions, but the set of possible values of velocity after the collision is continuous. Besides collisions of microparticles (molecules, atoms, electrons, etc.), this scheme also covers quantum transitions, and some pulse processes.

First, consider the following simple example of a univariate Poisson stepwise process $\xi(t)$. Let random times t_ν form a Poisson flow of events, so that the probability of the occurrence of n events in the interval $(0, t)$ is given by the Poisson law

$$P(n) = \frac{(n_1 t)^n}{n!} \exp(-n_1 t), \tag{5.137}$$

where $n_1 t$ is the mean number of events in $(0, t)$; n_1 is their mean in a unit time.

At time t_ν the process $\xi(t)$ in a step may get into the interval $(x, x+dx)$[10] with the probability $\varphi(x_\nu|x_{\nu-1})dx$ which depends on the preceding values of $x_{\nu-1}$ at time $t_{\nu-1}$. Thus, for the specified initial value $\xi(0) = x_0$, the total probability of a chain of n events with $x_1, \ldots, x_n$ at successive times of steps $t_1, \ldots, t_n$ $(t_0 = 0 < t_1 < \ldots < t_n)$, will be

$$\prod_{\nu=1}^{n} \varphi(x_\nu|x_{\nu-1})dx_\nu .$$

Let us find for $\xi(t)$ the conditional probability density $v(t, x|0, x_0)$ that can be written as $v(x|t, x_0)$. This is possible, because it is a priori obvious that by virtue of the time-independence of n_1 and $\varphi(x_\nu|x_{\nu-1})$, the process is homogeneous in t. We suppose first that a fixed number of n steps occurred in the interval $(0, t)$ and that we look for the conditional probability density $v(x|t, x_0, n)$ subject to this additional condition. It is clear that

$$v(x|t, x_0) = \sum_{n=0}^{\infty} v(x|t, x_0, n).$$

[10] ν is the *number of a step*. The set of possible x_ν at any ν is continuous.

As the transition $x_0 \rightarrow x_n = x$ occurs at any intermediate values of $x_1, \ldots, x_{n-1}$, we have

$$v(x|t, x_0, n) = P(n) p_n(x|x_0) \quad \text{where}$$

$$p_n(x|x_0) = \int \varphi(x|x_{n-1}) \ldots \varphi(x_1|x_0) dx_1 \ldots dx_{n-1}. \tag{5.138}$$

We note that, by (5.138),

$$p_n(x|x_0) = \int \varphi(x|x_{n-1}) p_{n-1}(x_{n-1}|x_0) dx_{n-1}. \tag{5.139}$$

In the special case of $n = 0$, i.e., in the absence of steps in $(0, t)$, x must be equal to x_0, so that $p_0(x|x_0) = \delta(x - x_0)$ and consequently

$$v(x|t, x_0, 0) = P(0)\delta(x - x_0).$$

We now add up $v(x|t, x_0, n)$ over all the mutually exclusive special cases $n = 0, 1, 2, \ldots$ to obtain the conditional probability density $v(x|t, x_0)$, which is not limited by a fixed n

$$\begin{aligned} v(x|t, x_0) &= \sum_{n=0}^{\infty} P(n) p_n(x|x_0) \\ &= P(0)\delta(x - x_0) + \sum_{n=1}^{\infty} P(n) p_n(x|x_0) \end{aligned} \tag{5.140}$$

or, after inserting $P(n)$ from (5.137), we get

$$v(x|t, x_0) = \exp(-n_1 t) \left\{ \delta(x - x_0) + \sum_{n=1}^{\infty} \frac{(n_1 t)^n}{n!} p_n(x|x_0) \right\}. \tag{5.141}$$

In the case of a very short time interval we substitute Δt for t, retaining only the terms up to the first order in Δt, i.e., neglecting multiple steps ($n = 2, 3, \ldots$), and obtaining thus for small times

$$v(x|\Delta t, x_0) \approx (1 - n_1 \Delta t)\delta(x - x_0) + n_1 \Delta t \varphi(x|x_0). \tag{5.142}$$

It is easily seen that the conditional probability density (5.141) obeys the Smoluchowski equation (see Exercise 5.13.17), i.e., it is the ***transition probability*** density defining the ***Markov*** process $\xi(t)$. Further, for $v(x|t, x_0)$, an integro-differential equation is readily obtained, whose solution for the initial condition

$$v(x|0, x_0) = \delta(x - x_0) \tag{5.143}$$

is given by (5.141). Indeed, differentiating (5.141) with respect to t, gives

$$\frac{\partial v}{\partial t} = -n_1 v + \sum_{n=1}^{\infty} \frac{(n_1 t)^{n-1}}{(n-1)!} n_1 \exp(-n_1 t) p_n(x|x_0).$$

Hence, by (5.139,140),

$$\frac{\partial v(x|t, x_0)}{\partial t} = -n_1 \Big[v(x|t, x_0) - \int \varphi(x|y) v(y|t, x_0) dy \Big]. \tag{5.144}$$

Equation (5.144) is a very special case of the Kolmogorov-Feller equation, i.e., the integro-differential equation satisfied by the transition probability for a Markov random process with purely discontinuous changes of state (steps). This general equation is still derived from the fundamental Smoluchowski equation, but it uses a generalized [as compared with (5.142)] expression for the transition probability in small time Δt. In fact, it is assumed that in the interval $(t, t+\Delta t)$ the process $\xi(t)$ either jumps (if there has been a step) from a value x at time t to the interval $(z, z+dz)$ with the conditional probability $\varphi(z|t, x)dz$, or retains the value x (no step). In the latter case the probability of the system being found in $(z, z+dz)$ is $\delta(z-x)dz$. Let the step, in turn, occur during a small time Δt with the probability $a(t, x)\Delta t$. Hence, to the *first order* in Δt

$$v(t+\Delta t, z|t, x) \approx [1 - a(t, x)\Delta t]\delta(z-x) + a(t, x)\Delta t \cdot \varphi(z|t, x). \tag{5.145}$$

Unlike (5.142), the temporal probability density of a step $a(t, x)$ is here not equal to the constant quantity n_1, but also depends on the initial time t, and on the value of x at that time. The conditional probability density φ of values z assumed as a result of a step is now dependent not only on the preceding value of x, but also on the time t, so that the process $\xi(t)$ is no longer stationary.

By virtue of (5.145) and certain additional requirements (continuity of a and φ as functions of t, and finiteness of a in any bounded interval of t), the Smoluchowski equation gives the following integro-differential equation for the transition probability density [5.12]:

$$\begin{aligned} \frac{\partial v(t, x|t_0, x_0)}{\partial t} = & - a(t, x) v(t, x|t_0, x_0) \\ & + \int a(t, y) \varphi(x|t, y) v(t, y|t_0, x_0) dy. \end{aligned} \tag{5.146}$$

This equation for v as a function of t and x is an analog of the Einstein-Fokker-Planck equation (5.61) for Markov processes with continuously varying states. Using the same assumptions and techniques, we can also derive a companion equation to (5.64), i.e., an equation for $v(t, x|t_0, x_0)$ as a function of t_0 and x_0:

$$\begin{aligned} \frac{\partial v(t, x|t_0, x_0)}{\partial t_0} = & \ a(t_0, x_0) \Big[v(t, x|t_0, x_0) \\ & - \int \varphi(y|t_0, x_0) v(t, x|t_0, y) dy \Big]. \end{aligned} \tag{5.147}$$

With the initial condition

$$v(t_0, x|t_0, x_0) = \delta(x - x_0) \tag{5.148}$$

both equations have a *unique* (and the *same*) solution.

Given a continuous set of states, how does the scheme of stepwise changes of states relate to the case of *discrete* possible states discussed in Sect. 5.3? Owing to their discrete nature, the states can change in steps only, as it were, willynilly.

For discrete states x_k, the conditional probability densities $v(t+\Delta t, z|t, x_i)$ and $\varphi(z|t, x_i)$ can be written in terms of delta-functions $\delta(z - x_k)$

$$v(t + \Delta t, z|t, x_i) = \sum_k p(t + \Delta t, x_k|t, x_i)\delta(z - x_k),$$

$$\varphi(z|t, x_i) = \sum_k \pi(x_k|t, x_i)\delta(z - x_k),$$

p and π are integral conditional probabilities of the state x_k (p is the transition probability independent of whether or not a step occurred, and π is the probability when the step occurred). Substituting these expressions into (5.145) gives

$$\sum_k p(t + \Delta t, x_k|t, x_i)\delta(z - x_k) \approx [1 - a(t, x_i)\Delta t]\delta(z - x_i)$$
$$+ a(t, x_i)\Delta t \sum_k \pi(x_k|t, x_i)\delta(z - x_k).$$

We next integrate this equation with respect to z (over the $\pm\varepsilon$-neighborhood of the state x_k), taking into account that

$$\int_{x_k-\varepsilon}^{x_k+\varepsilon} \delta(z - x_i)dz = \delta_{ik} \quad ,$$

and find the following expression for the transition probability in a short time Δt:

$$p(t + \Delta t, x_k|t, x_i) \approx [1 - a(t, x_i)\Delta t]\delta_{ik}$$
$$+ a(t, x_i)\Delta t \cdot \pi(x_k|t, x_i) = \delta_{ik} + a(t, x_i)[\pi(x_k|t, x_i) - \delta_{ik}]\Delta t.$$

But this relationship has the form of (5.12)

$$p(t + \Delta t, x_k|t, x_i) \approx \delta_{ik} + A_{ik}(t)\Delta t,$$

i.e., the form that was taken as an assumption in deriving the Kolmogorov equation for Markov processes with discrete possible states. Here

$$A_{ik}(t) = a(t, x_i)[\pi(x_k|t, x_i) - \delta_{ik}] \quad ,$$

which ensures that the condition (5.17) is met

$$\sum_k A_{ik}(t) = 0.$$

The Kolmogorov-Feller equations (5.146,147) for stepwise Markov processes thus treat both continuous and discrete possible states. For the integral transition probabilities $V(t, x|t_0, x_0)$ it is therefore advisable to write them using the Stieltjes integral

$$\frac{\partial V(t, x|t_0, x_0)}{\partial t} = -\int_{-\infty}^{x} a(t, y) d_y V(t, y|t_0, x_0) + \int_{-\infty}^{+\infty} a(t, y)\pi(x|t, y) d_y V(t, y|t_0, x_0), \tag{5.149}$$

$$\frac{\partial V(t, x|t_0, x_0)}{\partial t_0} = a(t_0, x_0)\Big[V(t, x|t_0, x_0) - \int_{-\infty}^{+\infty} V(t, x|t_0, y) d_y \pi(y|t_0, x_0)\Big]. \tag{5.150}$$

It is in this form that these equations have been derived by *Feller* [5.12].

The Kolmogorov-Feller equation (5.146) may be written in a more compact form, viz., as the *Boltzmann equation*, a principal equation in the kinetic theory of gases. Now let us introduce the probability density for a jump at a time t from a state y to a state x[11]

$$u(x|t, y) = a(t, y)\varphi(x|t, y). \tag{5.151}$$

The validity of this relationship is easily proved. In fact, $a(t, y)dt$ is the probability of a jump from y *somewhere* [i.e., the probability of the occurrence of a jump in $(t, t + dt)$], and $\varphi(x|t, y)dx$ is the conditional probability for a jump to bring the process into $(x, x + dx)$. The product of a and φ is the conditional probability density u, as defined above. If we integrate (5.151) over all x, taking into account

$$\int \varphi(x|t, y) dx = 1,$$

and interchange the variables $(x \rightleftarrows y)$, then we arrive at

$$\int u(y|t, x) dy = a(t, x). \tag{5.152}$$

If (5.152) is inserted into the first term on the right-hand-side of (5.146), and (5.151) into the second, then (5.146) will take the form

[11] Of course, this is only a shorthand notation. Rather, we should refer to $u(x|t, y)dx\,dt$ as the probability of finding the system in the interval $(x, x + dx)$ as a result of a jump from a state y in time $(t, t + dt)$.

$$\frac{\partial v(t,x|t_0,x_0)}{\partial t} = -v(t,x|t_0,x_0)\int u(y|t,x)dy + \int v(t,y|t_0,x_0)u(x|t,y)dy. \tag{5.153}$$

This is the Boltzmann equation (for a univariate process).

It is immediately evident that this is in essence the balance equation: the variation rate v (e.g., of particle concentration at a point x) is equal to the difference of two flows due to jumps: from x into some other states and from all the other states into x.

As already mentioned, the most general case is the mixed one. It is the superposition of x discrete and continuous "spectra" of the possible values of the process $\xi(t)$, when the state can change both in jumps and continuously. The integro-differential Kolmogorov-Feller equations for this general case are as follows[12]

$$\frac{\partial v(t,x|t_0,x_0)}{\partial t} = -\frac{\partial A(x,t)v}{\partial x} + \frac{1}{2}\frac{\partial^2 B(x,t)v}{\partial x^2} - a(t,x)v + \int_{-\infty}^{+\infty} a(t,y)v(t,y|t_0,x_0)\frac{\partial \pi(x|t,y)}{\partial x}dy, \tag{5.154}$$

$$\frac{\partial V(t,x|t_0,x_0)}{\partial t_0} = -A(x_0,t_0)\frac{\partial V}{\partial x_0} + \frac{B(x_0,t_0)}{2}\frac{\partial^2 V}{\partial x_0^2} + a(t_0,x_0) \times \left[V(t,x|t_0,x_0) - \int_{-\infty}^{+\infty} V(t,x|t_0,x_0)d_y\pi(y|t_0,x_0)\right], \tag{5.155}$$

where $A(x,t)$ and $B(x,t)$ are respectively, the coefficients of drift and diffusion defined in terms of the *continuous part* of V, in much the same manner as in deriving the Einstein-Fokker-Planck equation, see (5.58,59). Equation (5.155) follows from the Smoluchowski equation with the above-mentioned assumptions regarding A, B, a, and π. But, to derive (5.154) it suffices to suppose in addition the existence of the density $\varphi(x|t,y) = \partial\pi(x|t,y)/\partial x$, and the differentiability of $a(t,|y)$ and $\pi(x|t,y)$ with respect to t and y. Then, there exists the transition probability density $v = \partial V/\partial x$ described by (5.154).

If the jump probability is zero ($a = 0$), then we get from (5.154,155) the Einstein-Fokker-Planck equation (5.61) and the conjugate equation (5.64), assuming, of course, that the set of possible values is continuous. If $a \neq 0$, but the

[12] Kolmogorov, independently of Feller, suggested the existence of a class of Markov processes with jumps and in [5.3] gave without a proof the equations (5.149) and (5.154). He suggested that processes without probabilistic memory-effects should be referred to as "stochastically definite", but the term did not gain acceptance, and the name "Markov processes" established itself. *Feller* [5.12,23] gave an analysis of the conditions under which the integro-differential equations are derived, and also studied the questions of the existence and uniqueness of the solution to the problem with the initial condition (5.148).

drift and diffusion coefficients are zero ($A = 0$, $B = 0$), then we again return to the Kolmogorov-Feller equations (5.149,150) for purely stepwise Markov processes, *including* processes with discrete possible values.

It is hardly necessary to remind ourselves that all the equations are readily generalized to multivariate Markov processes $\xi(t) = \{\xi_1(t), \ldots, \xi_n(t)\}$.

5.12 First Passage Problem

This problem can also be formulated for non-Markov processes, but it is only for Markov processes that its exact solution can be worked out in the general form. This justifies the inclusion of this section in a chapter on Markov processes. Even in the special case of the stationary *normal* process $\xi(t)$ for which, it would seem, the answer to any problem could be expressed in one way or another in terms of the covariance $\psi(t_1, t_2)$, for the first passage problem it has as yet only been possible to obtain some particular results [5.24,25].

The problem is formulated as follows. Let a random process $\xi(t)$ assume, at an initial time $t_0 = 0$, a value x_0. What are the statistics (distribution functions, moments) of a *random time t* of a first passage for either a given one boundary $a<x_0$ or $b>x_0$, or for any one of them, if both are available? This means that, given the statistics of $\xi(t)$ "on the x-axis" (distributions w_n for values of the function itself), it is necessary to obtain the statistics "on the t-axis". Problems of this kind are among the most interesting in the theory of random processes. They also include issues concerning the statistics of intersections of a given level (threshold) with "spikes", of random processes [5.26].

The theory of first passage for a *Markov* process $\xi(t)$ was first developed by *Pontryaghin* et al. [5.27][13]. The solution of this problem found a wide variety of applications. In the case of Brownian motion, an example is the Brillouin experiment, in which a Brownian particle *sticks* to the wall of a vessel, i.e., this is the case of the *first* arrival of the particle at the absorbing boundary [5.7]. In radiophysics we may cite problems associated with the time of reaching the separatrix in fluctuations, in nonlinear dynamic systems. In particular, the time of passage through a potential barrier in a conservative system, or the time of spontaneous changes of self-oscillatory modes or quenching of oscillation in oscillators. The disruption of tracking in automatically-controlled servo systems, and other problems also come under this heading.

Following [5.27], we will look at a univariate (and continuous) Markov process[14] that is homogeneous in time. That is it features a transition probability density $v(x|t-t_0, x_0)$ and t-independent coefficients $A(x)$ and $B(x)$ in the

[13] This work, published in 1933 and discussed (with refinements) in the textbook [5.7], seems to have passed unnoticed in western countries, where the origin of the first passage problem studied for a Markov process is associated with the work of *Rice* (1944–1945) [5.8] and *Darling* and *Siegert* (1953) [5.28].

[14] The theory lends itself to a generalization to the multidimensional case [5.7].

Einstein-Fokker-Planck equation. The transition probability density v and the univariate density of the state probability $w_1(x)$ (if it exists) determine all the statistics of the Markov process "in x". But we will be concerned with another distribution different from w_n, viz., the probability of *time* t of the passage of either of the boundaries of the range of permissible variations of x

$$\varphi(t|x) = P\left\{ \begin{array}{c|c} x = a \text{ or } x = b & \text{If at } t_0 = 0 \\ \text{at least once in} & \text{a state } x \\ \text{time } t & \text{occurred} \end{array} \right\}$$

This is an *integral* probability [so that $\varphi(\infty|x) = 1$], but the probability density is $\partial\varphi/\partial t$, and hence the mean time of the passage of a or b is

$$\bar{t}(x) = \int_0^\infty t\frac{\partial\varphi}{\partial t}dt. \tag{5.156}$$

Initial and boundary conditions for $\varphi(t|x)$ are readily determined. Because $x{\neq}a$ and $x{\neq}b$ at $t = 0$, the initial condition will be

$$\varphi(0|x) = 0 \quad a<x<b. \tag{5.157}$$

But if at $t = 0$ either $x = a$, or $x = b$, i.e., the boundary has already been passed, then for any $t>0$ the following boundary condition will hold:

$$\varphi(t|a) = \varphi(t|b) = 1. \tag{5.158}$$

We now proceed to set up an equation for $\varphi(t|x)$ – a problem that is solved completely just for a Markov process using transition probability. But here we do not need the transition probability $v(y|t,x)dy$ from a point x at $t = 0$ to *any* interval $(y, y + dy)$ on the x-axis, but the probability $\hat{v}(y|t,x)dy$ for the same event, with y *belonging* to the interval (a, b). Of course, the initial conditions for v and $\hat{v}$ are the same

$$\hat{v}(y|0,x) = v(y|0,x) = \delta(y - x)$$

(therefore v and $\hat{v}$ practically coincide for very small t, when they are still very sharp functions of y). But for any $t>0$, it is obvious that $\hat{v}{\leq}v$ and the normalization conditions for v and $\hat{v}$ are also different. Whereas

$$\int_{-\infty}^{+\infty} v(y|t,x)dx = 1$$

we have for $\hat{v}$

$$\int_a^b \hat{v}(y|t,x)dy = 1 - \varphi(t|x). \tag{5.159}$$

In fact P_n be the probability to arrive at a time t at a certain point within the interval $(a,\ b)$, after n passages (touches, intersections) of the boundaries on the way. The complete system of possible events corresponds to $n = 0, 1, 2, \ldots$, so that $\sum_{n=0}^{\infty} P_n = 1$. But the left-hand side of (5.159) is P_0, [the probability of arriving at some point within $(a,\ b)$ without touching the boundaries], and $\varphi(t|x)$ is the probability of the same final result, but with the boundaries reached *at least once*, i.e.,

$$\varphi(t|x) = P_1 + P_2 + \ldots\ , \tag{5.160}$$

whence (5.159) follows.

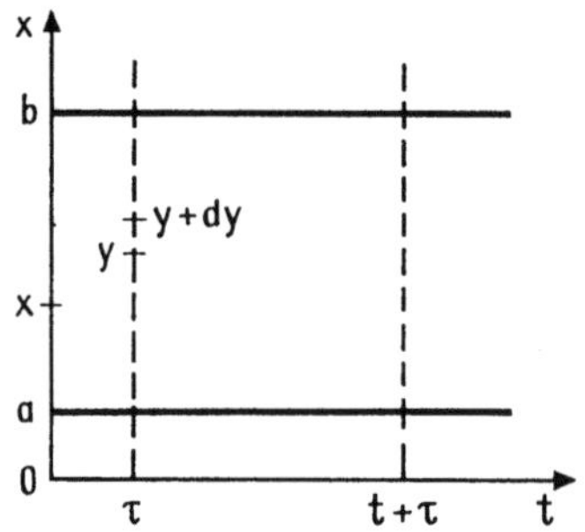

Fig. 5.16. The notation for the derivation of the integral equation (5.12.6) is specified

Using $\hat{v}$, we can write the following integral equation, see also Fig. 5.16,

$$\varphi(t+\tau|x) = \varphi(\tau|x) + \int_a^b \hat{v}(y|\tau, x)\varphi(t|y)dy. \tag{5.161}$$

It states that the probability of reaching a or b at least once within time $t+\tau$ (the left-hand-side) is equal to the probability of the same event in time τ [the term $\varphi(\tau|x)$ on the right-hand-side] plus the probability that the boundary was not touched within time τ, but during the rest of the time, i.e., t (second term on the right-hand-side). Really, $\hat{v}(y|\tau, x)dy$ is the probability of passing from x to y in time τ without touching the boundaries, and $\varphi(t|y)$ is the probability of at least one touching within the time t. Hence, $\hat{v}(y|\tau, x)\varphi(t|y)dy$ is the probability of at least one touching in the time interval $(0, t+\tau)$ with the passage through $(y, y+dy)$ at the time τ. The integral over y from a to b gives us the total probability that the boundaries were not touched in time τ, while they were touched during the interval $t+\tau$. We notice that, according to (5.161), the difference

$$\varphi(t+\tau|x) - \varphi(t|x) = \int_a^b \hat{v}(y|t, x)\varphi(\tau|y)dy$$

is the probability of reaching the boundary (at least once) in the interval $(t, t+\tau)$ after the "empty" interval $(0,\ t)$. Substituting dt for τ, we obtain that $(\partial\varphi/\partial t)dt$

represents, to the first order in dt, the probability of the *first* passage in the interval $(t, t+dt)$.

The limiting process $\tau \to 0$ in (5.161) gives in quite a rigorous way the (conjugate) Einstein-Fokker-Planck equation for $\varphi(t|x)$ [5.7]. But the same result is also obtainable using a physically plausible approximation of the initial integral equation (5.161), which was presented in [5.27]. Since τ is small (and further on $\tau \to 0$), we may from the very beginning ignore the term $\varphi(\tau|x)$ in (5.161), use $\hat{v} = v$ (see above) and extend the integration limits to $\pm\infty$, i.e., proceed from the equation

$$\varphi(t+\tau|x) = \int_{-\infty}^{+\infty} v(y|\tau, x)\varphi(t|y)dy. \tag{5.162}$$

Expanding $\varphi(t|y)$ in a power series of $y - x$,

$$\varphi(t|y) = \varphi(t|x) + \frac{\partial\varphi(t|x)}{\partial x}(y-x) + \frac{1}{2}\frac{\partial^2\varphi(t|x)}{\partial x^2}(y-x)^2 + 0[(y-x)^3] \quad ,$$

substituting this into (5.162) and dividing the equation by τ, we get (with the same assumptions as in Sect. 5.6 where we derived the Einstein-Fokker-Planck equation for the transition probability v) the equation

$$\frac{\partial\varphi(t|x)}{\partial t} = A(x)\frac{\partial\varphi}{\partial x} + \frac{B(x)}{2}\frac{\partial^2\varphi}{\partial x^2}, \tag{5.163}$$

where $A(x)$ and $B(x)$ are the same as earlier. This simplified derivation is somewhat unsatisfactory but, as mentioned above, (5.163) can also be obtained as the result of a rigorous derivation from (5.161).

We may either solve (5.163) with the initial and boundary conditions (5.157,158), or pass to the equation for the characteristic function

$$f_t(u|x) = \overline{e^{iut}} = \int_0^\infty e^{iut}\frac{\partial\varphi(t|x)}{\partial t}dt,$$

or set up ordinary differential equations for the moments $\overline{t^n}(x)$. We will take the last option.

We thus differentiate (5.163) with respect to t, multiply the result by t^n, and integrate with respect to t from 0 to ∞ to obtain

$$A(x)\frac{d\overline{t^n}}{dx} + \frac{B(x)}{2}\frac{d^2\overline{t^n}}{dx^2} = \int_0^\infty t^n\frac{\partial^2\varphi}{\partial t^2}dt = t^n\frac{\partial\varphi}{\partial t}\bigg|_0^\infty - n\overline{t^{n-1}}.$$

On the assumption that the probability density $\partial\varphi/\partial t$ is limited at zero and tends sufficiently fast to zero as $t \to \infty$, we get

$$\frac{B(x)}{2}\frac{d^2\overline{t^n}}{dx^2} + A(x)\frac{d\overline{t^n}}{dx} + n\overline{t^{n-1}} = 0. \tag{5.164}$$

In particular, at $n = 1$ and $n = 2$

$$\frac{B(x)}{2}\bar{t}'' + A(x)\bar{t}' + 1 = 0, \quad \text{and} \tag{5.165}$$

$$\frac{B(x)}{2}\overline{t^2}'' + A(x)\overline{t^2}' + 2\bar{t} = 0. \tag{5.166}$$

Evidently the solutions to such equations should be nonnegative and become zero every time the initial state x coincides with one of the boundaries

$$\overline{t^n}(a) = \overline{t^n}(b) = 0. \tag{5.167}$$

If there is only one boundary, say $b>x$, then we still have $\overline{t^n}(b) = 0$, and as $x \to -\infty$ it is natural to require that $\overline{t^n} \to \infty$.

Now we consider several examples.

Let $A(x) = 0$, and $B = \text{const}$. Equation (5.163) then takes the form

$$\dot{\varphi} = \frac{B}{2}\varphi'', \tag{5.168}$$

and its solution, subject to the conditions (5.157,158), will be

$$\varphi(t|x) = 1 + \frac{2}{\sqrt{\pi}}\left[\int\limits_{(b-x)/\sigma}^{(b-a)/\sigma} e^{-s^2}ds - \int\limits_{0}^{(x-a)/\sigma} e^{-s^2}ds\right], \quad \sigma = \sqrt{2Bt}. \tag{5.169}$$

The behavior of $\varphi(t|x)$ vs. x is shown in Fig. 5.17a for several values of t. For the mean time of the boundary passage we have, by (5.165), the equation

$$\frac{B}{2}\bar{t}'' = -1,$$

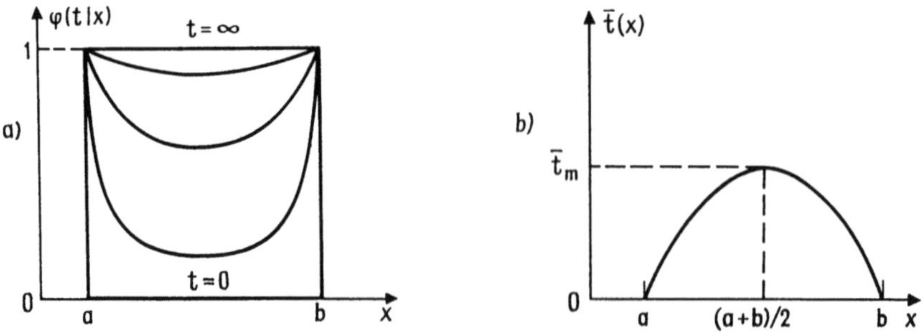

Fig. 5.17. (a) The probability $\varphi(t|x)$ in the interval (a, b) is given for five values of t ranging from 0 to ∞, (b) the mean time $\bar{t}$ of the first arrival at $x = a$ and $x = b$ is plotted as a function of the initial x within the interval (a, b)

whose solution (Fig. 5.17b), subject to (5.167), is

$$\bar{t}(x) = \frac{(b-x)(x-a)}{B}. \tag{5.170}$$

The longest time corresponds to the position mid-way between the boundaries $x = (a+b)/2$ and is equal to the time of diffusion (with a diffusion coefficient of $B/2$) through half of the segment (a, b)

$$2 \cdot \frac{B}{2} \bar{t}_m = \left(\frac{b-a}{2}\right)^2 .$$

We now remove the lower boundary to $-\infty$. Then, the solution to equation (5.168) should satisfy the conditions

$$\varphi(t|b) = 1, \quad \varphi(t| - \infty) = 0.$$

But this is exactly the solution derived from (5.169) as $a \to -\infty$

$$\varphi(t|x) = 1 - \frac{2}{\sqrt{\pi}} \int_0^{(b-x)/\sigma} e^{-s^2} ds, \quad \sigma = \sqrt{2Bt}.$$

The respective probability density will be

$$\frac{\partial \varphi}{\partial t} = \frac{2}{\sqrt{2\pi Bt}} \exp\left[-(b-x)^2/2Bt\right] .$$

It becomes $\delta(b-x)$ at $t = 0$, but the integral (5.156) diverges at the upper limit, as the density $\partial\varphi/\partial t$ decreases with t only as $1/\sqrt{t}$, i.e., the contribution of long times required to reach the boundary is too large. thus, $\bar{t}(b) = 0$, but for any $x<b$ we get $\bar{t}(x) = \infty$. This follows also from (5.170) as $a \to -\infty$.

In the general case of variable coefficients $A(x)$ and $B(x)$ and with *only one* boundary (at which $\bar{t} = 0$), we require that $\bar{t}(x)$ grows at the slowest rate as x moves further from the boundary, we will arrive at the following solution to (5.165),

$$\bar{t}_a(x) = 2 \int_a^x e^{-q(z)} dz \int_z^{\infty} e^{q(y)} \frac{dy}{B(y)} \quad \text{for} \quad a<x,$$

$$\bar{t}_b(x) = 2 \int_x^b e^{-q(z)} dz \int_{-\infty}^{z} e^{q(y)} \frac{dy}{B(y)} \quad \text{for} \quad b>x, \quad \text{where}$$

$$q(z) = 2 \int_0^z \frac{A(z)}{B(z)} dz.$$

5.13 Exercises

5.13.1. Show that the Smoluchowski equation follows from the definition (4.3) of the Markov process and the hierarchical dependence of finite-variate distributions.

Solution. According to (4.3), at $t_0<0<t$

$$w_3(x_0,t_0;y,\theta;x,t) = w_1(x_0,t_0)v(y,\theta|x_0,t_0)v(x,t|y,\theta),$$
$$w_2(x_0,t_0;x,t) = w_1(x_0,t_0)v(x,t|x_0,t_0).$$

The hierarchy condition yields

$$w_2(x_0,t_0;x,t) = \int w_3(x_0,t_0;y,\theta;x,t)dy.$$

Substituting w_2 and w_3 into this relation gives (5.3).

5.13.2. Let ξ_m $(m = 1,2,\dots)$ be a discrete sequence of independent random variables with the possible values $x_i = i = 0,1,2,\dots$ and the respective probabilities $p_i = P\{\xi_m = i\}$ with $\sum_{i=0}^{\infty} p_i = 1$. The transition probabilities (coinciding with the state probabilities) are independent of the number of tests, i.e., the sequence is homogeneous

$$p_{ij} \equiv p(x_j,m+1|x_i,m) = P\{\xi_{m+1} = j|\xi_m = i\} = P\{\xi_{m+1} = j\} = p_j.$$

Show that the sequence of variables

$$\eta_m = \xi_1 + \xi_2 + \dots + \xi_m$$

(by definition $\eta_0 = 0$) is Markovian, and find the matrix π_1 for the transition probabilities in one step.

Solution. The sequence η_m is Markovian since $\eta_{m+1} = \eta_m + \xi_{m+1}$, and hence, if the value of η_m is given, then the distribution of η_{m+1} is known. Denoting by $y_i = 1$ the possible values of η_m, we write the transition probability in one step as follows

$$\begin{aligned} p_{ij} &\equiv p\{y_j,m+1|y_i,m\} = P\{\eta_{m+1} = j|\eta_m = i\} \\ &= P\{\eta_m + \xi_{m+1} = j|\eta_m = i\} \\ &= P\{\xi_{m+1} = j-i\} = \begin{cases} p_{j-i} & \text{for } j\geq i, \\ 0 & \text{for } j<i. \end{cases} \end{aligned}$$

The matrix π_1 is thus

$$\pi_1 = \begin{pmatrix} p_0 & p_1 & p_2 & p_3 \cdots \\ 0 & p_0 & p_1 & p_2 \cdots \\ 0 & 0 & p_0 & p_1 \cdots \\ \dots & \dots & \dots & \dots\dots \end{pmatrix}.$$

5.13.3. The independent random variables ξ_m ($m = 1, 2, \ldots$) take on the values $+1$ or -1 with probabilities p and $q = 1 - p$. Investigate whether or not the sequence of variables $\eta_m = (\xi_m + \xi_{m+1})/2$ is Markovian.

Solution. The variable ξ_m may take on values $i = -1, 0$, and $+1$ with the probabilities $p_i = P\{\eta_m = i\}$ being equal respectively to q^2, $2pq$, and p^2, (the univariate distribution of η_m). For the two variables η_m and η_n, the joint probability (bivariate distribution) $p_{ij} = P\{\eta_m = i, \, \eta_n = j\}$ will be different, depending on the difference $n - m$. If $|n - m|>1$, then η_m and η_n will be independent, so that $p_{ij} = p_i p_j$. If $n = m - 1$, then η_m and η_{m-1} contain the same variable ξ_m and the matrix p_{ij} will be

$j \backslash i$	-1	0	1
-1	q^3	pq^2	0
0	pq^2	pq	p^2q
1	0	p^2q	p^3

Such a matrix is clearly also obtained at $n = m + 1$, when ξ_{m+1} enters into both, η_m and η_{m+1}. Finally, if $n = m$, then $p_{ij} = p_j \delta_{ij}$.

For the conditional probability

$$p(n, j|m, i) = P\{\eta_n = j|\eta_m = i\} = \frac{P\{\eta_m = i, \eta_n = j\}}{P\{\eta_m = i\}} = \frac{p_{ij}}{p_i}$$

we obtain the following matrices in the three above cases

$$p(n, j|m, i) = \begin{pmatrix} q^2 & 2pq & p^2 \\ q^2 & 2pq & p^2 \\ q^2 & 2pq & p^2 \end{pmatrix} \quad \text{for} \quad |n - m|>1, \tag{5.171}$$

$$= \begin{pmatrix} q & p & 0 \\ q/2 & 1/2 & p/2 \\ 0 & 1/2 & p \end{pmatrix} \quad \text{for} \quad |n - m| = 1, \tag{5.172}$$

$$= \begin{pmatrix} 1 & 0 & 0 \\ 0 & 1 & 0 \\ 0 & 0 & 1 \end{pmatrix} \quad \text{for} \quad n - m = 0. \tag{5.173}$$

If the η_m form a Markovian sequence, then the $p(n, j|m, i)$ should be the transition probabilities, i.e., satisfy the Smoluchowski equation

$$p(n, k|l, i) = \sum_j p(n, k|m, j) p(m, j|l, i), \quad l<m<n.$$

It is a straightforward exercise, however, to show that this is *not* the case. Suffice it to take the case of $n = m + 1$, $l = m - 1$, so that $n - l = 2$, and the left-hand side of the Smoluchowski equation to be the matrix (5.171). But the sum on the right-hand-side of the equation that constitutes the square of the matrix (5.172) is the matrix

$$\begin{pmatrix} q\left(q+\frac{p}{2}\right) & p\left(q+\frac{1}{2}\right) & \frac{p^2}{2} \\ \frac{q}{2}\left(q+\frac{1}{2}\right) & pq+\frac{1}{4} & \frac{p}{2}\left(p+\frac{1}{2}\right) \\ \frac{q^2}{2} & q\left(p+\frac{1}{2}\right) & p\left(p+\frac{q}{2}\right) \end{pmatrix} .$$

5.13.4. Write the matrix for transition probabilities in one step for the example of the random walk (Fig. 5.2) considered in Sect. 5.2, but this time for the case of absorbing walls.

Solution. It is obvious that for $1<i<N$ the transition probabilities p_{ik} remain the same as in the case of reflecting walls. But a particle striking the wall ($i=1$ or $i=N$) stops, i.e.,

$$p_{1k}=\begin{cases}1 & (k=1)\\ 0 & (k\neq 1)\end{cases}, \quad p_{Nk}=\begin{cases}1 & (k=N)\\ 0 & (k\neq N)\end{cases} .$$

Thus

$$\pi_1=(p_{ik})=\begin{pmatrix} 1 & 0 & 0 & 0 & \cdots & 0 & 0 & 0\\ q & 0 & p & 0 & \cdots & 0 & 0 & 0\\ 0 & q & 0 & p & \cdots & 0 & 0 & 0\\ \cdots & \cdots & \cdots & \cdots & \cdots & \cdots & \cdots & \cdots\\ 0 & 0 & 0 & 0 & \cdots & q & 0 & p\\ 0 & 0 & 0 & 0 & \cdots & 0 & 0 & 1 \end{pmatrix} .$$

5.13.5. Show that for a normal stationary Markov process the correlation coefficient is

$$K(\tau)=e^{-\alpha|\tau|} \quad ,$$

where $\alpha>0$ (the Doob theorem, see Ref. [5.29, Sect. 8] and [5.30]).

Solution. According to (2.28), the conditional probability for a normal process with $\overline{x}=0$ and $\overline{x^2}=1$ is

$$v(x_2|x_1)=\frac{w_2(x_1,x_2)}{w_1w_2}=\frac{1}{\sqrt{2\pi(1-K^2)}}\exp\left\{-\frac{(Kx_1-x_2)^2}{2(1-K^2)}\right\}, \tag{5.174}$$

were $K=\overline{x_1x_2}=K(t_2-t_1)$ is the correlation coefficient (as already stated, the process is stationary and $t_2>t_1$). The process being Markovian, v should obey the Smoluchowski equation

$$v(x_2|x_1)=\int\limits_{-\infty}^{+\infty} v(x_2|y)v(y|x_1)dy.$$

Inserting an expression of the form (5.171) for $v(x_2|y)$ with $K_2 = \overline{x_2 y} = K(t_2 - \theta)$, and for $v(y|x_1)$ with $K_1 = \overline{y x_1} = K(\theta - t_1)$ and taking the integral with respect to y gives

$$\int_{-\infty}^{+\infty} v(x_2|y)v(y|x_1)dy = \frac{1}{\sqrt{2\pi(1 - K_2^2 K_1^2)}} \exp\left\{-\frac{(K_2 K_1 x_1 - x_2)^2}{2(1 - K_2^2 K_1^2)}\right\},$$

which is to be equal to (5.171). Hence, $K = K_2 K_1$, or

$$K(t_2 - t_1) = K(t_2 - \theta)K(\theta - t_1), \quad t_2 > \theta > t_1.$$

For positive arguments the solution to this functional equation is

$$K(\tau) = e^{-\alpha\tau} \quad , \quad \tau \geq 0,$$

where α is a complex number. As K should meet the conditions $|K| \leq 1$, and $K(-\tau) = K^*(\tau)$, it is necessary that α be real and positive. Further, it follows from these conditions that $K(\tau)$ must be even when, instead of $\tau > 0$, we take $\tau < 0$.

5.13.6. Find the relation between the transition probability $p_{N,m} = p(N, m|0, 0)$ to go in N steps from the point 0 to a point m and the binomial probability $P_N(n)$ that, out of the total of N independent steps, n steps have been made to the right.

Solution. $P_N(n)$ is the probability that a shift to the right by $2n - N$ steps (from any initial point) will occur. If we are interested in the shift to the right by m steps, i.e., $m = 2n - N$, then $n = (m + N)/2$ steps should be made to the right. This is of course possible with the same parity of N and m. Thus

$$p_{N,m} = P_N\left(\frac{m + N}{2}\right) \quad \text{where} \quad m + N \quad \text{is odd.}$$

It is easily verified by substitution that this relation satisfies both the Markov equation

$$p_{N,m} = p \cdot p_{N-1,m-1} + q \cdot p_{N-1,m+1} \quad ,$$

and the normalization condition $\sum_{m-N}^{N} p_{N,m} = 1$ (the sum over the values of m having the same parity as N).

5.13.7. Find the transition probability $p(x_k|t, x_i) \equiv p_{ik}(t)$ for a Markov process homogeneous in t, with discrete states, presuming the (constant) coefficients A_{jk} to take on the values

$$A_{j,j+1} = n_1, \quad A_{jj} = -n_1, \quad A_{jk} = 0 \quad \text{for other}\, k,$$

and for the initial conditions $p_{ik}(0) = \delta_{ik}$ (*Kolmogorov* [5.3]).

Solution.

$$p_{ik}(t) = \begin{cases} \dfrac{(n_1 t)^{k-i}}{(k-i)!} e^{-n_1 t} & \text{for } k \geq i, \\ 0 & \text{for } k < i. \end{cases}$$

5.13.8. Find the stationary state probabilities P_k for a Markov process homogeneous in t, with the following (constant) values of the coefficients $A_{jk}(j,k = 1,2,\ldots)$

$$A_{j,j+1} = \alpha, \quad A_{j,j-1} = \beta > \alpha, \quad A_{11} = -\alpha, \quad A_{jj} = -(\alpha+\beta) \quad \text{for} \quad j > 1,$$
$$A_{jk} = 0 \quad \text{for other } k$$

(*Kolmogorov* [5.3]).

Solution. In this case (5.24) has the form

$$\beta P_2 - \alpha P_1 = 0 \qquad (k = 1),$$
$$\beta P_{k+1} - \alpha P_k = \beta P_k - \alpha P_{k-1} \quad (k > 1),$$

i.e., $\beta P_k = \alpha P_{k-1}$ for all k. Consequently, $P_k = P_1(\alpha/\beta)^{k-1}$ and from the condition $\sum_{k=1}^{\infty} P_k = 1$ it follows that $P_1 = 1 - \alpha/\beta$. Finally, $P_k = (1 - \alpha/\beta)(\alpha/\beta)^{k-1}$.

5.13.9. Solve the preceding problem, but with the following values for the coefficients $A_{jk}(j,k = 1,2,\ldots)$

$$A_{j,j+1} = \alpha, \quad A_{j,j-1} = (j-1)\beta, \quad A_{jj} = -\alpha - (j-1)\beta,$$
$$A_{jk} = 0 \quad \text{for other } k$$

(*Kolmogorov* [5.3]).

Solution.

$$P_{k+1} = \frac{\exp(-\alpha/\beta)}{k!} \left(\frac{\alpha}{\beta}\right)^k.$$

5.13.10. *Fuctuations of photographic density* [5.31]. Let an area of photoemulsion corresponding to the least resolvable point object (cell, for short) contain at the time $t = 0$ of commencement of exposure N undeveloped grains. Each of them may only be in one of two states - developed or undeveloped. If by a time t we have m undeveloped grains ($m = 1,2,\ldots,N$), then it is supposed that the probability of any one of them being developed in the interval $(t, t+dt)$ is $A_{m,m-1}(t)dt = m\alpha(t)dt$, and the probabilities of two or more grains being developed are of higher order in dt.

Write and solve a set of Kolmogorov equations for the transition probability $p(t,m|0,N) \equiv p_m(t)$ $(m = 1,2,\ldots,N)$ from N at $t = 0$ undeveloped grains to m such grains at time t. Find $\overline{m}$ and $D[m]$.

Solution. For m undeveloped grains at time t during a next interval dt there may be either a transition to $m-1$ undeveloped grains with the probability $m\alpha(t)dt$, or the retention of the number of undeveloped grains with the probability $1 - A_{m,m}(t)dt$; but $A_{m,m} + A_{m,m-1} = 0$, so that $A_{m,m}(t) = -m\alpha(t)$. Consequently, at $m<N$ the equation for $p_m(t)$ will be

$$\frac{dp_m}{dt} = A_{m,m}p_m + A_{m+1,m}p_{m+1} = -m\alpha(t)p_m + (m+1)\alpha(t)p_{m+1},$$

and at $m = N$, as $p_{N+1} = 0$,

$$\frac{dp_N}{dt} = -N\alpha(t)p_N.$$

The solution of these equations subject to the initial condition $p_m(0) = \delta_{Nm}$ is given by the binomial law

$$p_m(t) = \frac{N!}{(N-m)!m!}p^m(t)q^{N-m}(t), \quad \text{where}$$

$$p(t) = e^{-\varrho(t)} \quad , \; q(t) = 1 - p(t) = 1 - e^{-\varrho(t)} \quad , \; \varrho(t) = \int\limits_0^t \alpha(\theta)d\theta.$$

Clearly, $p(t)$ is the probability that in time t a grain will remain undeveloped. The mean and variance for m are thus

$$\overline{m} = Np(t) = Ne^{-\varrho(t)}, \; \sigma^2 \equiv D[m] = Np(t)q(t) = Ne^{-\varrho(t)}[1 - e^{-\varrho(t)}] \quad .$$

The process is nonstationary, and $D[m]$ varies with $\overline{m}$

$$\sigma^2 = \overline{m}\left(1 - \frac{\overline{m}}{N}\right) = \overline{n}\left(1 - \frac{\overline{n}}{N}\right),$$

where $\overline{n} = N - \overline{m}$ is the mean number of the developed grains.

At a fixed $\overline{m}$, the binomial distribution changes into the Poisson distribution with increasing N. Therefore in problem (5.13.7), where it was assumed from the very beginning that $N = \infty$, we arrived at the stationary Poisson law (as the A_{jk} were taken to be constant).

According to [5.31], a good approximation for $\alpha(t)$ is a linear function of the exposure time t

$$\alpha(t) = t/\vartheta^2.$$

Then

$$\varrho(t) = t^2/2\vartheta^2$$

and

$$\bar{n} = N - \bar{m} = N[1 - e^{-\varrho(t)}] = N(1 - e^{-t^2/2\vartheta^2})$$

The density, proportional to $\bar{n}$, grows with t following the same law. Accordingly, ϑ is an exposure such that the blackening rate is maximal. Clearly, ϑ varies with the luminescence of the emulsion. Unlike the blackening that grows monotonically with t, the photographic "noise" (i.e., σ^2) has a maximum $\sigma^2 = N/4$ at $t = \vartheta\sqrt{\ln 4}$. At $t = 0$ and $t = \infty$, σ^2 vanishes. The noise is all the stronger, the more grains there are in a cell, since $\sigma^2 \sim N$.

5.13.11. Using the Rayleigh method (Sect. 5.4), derive the Einstein-Fokker-Planck equation for the transition probability $v(t,x) \equiv v(t,x|0,0)$ for a particle that may in the time Δt make $+ka$ and $-ka$ steps with the respective probabilities p_k and q_k ($k = 1,2,\ldots,n$; so-called n-step walk). Of course, in this case

$$\sum_{k=1}^{n} (p_k + q_k) = 1.$$

Solution. Putting $t = N\Delta t$, and $x = ma$, we obtain for the transition probability $p_{N,m}$ the Markov equation

$$p_{N,m} = \sum_{k=1}^{n} (p_k p_{N-1,m-k} + q_k p_{N-1,m+k}), \qquad \text{or}$$

$$v(t,x) = \sum_{k=1}^{n} [p_k v(t - \Delta t, x - ka) + q_k v(t - \Delta t, x + ka)] \quad .$$

Expanding the right-hand side in a power series of a gives

$$\begin{aligned} v(t,x) - v(t - \Delta t, x) = & -a \sum_{k=1}^{n} (p_k - q_k) k \frac{\partial v}{\partial x} \\ & + \frac{a^2}{2} \sum_{k=1}^{n} (p_k + q_k) k^2 \frac{\partial^2 v}{\partial x^2} - \ldots \end{aligned}$$

On the assumption that a and $\sum_{k=1}^{n}(p_k - q_k)k$ are of the order of $\sqrt{\Delta t}$ as $\Delta t \to 0$, in the limit the only nonzero terms will be those that contain $\partial v/\partial x$ and $\partial^2 v/\partial x^2$,i.e., we obtain the differential equation

$$\frac{\partial v(t,x)}{\partial t} = -A\frac{\partial v}{\partial x} + \frac{B}{2}\frac{\partial^2 v}{\partial x^2},$$

where

$$A = \lim_{\Delta t \to 0} \frac{a}{\Delta t} \sum_{k=1}^{n} (p_k - q_k)k, \quad B = \lim_{\Delta t \to 0} \frac{a^2}{\Delta t} \sum_{k=1}^{n} (p_k + q_k)k^2.$$

Let, for example,

$$p_k = p_1 e^{-(k-1)\alpha} \quad , \quad q_k = q_1 e^{-(k-1)\alpha}$$

($\alpha > 0$ and $k = 1, 2, ...$, i.e., $n = \infty$). Then the normalization condition yields

$$\sum_{k=1}^{\infty} (p_k + q_k) = \frac{p_1 + q_1}{1 - e^{-\alpha}} = 1, \quad \text{i.e.,} \quad p_1 + q_1 = 1 - e^{-\alpha} \quad .$$

Further, we have

$$\sum_{k=1}^{\infty} p_k k = -p_1 e^{\alpha} \frac{\partial}{\partial \alpha} \sum_{k=1}^{\infty} e^{-k\alpha} = \frac{p_1}{(1 - e^{-\alpha})^2}, \quad \sum_{k=1}^{\infty} q_k k = \frac{q_1}{(1 - e^{-\alpha})^2}.$$

In much the same way we get

$$\sum_{k=1}^{\infty} p_k k^2 = p_1 \frac{1 + e^{-\alpha}}{(1 - e^{-\alpha})^3}, \quad \sum_{k=1}^{\infty} q_k k^2 = q_1 \frac{1 + e^{-\alpha}}{(1 - e^{-\alpha})^3}.$$

Thus

$$A = \frac{A_0}{(1 - e^{-\alpha})^2}, \quad B = \frac{B_0(1 + e^{-\alpha})}{(1 - e^{-\alpha})^2}, \quad \text{where}$$

$$A_0 = \lim_{\delta t \to 0} \frac{(p_1 - q_1)a}{\Delta t}, \quad B_0 = \lim_{\Delta t \to 0} \frac{a^2}{\Delta t}.$$

As $\alpha \to \infty$, the only nonzero terms are p_1 and q_1, and we come back to the case of the one-step walk considered in Sect. 5.4, with $A = A_0$ and $B = B_0$. As $\alpha \to 0$, the probabilities of multistep jumps increase, thus speeding up both the systematic motion and the diffusion

$$\bar{s} = At \approx \frac{A_0}{\alpha^2}, \quad \overline{(\Delta s)^2} = 2Bt \approx \frac{4B_0}{\alpha^2} t.$$

5.13.12. Use equation (5.44) to calculate the distribution of the magnitude of a vector $\boldsymbol{r}$, whose components x and y are independent and uniformly distributed over the interval $(-A, A)$, i.e.,

$$w(x) = \begin{cases} 1/2A & \text{for } \; -A \leq x \leq A, \\ 0 & \text{for } \; |x| > A. \end{cases}$$

Solution.

$$w_r(r) = \begin{cases} \pi r/2A^2 & \text{for } r\leq A, \\ \dfrac{\pi r}{2A^2}\left(1 - \dfrac{4}{\pi}\text{arc cos}\,\dfrac{A}{r}\right) & \text{for } A<r\leq\sqrt{2}A, \\ 0 & \text{for } \sqrt{2}A<r. \end{cases}$$

5.13.13. In (5.59) the coefficient B is defined in terms of $\overline{(x-y)^2}$, i.e., in terms of the mean square deviation of a random final point x at time t from a given starting point y at time $t-\tau$. The passage from y to x is due both to the chaotic motion (diffusion) and to the systematic drift with the velocity

$$A(y,t) = \lim_{\tau\to 0}\frac{\overline{x-y}}{\tau} = \lim_{\tau\to 0}\frac{\bar{x}-y}{\tau},$$

where $\bar{x}$ is the mean x at the time t. It would seem natural to introduce as a measure of a purely random (diffusion) displacement not the quantity B, but

$$\tilde{B} = \lim_{\tau\to 0}\frac{\overline{(x-\bar{x})^2}}{\tau}.$$

Show that $\tilde{B} = B$, and discuss the result.

Solution. To the first order in τ we have

$$\bar{x} - y\approx A(y,t)\tau.$$

Therefore,

$$\begin{aligned} \tilde{B}(y,t) &= \lim_{\tau\to 0}\frac{\overline{(x-\bar{x})^2}}{\tau} \\ &= \lim_{\tau\to 0}\left\{\frac{1}{\tau}\overline{(x-y-A\tau)^2}\right\} \\ &= \lim_{\tau\to 0}\left\{\frac{1}{\tau}[\overline{(x-y)^2} - 2A\tau\overline{(x-y)} + A^2\tau^2]\right\} \\ &= \lim_{\tau\to 0}\left\{\frac{1}{\tau}[\overline{(x-y)^2} - A^2\tau^2]\right\} = B(y,t). \end{aligned}$$

This might be explained by the fact that the total deviation $x - y$ and the purely random deviation $x - \bar{x}$ both increase according to a diffusion law (for both the standard deviations are of the order of $\sqrt{\tau}$), and differ only in a higher (first) order in τ

$$(x-y) - (x-\bar{x}) = \bar{x} - y\approx A\tau.$$

5.13.14. The Kolmogorov equation (5.18) for a random process with discrete states was obtained from the Smoluchowski equation on the assumption that transition probabilities in a unit time, $A_{jk}(t)$, $(k\neq j)$, with $A_{jj}(t) = -\sum_{k\neq j} A_{jk}(t)$ exist. Given this assumption, (5.18) can be derived from the Markov equation. Show this for the one-step random walk, in which the only possible displacements are one step to the right or left, or no step (with the probabilities p, q, and $\tilde{p}$, respectively).

Solution. The Markov equation for the transition probability $p_{N,n}$ will be

$$p_{N+1,m} = p\cdot p_{N,m-1} + q\cdot p_{N,m+1} + \tilde{p}\cdot p_{N,m}, \quad p+q+\tilde{p} = 1.$$

Setting $N\Delta t = t$, we rewrite this equation as

$$p_m(t+\Delta t) = p\cdot p_{m-1}(t) + q\cdot p_{m+1}(t) + \tilde{p}\cdot p_m(t).$$

If

$$p = A_{m-1,m}\Delta t + o(\Delta t), \quad q = A_{m+1,m}\Delta t + o(\Delta t),$$
$$\tilde{p} = 1 + A_{m,m}\Delta t + o(\Delta t),$$

where

$$A_{m,m} = -(A_{m-1,m} + A_{m+1,m}),$$

then in the $\Delta t\to 0$ limit we obtain the Kolmogorov equation

$$\dot{p}_m(t) = A_{m-1,m}p_{m-1}(t) + A_{m+1,m}p_{m+1}(t) - (A_{m-1,m} + A_{m+1,m})p_m(t).$$

In particular, if the walk is symmetric (i.e., $A_{m-1,m} = A_{m+1,m}$), then selecting a scale for t such that these temporal probability densities be unity, we get

$$\dot{p}_m = p_{m-1} + p_{m+1} - 2p_m.$$

We can, of course, think of p_m as the transition probability $p(t,ma|0,0)$, or as the state probability $p(t,ma)$. It should be noted that in order for $A_{j,m}$ to exist, the introduction of the "no step" state is necessary.

5.13.15. In Sect. 5.4 the Einstein-Fokker-Planck equation for the process with continuous possible states was derived from the Markov equation with a discrete sequence. It has been shown in the preceding problem that the limiting process $\Delta t\to 0$ under appropriate conditions enables the Kolmogorov equation to be obtained from the Markov equation for the process with discrete states. It is only natural to expect a subsequent convergence of these states (e.g., an unlimited shortening of the step length, a, in the random walk) to facilitate the derivation of the Einstein-Fokker-Planck equation from the Kolmogorov equation. Show this, referring to the symmetric one-step walk with the Kolmogorov equation

$$\dot{p}_m = p_{m+1} + p_{m-1} - 2p_m.$$

Solution. Setting $x = ma$ and introducing the probability density $v(t,x) = p(t,ma)/a = p_m/a$, we reduce the Kolmogorov equation to the form

$$\dot{v}(t,x) = v(t,x+a) + v(t,x-a) - 2v(t,x).$$

Expanding $v(t, x \pm a)$ in the powers of a gives

$$\dot{v}(t,x) = a^2 \frac{\partial^2 v}{\partial x^2} + \frac{a^4}{12} \frac{\partial^4 v}{\partial x^4} + \dots .$$

If we substitute $\tau = a^2 t$ for t and expand $v(\tau, x, a^2)$ in the powers of a^2

$$v(\tau, x, a^2) = v^{(0)}(\tau, x) + a^2 v^{(1)}(\tau, x) + \dots,$$

we arrive at the sequence of equations of successive approximations

$$\frac{\partial v^{(0)}}{\partial \tau} = \frac{\partial^2 v^{(0)}}{\partial x^2}, \quad \frac{\partial v^{(1)}}{\partial \tau} = \frac{\partial^2 v^{(1)}}{\partial x^2} + \frac{1}{12} \frac{\partial^4 v^{(0)}}{\partial x^4}, \dots .$$

The zero-order approximation, i.e., the equation for $v^{(0)}$, constitutes the Einstein-Fokker-Planck equation.

"Corrections for discreteness" of $v^{(1)}$, $v^{(2)}$, ... are of interest in many problems, where a description of the discrete random structure of a process is replaced by a "smoothed" description using partial differential equations in continuous variables. It is sometimes important to know how such equations depend on the multistep nature of an initial discrete problem, and what boundary conditions for continuous variables (in various orders in a^2) are obtained using different ways of limiting the range of discrete variables. These questions have been studied in detail in [5.32], based on the models of a one-step and multistep random walk.

5.13.16. Use the characteristic function of the conditional distribution $\tilde{v}(t, x|t-\tau, y;\, T, X)$ to show that an equation of the form (5.74) for the conditional probability $v(t, x|T, X)$ follows from the integral relation (5.73) and that the coefficients A_n in (5.74) are determined from (5.75) [5.13].

Solution. Write the characteristic function as

$$\begin{aligned} \tilde{\varphi}(t, u|t-\tau, y; T, X) &= \int_{-\infty}^{+\infty} e^{iu(x-y)} \tilde{v}(t, x|t-\tau, y; T, X) dx \\ &= \sum_{n=0}^{\infty} \frac{(iu)^n}{n!} a_n(t-\tau, y; T, X), \end{aligned}$$

where

$$a_n(t-\tau,y;T,X) = \int_{-\infty}^{+\infty} (x-y)^n \tilde{v}(t,x|t-\tau,y;T,X)dx.$$

The inverse Fourier transformation yields

$$\begin{aligned}\tilde{v}(t,x|t-\tau,y;T,X) &= \frac{1}{2\pi}\int_{-\infty}^{+\infty} e^{-iu(x-y)}\tilde{\varphi}(t,u|t-\tau,y;T,X)du \\ &= \sum_{n=0}^{\infty}\frac{a_n}{2\pi n!}\int_{-\infty}^{+\infty} e^{-iu(x-y)}(iu)^n du \\ &= \sum_{n=0}^{\infty}\frac{(-1)^n}{n!}a_n\frac{\partial^n\delta(x-y)}{\partial x^n}.\end{aligned}$$

Substituting this for v in the integral (5.73) gives

$$\begin{aligned}v(t,x|,T,X) &= \sum_{n=0}^{\infty}\frac{(-1)^n}{n!}\int_{-\infty}^{+\infty} a_n(y,t-\tau;T,X) \\ &\quad\times v(t-\tau,y|T,X)\frac{\partial^n\delta(x-y)}{\partial x^n}dy \\ &= \sum_{n=0}^{\infty}\frac{(-1)^n}{n!}\frac{\partial^n}{\partial x^n}[a_n(x,t-\tau;T,X)v(t-\tau,x|T,X)].\end{aligned}$$

The summand with $n=0$ is simply $v(t-\tau,x|T,X)$. We transpose it to the left-hand-side, divide the equation by τ and pass to the limit as $\tau\to 0$. Since, by (5.75),

$$\lim_{\tau\to 0}\frac{a_n(x,t-\tau;T,X)}{\tau} = A_n(x,t|T,X),$$

we obtain

$$\frac{\partial v(t,x|T,X)}{\partial t} = \sum_{n=1}^{\infty}\frac{(-1)^n}{n!}\frac{\partial^n}{\partial x^n}[A_n(x,t|T,X)v(t,x|T,X)].$$

The above arguments also seem to apply when $\tilde{v}=v$, and the conditions (T,X) are reduced to the preceding state $x_0=x(t_0)$ of the process $x(t)$, i.e., to a Markov process. We arrive then at equation (5.61) for the transition probability $v(t,x|t_0,x_0)$, as only A_1 and A_2 are nonzero.

5.13.17. Show that the probability density $v(x|t,x_0)$, expressed by (5.141), obeys the Smoluchowski equation

$$v(x|t,x_0) = \int_{-\infty}^{+\infty} v(x|t-\theta,y)v(y|\theta,x_0)dy, \qquad 0<\theta<t.$$

Solution. Using the binomial expansion for $t^n = [(t-\theta)+\theta]^n$, it is easily seen that for the Poisson law $P(n) \equiv P_n(t) = (n_1 t)^n \exp(-n_1 T)/n!$ the following relation holds:

$$P_n(t) = \sum_{k=0}^{n} P_{n-k}(t-\theta)P_k(\theta). \tag{5.175}$$

Further, from the expression (5.138) for $p_n(x|x_0)$, we obtain the equation

$$\int_{-\infty}^{+\infty} p_m(x|y)p_k(y|x_0)dy = p_{m+k}(x|x_0), \tag{5.176}$$

which is also valid for vanishing m and/or k.

Presenting $v(x|t, x_0)$ in the form (5.140)

$$v(x|t,x_0) = \sum_{n=0}^{\infty} P_n(t)p_n(x|x_0),$$

gives

$$v(x|t-\theta, y)v(y|\theta, x_0) = \sum_{k,m=0}^{\infty} P_m(t-\theta)P_k(\theta)p_m(x|y)p_k(y|x_0).$$

Integrating this expression with respect to y and taking (5.176) into consideration, we get

$$\begin{aligned}&\int_{-\infty}^{+\infty} v(x|t-\theta,y)v(y|\theta,x_0)dy\\ &= \sum_{k,m=0}^{\infty} P_m(t-\theta)P_k(\theta)p_{m+k}(x|x_0)\\ &= \sum_{n=0}^{\infty} p_n(x|x_0)\sum_{k=0}^{n} P_{n-k}(t-\theta)P_k(\theta).\end{aligned}$$

By (5.175), this is equal to $v(x|t, x_0)$.

5.13.18. Find the transition probability, univariate probability density and covariance for the special case of the Poisson process (5.141) where the posterior (after a jump) probability density $\varphi(x_\nu/x_{\nu-1})dx_\nu$ is independent of $x_{\nu-1}$, i.e., equal to $\varphi(x_\nu)dx_\nu$. If x_ν is the velocity of a particle in a gas, then this case refers to a light particle (e.g., an electron) colliding with heavy particles. At a collision the light particle essentially "forgets" its velocity before the collision.

Solution. According to (5.138), we have $p_n(x|x_0) = \varphi(x)$. Therefore

$$\begin{aligned}v(x|t,x_0) &= e^{-n_1 t}\left\{\delta(x-x_0) + \varphi(x)\sum_{n=1}^{\infty}\frac{(n_1 t)^n}{n!}\right\}\\ &= e^{-n_1 t}\delta(x-x_0) + (1-e^{-n_1 t})\varphi(x),\end{aligned} \tag{5.177}$$

i.e., $x = x_0$ with the probability $\exp(-n_1 t)$ that in $(0,t)$ there will be no collisions, or x will have the distribution $\varphi(x)$ with the probability $1 - \exp(-n_1 t)$ that in $(0,t)$ there will be at least one collision. Such an evolution of the distribution (5.175) describes, for example, the behavior of velocities in a beam of light particles traveling through a gas composed of heavy particles. The term with $\delta(x - x_0)$ describes those particles that did not undergo collisions up to the time t and, therefore, still belong to the beam, and the second term describes the particles scattered by the gas.

As $t \to \infty$, the distribution $w_1(x) = \varphi(x)$ becomes stationary, and hence the bivariate distribution will be

$$\begin{aligned} w_2(x,t;x_0,0) &= w_1(x_0)v(x|t,x_0) \\ &= \varphi(x_0)[e^{-n_1 t}\delta(x - x_0) + (1 - e^{-n_1 t})\varphi(x)]. \end{aligned}$$

Using w_2 we calculate the covariance to get

$$\psi(t) = \overline{x x_0} - \overline{x} \cdot \overline{x}_0 = \int x x_0 w_2(x,t;x_0,0)\,dx\,dx_0 - \overline{x}^2 = \sigma^2 e^{-n_1 t} \quad ,$$

where $\sigma^2 = \overline{x^2} - \overline{x}^2$. The covariance is the same as for a normal Markov process (see Exercise 5.13.5), though the distribution w_1, even for the Gaussian distribution $\varphi(x)$, is not normal.

5.13.19. Find the mean time required to reach boundaries with a constant systematic velocity A directed to the upper boundary $b > x$, (i.e., $A > 0$) when (a) there is a lower boundary $a < x$, and (b) there is no lower boundary. The diffusion coefficient is constant.

Solution. With constant A and B equation (5.165)

$$\frac{B}{2}\overline{t}'' + A\overline{t}' + 1 = 0$$

has the general solution

$$\overline{t}(x) = \frac{1}{A}\left(C_1 - x + \frac{1}{\lambda}\right) + C_2 e^{-\lambda x} \quad , \quad \lambda = \frac{2A}{B}. \tag{5.178}$$

With two boundaries, finding C_1 and C_2 from the condition that $\overline{t}(a) = \overline{t}(b) = 0$, we arrive at

$$\overline{t}(x) = \frac{1}{A}\left[\frac{a e^{-\lambda b} - b e^{\lambda a} + (b - a)e^{-\lambda x}}{e^{-\lambda b} - e^{-\lambda a}} - x\right].$$

At $A = 0$ (and hence $\lambda = 0$) we return to (5.151).

If for $A \neq 0$ the lower boundary is placed at infinity ($a \to -\infty$), we have

$$\overline{t}(x) = \frac{b - x}{A},$$

i.e., the mean time required to reach the upper boundary b in this case is simply the time taken to cover the distance to the boundary with systematic velocity A. This also follows from the general solution (5.175) if we impose only one condition $\bar{t}(b) = 0$ and require $\bar{t}(x)$ to be nonnegative. This last requirement is seen to reduce to the condition that $\bar{t}(x)$ grows in the slowest way as $x \to -\infty$ (i.e., as the initial point x is displaced still further from the boundary b), which occurs if (5.178) has no exponential term ($C_2 = 0$).

6. Stochastic Differential Equations

The previous chapter was concerned with differential, integral, and mixed equations for some *distribution functions* of random processes, namely, state and transition probabilities. However, there exists another approach that involves equations for random functions themselves which are in general multicomponent, i.e., $\xi(t)$. In addition to $\xi(t)$ and its derivatives with respect to t, the equations also contain other deterministic and random functions $f(t)$ with given stochastic properties of the latter. Such equations are called *stochastic* differential equations (SDE). Their significance becomes especially clear if they are viewed as *deterministic* equations for *random* realizations of the desired solution $\xi(t)$ for known realizations of $f(t)$. This simple interpretation enables us to make extensive use of the well-developed mathematical techniques of conventional analysis for the construction and solution of SDE's. At the same time, it suggests that the statistics of the solution $\xi(t)$ must be completely determined by the given stochastic behavior of the random functions $f(t)$.

This chapter contains some basic information on SDE's for univariate random functions (processes), i.e., it deals with *ordinary* differential equations (SDE's for random fields that depend not only on the time t, but also on the spatial coordinates of the point $r = (x, y, z)$ i.e., partial differential equations are discussed in [6.1,2].)

Section 6.1 gives a clear formulation of the problem. In particular, it is stressed that the applicability of SDE's is not restricted by the condition that the response of the system $\xi(t)$ be Markovian. In Sect. 6.2 we introduce a further type of random process (processes with *independent increments*) and the concept of *delta-correlated* processes. Both play an important role in the theory of SDE's. In Sect. 6.3 they find their immediate applications: as an illustration of the main ideas the response $\xi(t)$ to the random force $f(t)$ of a linear system with 1/2 degrees of freedom is considered. The solution of the appropriate first-order SDE is trivial. But a comparison of this solution with that of the Einstein-Fokker-Planck equation shows that for the delta-correlated force $f(t)$, the response $\xi(t)$ is a diffusion Markov process. We then pass to problems of a more general nature.

Assuming that the delta-correlation of random forces ensures a Markovian response, we examine in Sect. 6.4 first-order SDE's for an arbitrary nonlinear and nonautonomous system. Further, we formulate a general theorem for the Markov approxmation for a dynamic system with an arbitrary number of degrees of freedom and Gaussian random forces with sufficiently short correlation times. The proof of this theorem, which calls for more powerful mathematical tools, is given in [Ref. 6.1, Chap. 1].

Section 6.5 examines two examples showing the advantages of SDE's in problems where there is an arbitrary correlation between the external actions. The first illustrative example deals with a linear system with 1/2 degrees of freedom and *Gaussian parametric* force. The second example is of more general interest, to radiophysicists amongst others. It is concerned with deterministic linear systems (specifically, electric circuits) under the influence of random forces (noise) with arbitrary statistics.

6.1 Statement of the Problem

It has already been noted that Markov processes may be treated as a direct generalization of deterministic processes, viz., as a response of a dynamic system to a random force. This by no means implies that the response of any dynamic system to an arbitrary random force is bound to be a process without stochastic memory effects. On the contrary, it may be seen a priori that this requires certain conditions to be met. However, the purely stochastic treatment given in the previous chapter, i.e., the derivation of equations for distribution functions (state probabilities, transition probabilities) in essence ignored both the dynamic behavior of the system, and the stochastic properties of the forces to which it was subjected. Therefore, in using the theory of Markov processes (e.g., to describe the oscillator) we were forced to rely on various qualitative considerations pertaining to the Markovian nature of the response. Indeed, the coefficients $A_{jk}(t)$ in the Kolmogorov equations for Markov processes with discrete states (Sect. 5.3), the coefficients $A(x,t)$ and $B(x,t)$ in the Einstein-Fokker-Planck equation for diffusion Markov processes (Sect. 5.6,7), and also the kernel $\varphi(x|t,x_0)$ and the probability $a(t,x)$ in the Kolmogorov-Feller equation for the stepwise Markov process (Sect. 5.11) are all the result of certain assumptions as to the behavior of the desired transition probabilities during infinitesimal time intervals. No relation of these functions to the dynamic characteristics of the system or to the stochastic properties of random forces was revealed.

We now turn to another treatment of random processes in dynamic systems which does not use equations for the transition probability (or other distribution functions) but for the *random response $\xi(t)$ itself.* This approach has been introduced by *Langevin* [6.3] to describe Brownian motion, but its significance and potentialities are much wider. Suffice it to indicate that the approach is not limited by the requirement that the response $\xi(t)$ be Markovian.

As a concrete example we take the case of a linear dynamic system (which again is not obligatory) described, for instance, by the differential equation

$$\hat{L}\xi(t) = f(t). \tag{6.1}$$

Here $f(t)$ is the external force, and $\hat{L}$ is a linear differential operator that in the general case also varies explicitly with time t through the coefficients $a_k(t)$ in

$$\hat{L} = \sum_{k=0}^{n} a_k(t) \frac{d^k}{dt^k}$$

(*parametric action*). Let the dynamic problem described by (6.1) with specified initial conditions $\xi^{(k)}(0)$ $(k = 0, 1, \ldots, n-1)$ have a unique solution $\xi(t)$. Suppose next that the force $f(t)$, the coefficients $a_k(t)$ and the initial values $\xi^{(k)}(0)$ are *random*, and that for almost all the realizations of these random functions and variables (i.e., almost certainly) the uniqueness conditions still hold for the solution $\xi(t)$. Equation (6.1) then gives rise to a set of realizations $\xi(t)$ and it is natural to expect that all the stochastic properties of this (now random) process $\xi(t)$ will be completely defined by the statistics of initial random functions and variables. This being so, we may also expect that in the special case where the response $\xi(t)$ is a diffusion Markov process, the statistics of the initial functions and variables will define the coefficients $A(x,t)$ and $B(x,t)$ in the Einstein-Fokker-Planck equation for the transition probability $v(x,t|x_0,t_0)$ for the random process $\xi(t)$.

This, of course, gives rise to a number of purely mathematical questions. For example, when can we derive differential equations for w_n, or at least for any moments $\xi(t)$? When is the process $\xi(t)$ necessarily Markovian (which, as stated above, is not obligatory in the case of the problem we have just discussed)? *Et cetera.* A full description of these questions (not all of which have known answers) lies beyond the scope of the book. Therefore, we will confine ourselves to some results of direct practical importance.

To begin with, we should like to discuss one more class of random functions that plays a prominent role just in those cases where the response $\xi(t)$ of the dynamic system appears to be a Markov process.

6.2 Random Functions with Independent Increments

We have already several times dealt with such functions though the term has not been used. Examples discussed above are: translational (Sects. 2.7 and 5.4) and rotational (Sect. 5.10) Brownian motion; fluctuational phase shift in the vacuum-tube oscillator (Sect. 5.9); and the charge transported by the electron flux to the tube anode when there is no depression of the shot noise (Sects. 2.5 and 3.3). All these examples deal with such random functions $\xi(t)$ whose *increments* in *nonoverlapping* time intervals are *independent*.

If we break down the interval (0, t) into an arbitrary number n of sequential intervals $\Delta t_i = t_{i+1}$ $(t_0 = 0, t_n = t)$, we can always represent $\xi(t)$ as the sum of *independent* summands given by the value $\xi(0)$ and the increments $\Delta_i \xi = \xi(t_{i+1}) - \xi(t_i)$ of the function $\xi(t)$ in these sequential intervals

$$\xi(t) = \xi(0) + \sum_{i=0}^{n-1} \Delta_i \xi. \tag{6.2}$$

This is valid for the path covered by a Brownian particle in time t, the number of radioactive nuclei decayed in time t (on the assumption that the loss of the nuclei through decay can be ignored), the fluctuational phase shift in the vacuum-tube generator, and so forth. The terms in (6.2) being independent, the variance of $\xi(t)$ is

$$D[\xi(t)] = D[\xi(0)] + \sum_{i=0}^{n-1} D[\Delta_i \xi]. \tag{6.3}$$

In particular, if the above function $\xi(t)$ is almost certainly *continuous*, then the independent summands $\Delta_i\xi$ can be made arbitrarily small by increasing n. In the limit $n \to \infty$ we conclude from the central limit theorem that the function $\xi(t) - \xi(0)$, where $\xi(t)$ is a continuous function with independent increments, is normally distributed with the variance $\int_0^t B(t)dt$, where $B(t) \geq 0$.

Assume now a process of the type considered to be *homogeneous in t*. This implies that the distribution of $\Delta_i\xi$ varies only with the duration of the interval $\Delta t_i = t_{i+1} - t_i$ (but continues to be independent of the values assumed by the function at times $0 \leq t_i$). Let us make $\xi(0) = 0$. Then $\xi(t)$ is equal to the sum of its increments in the interval $(0, t)$

$$\Delta_i \xi = \xi(t_{i+1}) - \xi(t_i) \equiv \eta(t_{i+1} - t_i) = \eta(\Delta t_i)$$

where $\xi(t) = \eta(t-0) = \eta(t)$. Equations (6.2,3) will take the form

$$\xi(t) = \eta(t) = \sum_{i=0}^{n-1} \eta(\Delta t_i), \quad \text{and} \tag{6.4}$$

$$D[\eta(t)] = \sum_{i=0}^{n-1} D[\eta(\Delta t_i)]. \tag{6.5}$$

The only nonnegative solution to the functional equation (6.5) is the linear homogeneous function

$$D[\xi(t)] = Bt \quad (B > 0), \tag{6.6}$$

i.e., the diffusion law. Thus, a *homogeneous continuous function with independent increments is normally distributed with a variance proportional to t.* We saw this in Sect. 5.4, where we passed from a random function with independent increments and discrete possible values (jumps by $\pm a$ with probabilities independent of the jump number), to its continuous counterpart ($a \to 0$). If from the very beginning the problem is formulated for the continuous function, then the same result is obtained assuming homogeneous random impacts (B =const, Sect. 5.7,9).

We now turn our attention to the following significant fact: a continuous function with independent increments is *not differentiable*[1]. Indeed, the variance of the increment $\Delta\xi = \xi(t+\Delta t) - \xi(t)$ is a quantity of the order Δt

$$D[\Delta\xi] \sim \Delta t$$

(for a homogeneous function we have simply $D[\Delta\xi] = B\Delta t$, where B =const.). Consequently,

$$D\left[\frac{\Delta\xi}{\Delta t}\right] = \frac{1}{(\Delta t)^2} D[\Delta\xi] \sim \frac{1}{\Delta t}$$

i.e., at any instant of time t the variance of a normally distributed variable $\Delta\xi/\Delta t$ grows indefinitely as $\Delta t \to 0$. This implies precisely that the derivative $d\xi/dt$ does not exist at any point t (Sect. 4.6). It thus turns out that if the path $s(t)$ covered by a Brownian particle, or a charge $q(t)$ that came to the anode of a vacuum tube, or the fluctuational phase shift $\varphi(t)$ in an oscillator are treated as continuous functions with independent increments, then neither the instantaneous velocity $u = ds/dt$, nor the instantaneous current $I = dq/dt$, nor the instantaneous frequency $\omega = d\varphi/dt$ exists. But physicists and engineers make extensive use of the random functions $u(t)$, $I(t)$, $\omega(t)$, etc., and obtain satisfactory results. It is therefore worthwhile to dwell a bit on this issue.

Consider for the sake of the argument the one-dimensional translational motion of a Brownian particle, assuming that $\xi(t) \equiv s(t)$ is a homogeneous function with independent increments. Suppose for simplicity that $\overline{s(t)} = 0$.

Taking in (5.77) the limit $n \to \infty$ (Sect. 5.6), enables $s(t)$ to be expressed in the form of the integral

$$s(t) = \int_0^t ds(t'),$$

where $ds(t')$ is the increment of the path in the interval $(t', t' + dt')$. Hence,

$$D[s(t)] = \overline{s^2(t)} = \int_0^t \int_0^t \overline{ds(t')ds(t'')}.$$

According to (6.6), the double integral on the right-hand side must be equal to Bt, which is only possible if the covariance for $ds(t)$ has the form

$$\overline{ds(t')ds(t'')} = B\delta(t' - t'')dt'\,dt''. \tag{6.7}$$

Evidently, this is a direct consequence of the mutual independence of increments in nonoverlapping intervals.

If we formally use the concept of the instantaneous velocity $u(t) = ds/dt$, then

---

[1] In neither of the stochastic definitions of the differentiability, see Sect. 4.6.

$$\overline{ds(t')ds(t'')} = \overline{u(t')u(t'')}dt'\,dt''.$$

Comparing this with (6.7) indicates that $u(t)$ must here be regarded as the so-called *delta-correlated* random process

$$\overline{u(t')u(t'')} = B\delta(t' - t''). \tag{6.8}$$

Accordingly, the mean square of $u(t)$ at any t is infinite. We can then readily find the location of such a singularity.

The assumption that increments Δs in nonoverlapping intervals Δt are independent holds good given that in time Δt a particle is subject to very many random pushes, i.e., $\Delta t \gg \vartheta$, where ϑ is the mean time between pushes. But the definition of continuity of $s(t)$ requires that we have $\Delta s \to 0$ (if only in probability) as Δt *decreases without bound.* Both conditions are only rigorously compatible at $\vartheta = 0$, i.e., with *infinitely frequent* pushes, which excludes the existence of an instantaneous velocity $u = ds/dt$.

If $\vartheta \neq 0$, as is often the case, then it may be that in the time $t < \vartheta$ not a single random push occurs, and acordingly, $s(t)$ will not be a function with independent increments. The instantaneous velocity $u(t)$ does then exist and has a correlation time of the order ϑ.

Let, for example,

$$\overline{u(t')u(t'')} = \frac{B}{2\vartheta} e^{-|t'-t''|/\vartheta} \quad \left(\overline{u^2} = \frac{B}{2\vartheta}\right). \tag{6.9}$$

Then, as $\Delta s = \int_t^{t+\Delta t} u(t')dt'$, we have

$$\begin{aligned} D[\Delta s] &= \overline{(\Delta s)^2} = \iint\limits_t^{t+\Delta t} \overline{u(t')u(t'')}dt'\,dt'' \\ &= 2\int\limits_t^{t+\Delta t} e^{-t'/\vartheta}dt' \int\limits_t^{t'} \frac{B}{2\vartheta} e^{t''/\vartheta} dt'' \\ &= B[\Delta t - \vartheta(1 - e^{-\Delta t/\vartheta})]. \end{aligned} \tag{6.10}$$

For $\Delta t \ll \vartheta$ we get

$$\overline{(\Delta s)^2} \approx \frac{B}{2\vartheta}(\Delta t)^2 = \overline{u^2}(\Delta t)^2, \tag{6.11}$$

so that for any $\Delta t > 0$, $\overline{(\Delta s)^2} \to \infty$ as ϑ tends to zero. But for $\Delta t \gg \vartheta$ the function $s(t)$ behaves *as if* its increments were independent, and as it follows from (6.10), the diffusion law comes to the fore

$$\overline{(\Delta s)^2} = B\Delta t.$$

These conclusions are in no way associated with the selection of the *exponential* covariance (6.9) for $u(t)$. What matters is only whether or not this covariance is sufficiently sharp for the time scale of the interval Δt (i.e., $\vartheta \ll \Delta t$), otherwise this interval may be too small ($\Delta t \ll \vartheta$).

When taking into consideration the nonzero correlation time ϑ we cannot in (6.3) pass to the limit $n \to \infty$, still assuming the increments $\Delta_i \xi$ in the time intervals Δt_i to remain independent. The increasing of n is constrained by the condition that intervals $\Delta t_i \sim t/n$ are still long as compared with ϑ, i.e., n is also limited from above ($1 \ll n \ll t/\vartheta$), thereby the distribution of the function $\xi(t) - \xi(0)$ is only approximately normal, but *approaches it more accureately, the shorter* ϑ.

Note now that the condition $\Delta t \gg \vartheta$ for a Brownian particle means that $\Delta t \gg 10^{-21} s$. Thus, for periods of time that on the macroscopic scale can fairly well be considered infinitesimal (say, 10^{-12}s), the number of random pushes is still enormous (about 10^9). Every time we encounter such a situation, i.e., when times of the order ϑ are in practice intangible, and in no way manifest themselves in phenomena observed, practically everything happens as if $\vartheta = 0$ [2]. A graphic and convenient idealization of $\vartheta = 0$ enables $s(t)$ to be regarded as a continuous function with independent increments. At the same time – recalling that *in fact there exists a certain sufficiently small* $\vartheta \neq 0$ it enables us to use the instantaneous velocity $u = ds/dt$, assigning to it the delta-correlation (6.8). To all intents and purposes, this does not cause any mathematical difficulties since the δ-function is only meaningful when it appears under the integral sign.

6.3 Simple Example of a Stochastic Differential Equation

We now return to the dynamic equation of first order (for a system with 1/2 degrees of freedom), an example of which is the equation for small amplitude fluctuations in the self-excited oscillator [the first of the equations (5.111)], i.e., to an equation of the form

$$\dot{x} = -\beta x. \tag{6.12}$$

The same equation occurs in problems concerned with: the velocity u of the one-dimensional motion of a particle of mass m in a medium with viscous friction $F = -hu$ with $\beta = h/m$; the displacement s of a particle devoid of mass and tied up to a spring with the constant k ($\beta = k/h$); the voltage V across the capacitor of aRC-circuit ($\beta = 1/RC$); or the currentI through an IR-circuit ($\beta = R/L$), and so on.

As stated in Sect. 5.8, we assume that in the dynamic system (6.12) exposed to sufficiently frequent (as compared with the relaxation time $1/\beta$) homo-

[2] From the spectral point of view this means that the upper boundary of the frequencies used lies much lower than $1/\vartheta$ [Ref. 6.4, Sect. 3.1].

geneous impacts, the response $x(t)$ will be a continuous homogeneous Markov process with the transition probability $v(x|t, x_0)$ satisfying the Einstein-Fokker-Planck equation

$$\frac{\partial v}{\partial t} = \beta \frac{\partial x v}{\partial x} + \frac{B}{2} \frac{\partial^2 v}{\partial x^2}, \tag{6.13}$$

i.e., the equation (5.112), but only in the univariate case, where v is only dependent on one spatial variable. Using the technique discussed in Sect. 5.8, we equate the coefficient $A(x,t)$ to the expression for $\dot{x}$, i.e., to the right-hand-side of (6.12): $A = \beta x$.

Given the initial condition

$$v(x|0, x_0) = \delta(x - x_0),$$

the solution to (6.13) is expressed by Gauss' law

$$v(x|t, x_0) = \frac{1}{\sqrt{2\pi}\sigma} e^{-(x-\overline{x})^2/2\sigma^2} \quad , \tag{6.14}$$

where

$$\overline{x} = x_0 e^{-\beta t} \quad , \quad \sigma^2 = \frac{B}{2\beta}(1 - e^{-2\beta t}), \tag{6.15}$$

see (5.115,116). In the limit $t \to \infty$, i.e., for $t \gg 1/\beta$, (6.14) becomes the stationary distribution $w_1(x)$ independent of x_0 with $\overline{x} = 0$ and $\sigma_\infty^2 = B/2\beta$. In the problem concerned with the velocity u of the particle in a viscous medium, where $x \equiv u$ and $\beta = h/m$, the distribution $w_1(u)$ must be Maxwellian

$$w_1(u) = \sqrt{\frac{m}{2\pi kT}} e^{-mu^2/2kT} \quad ,$$

so that $\sigma_\infty^2 = kT/m$ may also hold for the other problems mentioned above. Just as a corollary to the theorem describing the equipartition of energy in degrees of freedom: the average energy of a system with 1/2 degrees of freedom must be $kT/2$ (here $\overline{mu^2}/2 = m\sigma_\infty^2/2 = kT/2$).

Such is the purely stochastic method for solving the fluctuation problem given these assumptions. Now we will proceed in a different manner. We *introduce* into (6.12) a random (or fluctuational) force $f(t)$ to obtain

$$\dot{x} = -\beta x + f(t). \tag{6.16}$$

If for definiteness we consider the problem of the motion of a particle in an infinite viscous medium, then we obtain the equation of motion

$$m\dot{u} = -hu + \tilde{F}(t), \tag{6.17}$$

where the force on the particle is broken down into two parts: the *systematic* force of friction $F = -hu$ and the *random* force $\tilde{F}(t)$ [6.3].

Supposing that the systematic force is expressed by the Stokes law (for the spherical particle of radius a we have $h = 6\pi a\eta$, where η is the viscosity of the liquid) we make in fact two assumptions:

First, the particle must be in a laminar flow, i.e., the Reynolds number

$$\mathrm{Re} = \frac{\varrho a u}{\eta},$$

where ϱ is the density of the liquid, must be small. If u is assumed to be the root-mean-square velocity of thermal motion $[u \sim (kT/\varrho_1 a^3)^{1/2}$, where ϱ_1 is the density of the particle], i.e., considering the fastest vibrations of the particle, then

$$\mathrm{Re} = \frac{\varrho}{\eta}\left(\frac{kT}{\varrho_1 a}\right)^{1/2}.$$

At $\eta/\varrho \sim 10^{-2}$, $\varrho \sim \varrho_1 \sim 1$, $kT \sim 10^{-13}$, we have $\mathrm{Re} \sim 3.10^{-5} a^{-1/2}$, which even for molecular dimensions $a \sim 10^{-8}$ gives the value $\mathrm{Re} \sim 0.3$. Thus the condition of laminarity is satisfied.

Second, the *total* systematic force acting on a ball moving in a viscous incompressible liquid is, according to Boussinesq,

$$F = -hu - m'\dot{u} - 6\sqrt{\frac{hm'}{\pi}}\int_0^\infty \dot{u}(t - \alpha^2)d\alpha,$$

where $m' = 2\pi a^3 \varrho/3$ is the associated mass which equals one half of the mass of the liquid displaced by the particle. In (6.17) only the first term for the complete force F is retained. But at $\varrho \sim \varrho_1$, the second and third terms are of the same order as $m\dot{u}$. As far as the term $m'\dot{u}$ is concerned, this is of no importance, as the role of this term is reduced to changing the effective mass of the particle. More important is the third term which describes the viscous hydrodynamic memory-effect (see Sects. 4.2 and 5.1), which, if included, imparts to the system an infinite set of degrees of freedom.

For the case of a viscous (and thereby stochastic) memory-effect, the mean square of the particle displacement has been found by *Vladimirskii* and *Terletskii* [6.5]. The usual expression holds only for intervals of time t that are sufficiently long as compared with the relaxation time, $t \gg (m + m')/h$. We confine ourselves to a simplified problem statement based on (6.16).

We will treat this stochastic equation as if it were a conventional differential equation. Integrating it with the initial condition $x(0) = x_0$, gives[3]

[3] In what follows we will regard the initial conditions as deterministic. Otherwise x_0 would be a random variable specified by its distribution $w(x_0)dx_0$, and the conditional moments calculated later (i.e., moments with x_0 fixed) should be averaged over this distribution as well.

$$x(t) = x_0 e^{-\beta t} + e^{-\beta t} \int_0^t e^{\beta\theta} f(\theta) d\theta .$$

As it has been assumed that $\overline{f(t)} = 0$, the averaging of (6.18) over an ensemble of random forces gives

$$\bar{x} = x_0 e^{-\beta t} , \tag{6.19}$$

i.e., for $\bar{x}$ both (6.12) and the Einstein-Fokker-Planck equation (6.13) yield the same dynamic law.

We now evaluate the variance of $x(t)$. According to (6.18,19),

$$D[x(t)] = \overline{(x - \bar{x})^2} = e^{-2\beta t} \iint_0^t e^{\beta(\theta+\theta')} \langle f(\theta) f(\theta')\rangle d\theta \, d\theta' , \tag{6.20}$$

and hence, to obtain $D[x]$ it is necessary to specify the covariance of the random force, $\psi_f(\theta, \theta') = \langle f(\theta) f(\theta')\rangle$. We could take *any* covariance (subject however to general limitations of its form), but we will make a special assumption, namely: suppose that $f(t)$ is a *stationary delta-correlated* process,

$$\psi_f(\theta - \theta') = \langle f(\theta) f(\theta')\rangle = C\delta(\theta - \theta') , \tag{6.21}$$

where C is constant. Note that thereby the momentum

$$p(t) = \int_0^t f(\theta) d\theta$$

will be a continuous random function with independent increments, and hence distributed normally at any t (Sect. 6.2).

Substituting (6.21) into (6.20) gives

$$\begin{aligned} \sigma^2 &= D[x] = C\, e^{-2\beta t} \iint_0^t e^{\beta(\theta+\theta')} \delta(\theta - \theta') d\theta \, d\theta' \\ &= C\, e^{-2\beta t} \int_0^t e^{2\beta\theta} d\theta = \frac{C}{2\beta}(1 - e^{-2\beta t}) . \end{aligned}$$

If we set $C = B$, then this will coincide with (6.15) for σ^2, obtained from the Einstein-Fokker-Planck equation (6.13).

We have so far only found the moments $\bar{x}$ and $D[x]$, but that is not all one can do. The momentum increment $\Delta p = f(t)\Delta t$ is distributed normally with any Δt, since the difference $x - \bar{x}$ is, from (6.18), the sum (or rather the limit of the sum) of normally distributed variables. Accordingly, the distribution $x - \bar{x}$

is also given by the Gaussian law with the variance (6.22). For $C = B$ this *conditional* distribution [given that $x(0) = x_0$], simply coincides with (6.14). Further, it is readily proven by direct substitution that conditional probabilities of this kind satisfy the Smoluchowski equation (i.e. are transition probabilities), thus suggesting that the process $x(t)$ is Markovian. Therefore, if the random force $f(t)$ in the stochastic differential equation (6.16) is stationary and delta-correlated, see (6.21), then the response $x(t)$ is a diffusion Markov process, whose transition probability obeys the Einstein-Fokker-Planck equation with $B = C$.

Both treatments – using the Einstein-Fokker-Planck equation or a stochastic differential equation for the random function $x(t)$ – are equivalent for the problem we have considered. This is not to say that they are identical beyond the confines of the problem. The Einstein-Fokker-Planck equation has, for example, apparent advantages in the cases where some constraints are imposed on the set of possible values of the random function $x(t)$ (say, the presence of reflecting or absorbing walls, etc.). These constraints are taken care of by simply introducing appropriate boundary conditions. In the Langevin treatment the introduction of such constraints is fairly complicated. On the other hand, as stressed above, the Langevin technique does not require the force to be delta-correlated.

It should be noted that it is exactly in the case of delta-correlated forces that the handling of the differential equation (6.16) is, in a sense, symbolic. This equation is written not in $\overline{x}$ but in the instantaneous value $x(t)$. But with infinitely frequent pushes the response $x(t)$ is a *nondifferentiable* function, i.e., $\dot{x}(t)$ does not exist (in either of the stochastic senses of the notion of a derivative). Therefore the entire "differential" equation has only a *symbolic* sense. This is to be understood as follows.

By formally integrating (6.16) we obtain the solution (6.18) for $x(t)$, which presents no problems as only a delta-correlated force is contained under the integral. Put another way, (6.16) is (in the case of a delta-correlated force) a mathematically incorrect form for the subsequent *solution* which is quite important and, in the final analysis, the only one of interest to us. This approach is justified by the well-known advantages of dealing with differential equations in *formulating* a problem, such as the possibility of using general dynamic laws, the whole arsenal of current mathematical techniques in seeking the solution, and so forth. Finally, with a non-delta-correlated $f(t)$ all these reservations become irrelevant: stochastic differential equations for random functions then themselves acquire a definite mathematical content and, in addition, allow us to go beyond the confines of the class of Markov processes.

The constant C in the covariance (6.21) seems to characterize the intensity of random impacts. Let us look once more at variables in which the force and response are *energy conjugate*, i.e., the product of the force by the derivative of the response is the power transferred to the system. This is valid, say, for the force $\tilde{F}(t)$ in (6.17), as the power transferred to the particle is $\tilde{F}(t)\dot{u}$. Dividing (6.17) by the particle mass m reduces it to (6.16). Consequently, $f(t) = \tilde{F}(t)/m$, so that the covariance of the *real* force $F(t)$, according to (6.21), is

$$\langle\tilde{F}(t)\tilde{F}(t')\rangle = m^2\langle f(t)f(t')\rangle = m^2 C\delta(t-t') \equiv \tilde{C}\delta(t-t').$$

In the problem on the velocity of a Brownian particle we have set $C = B$ and $B = 2\,kTh/m^2$. Accordingly, the constant C in the covariance of the force $\tilde{F}(t)$ is

$$\tilde{C} = m^2 C = m^2 B = 2\,kTh, \tag{6.23}$$

i.e., it is connected *only with the coefficient of systematic friction h*. In the problem of a current in the LR-circuit, $\tilde{F}(t)$ refers to the random thermal e.m.f. $e(t)$ (Sect. 5.8), and h to the resistance of the circuit R so that the correlation constant for $e(t)$ will be $\tilde{C} = 2\,kTR$

$$\langle e(t)e(t')\rangle = 2\,kTR\delta(t-t'). \tag{6.24}$$

The use of stochastic differential equations enables us to go a step further. Solution (6.18) for x is the velocity u of a Brownian particle, or the current I through the LR-circuit. But $u = \dot{s}$, and $I = \dot{q}$, where s is the path covered by the particle, and q is the charge passing through the cross-sectional area. Consequently, we can also obtain the statistics for these variables. Let us now consider s.

Integrating (6.18) for $u = ds/dt$ with the (deterministic) initial condition $s(0) = s_0$ gives

$$s - s_0 = \frac{u_0}{\beta}(1 - e^{-\beta t}) + \frac{1}{\beta}\int_0^t [1 - e^{-\beta(t-\theta)}]f(\theta)d\theta. \tag{6.25}$$

Hence

$$\bar{s} = s_0 + \frac{u_0}{\beta}(1 - e^{-\beta t}). \tag{6.26}$$

For the variable

$$s - \bar{s} = \frac{1}{\beta}\int_0^t [1 - e^{-\beta(t-\theta)}]f(\theta)d\theta \tag{6.27}$$

we will use the reasoning followed earlier in drawing the conclusion that $x - \bar{x}$, i.e., in this case $u - \bar{u}$ is distributed normally. The transition probability for $s(t)$ is thus

$$v(s|t, s_0) = \frac{1}{\sqrt{2\pi D[s]}} \exp\left\{-\frac{(s-\bar{s})^2}{2D[s]}\right\} \quad .$$

The variance of $s(t)$ is worked out in the same way as for $u(t)$. In both cases we deal with a random function of the form

$$R(t) - \overline{R} = \int_0^t H(t,\theta) f(\theta) d\theta. \tag{6.28}$$

Hence, by (6.21), we have

$$\begin{aligned} D[R] &= \overline{(R-\overline{R})^2} = \int_0^t \int_0^t H(t,\theta) H(t,\theta') \overline{f(\theta) f(\theta')} d\theta\, d\theta' \\ &= C \int_0^t H^2(t,\theta) d\theta. \end{aligned}$$

If H varies with $t - \theta$ only, it is easily seen that

$$D[R] = C \int_0^t H^2(\theta) d\theta. \tag{6.29}$$

For $s(t)$ the function H is, by (6.27),

$$H(\theta) = \frac{1}{\beta}(1 - e^{-\beta\theta}).$$

Substituting this into (6.29) and using (6.23) gives

$$D[s] = \frac{kT}{\beta^2 m}(2\beta t - 3 + 4e^{-\beta t} - e^{-2\beta t}).$$

For $\beta t \ll 1$ this yields

$$D[s] = \frac{2\,kT}{3} \frac{\beta t^3}{m} = \frac{2\,kTh}{3m^2} t^3.$$

Strictly speaking, for a Brownian particle we may not consider times $t \ll 1/\beta$ (relaxation times) because the problem is stated without any allowance for viscous memory-effects, but in the problem of the LR-circuit this constraint is disregarded. In the opposite case of $\beta t \gg 1$ we have

$$D[s] = \frac{2\,kT}{\beta m} t = \frac{2\,kT}{h} t. \tag{6.30}$$

This relation is due to Einstein. It indicates that Brownian motion becomes more intensive with temperature and decays when the coefficient of viscous friction h is increased. The mass m does not figure in (6.30), which suggests that ignoring inertia, this formula can be derived from the equation

$$h\dot{s} = \tilde{F}(t),$$

for $s(t)$, hence

$$s - s_0 = \frac{1}{h} \int_0^t \tilde{F}(t) dt.$$

6.4 General Case of First-Order Equations and a Set of Such Equations for Gaussian Delta-Correlated Action

We now turn to the more general case of a *nonlinear and open* system with 1/2 degrees of freedom (first-order system) described by the differential equation

$$\dot{x} = \Phi[x, t; F(x,t)], \tag{6.31}$$

where Φ is a deterministic function of x, t, and the external action $F(x,t)$. The latter is assumed to be a random and, in the general case, nonstationary process that also depends on the state x of the system itself.

It is always possible to present the right-hand side of (6.31) as the sum of its mean [over the ensemble of random actions $F(x,t)$] and fluctuating parts, i.e., the random deviation from this mean:

$$\langle \Phi[x,t; F(x,t)] \rangle = X(x,t), \quad \Phi = \langle \Phi \rangle + f(x,t).$$

So, we can always write (6.31) as an equation with the *additive* random action

$$\dot{x} = X(x,t) + f(x,t), \tag{6.32}$$

where X is a deterministic function, and $\langle f(x,t) \rangle = 0$.

Clarifying the conditions under which the solution $x(t)$ will in this more general case be a continuous Markov process, requires strictly speaking an individual study, but we shall confine our discussion here to certain graphic considerations and to some facts pertaining to the linear equation (6.16).

The solution to (6.16) will be (approximately) Markovian provided $\vartheta_x \gg \vartheta_f$, where $\vartheta_x = 1/\beta$ is the relaxation time of the system and ϑ_f is the correlation time for the stationary force $f(t)$. If we treat $f(t)$ as delta-correlated, i.e., set $\vartheta = 0$, then the above condition is met for an arbitrarily small ϑ_x and the process $x(t)$ will now be fairly rigorously Markovian. Physically, it is quite plausible that for the solution to (6.32) to be Markovian, a correlation $f(x,t)$ is required that is short as compared with the time-scale of the process $x(t)$. Of course, the concepts of the correlation time ϑ_f for the *nonstationary* process $f(x,t)$ and of the relaxation time ϑ_x for the nonlinear dynamic system are both somewhat obscure. It is clear, however, that we are dealing here with *instantaneous* rates of variation of $x(t)$ and moments of the function $f(x,t)$. Our condition is as follows: at any t we should be able to select a time interval τ such that

$$\vartheta_x \gg \tau \gg \vartheta_f .$$

Thus, in a macroscopically infinitesimal time τ, the variable $x(t)$ remains practically unchanged, whereas the covariance of the process $f(x,t)$, which varies with t and $t' = t + \Delta t$, becomes zero already at Δt of the order of $\vartheta_f \ll \tau$. If $\vartheta_f = 0$, i.e., the covariance of the process $f(x,t)$ is proportional to $\delta(t-t')$,

$$\langle f(x,t)f(x',t')\rangle = C(x,x',t)\delta(t-t'), \tag{6.33}$$

then the condition imposed will be satisfied even as $\tau \to 0$ [where the rate of variation of $x(t)$ is finite, i.e., $\vartheta_x \neq 0$].

Assuming that this ensures the Markovian character of the process $x(t)$, i.e., that it has a transition probability $v(t,x|t_0,x_0)$ satisfying the Einstein-Fokker-Planck equation

$$\frac{\partial v}{\partial t} = -\frac{\partial A(x,t)v}{\partial x} + \frac{1}{2}\frac{\partial^2 B(x,t)v}{\partial x^2}, \tag{6.34}$$

we compute the variables $A(x,t)$ and $B(x,t)$ in accordance with their definitions (5.58,59):

$$A(x,t) = \lim_{\Delta t \to 0} \frac{\overline{\Delta x}}{\Delta t} \quad \text{and} \quad B(x,t) = \lim_{\Delta t \to 0} \frac{\overline{(\Delta x)^2}}{\Delta t}. \tag{6.35}$$

The averaging in these relations was ealier taken to be the averaging in transition probability $v(t+\Delta t, x+\Delta x|t,x)$. But now it is expected that the stochastic equation (6.32) defines *all* the stochastic properties of $x(t)$ [in particular all the moments of $x(t)$] through specified probabilistic characteristics of the force $f(x,t)$. It is therefore only natural, to perform the averaging over the ensemble of realizations of $f(x,t)$.

According to (6.32) the increment $x(t)$ in small time τ is

$$\Delta x(t,t+\tau) = \int_t^{t+\tau} \{X[x(\theta),\theta] + f[x(\theta),\theta]\}d\theta. \tag{6.36}$$

Both for X and for f we may write

$$X[x(\theta),\theta] = X[x(t)+\Delta x(t,\theta),\theta] \approx X[x(t),\theta] + X'_x[x(t),\theta]\cdot\Delta x(t,\theta).$$

As far as $\Delta x(t,\theta)$ is concerned, a first approximation will suffice, i.e., in (6.36) we can put $x(\theta) = x(t)$

$$\Delta x(t,\theta) \approx \int_t^{\theta} \{X[x(t)\theta'] + f[x(t),\theta']\}d\theta'.$$

Inserting this into the preceding equality, and the result into (6.36) gives

$$\Delta x(t,t+\tau) = \int_0^{t+\tau} d\theta\{X(x,\theta) + f(x,\theta)$$
$$+ \int_t^{\theta} [X'_x(x,\theta) + f'_x(x,\theta)][X(x,\theta') + f(x,\theta')]d\theta'\} \quad .$$

In averaging this expression, it should be taken into account that for any $x = x(t)$ and θ we have $\langle f(x,\theta)\rangle = 0$, and hence $\langle f'_x(x,\theta)\rangle = 0$, so that

$$\langle \Delta x(t, t+\tau)\rangle = \int_t^{t+\tau} d\theta \Big\{ X(x,\theta)$$

$$+ \int_t^{\theta} [X'_x(x,\theta)X(x,\theta') + \langle f'_x(x,\theta)f(x,\theta')\rangle] d\theta' \Big\}. \tag{6.37}$$

Further, it follows from (6.33) that

$$\begin{aligned} \langle f'_x(x,\theta)f(x,\theta')\rangle &= \left[\frac{\partial}{\partial x'}\langle f(x',\theta')f(x,\theta)\rangle\right]_{x'=x} \\ &= \left.\frac{\partial C(x,x',\theta)}{\partial x'}\right|_{x'=x} \delta(\theta-\theta'). \end{aligned}$$

As the integral of the delta-function at the boundary of the interval gives 1/2, (6.37) becomes

$$\begin{aligned} &\langle \Delta x(t,t+\tau)\rangle \\ &\approx \int_t^{t+\tau} d\theta \Big\{ X(x,\theta) + X'_x(x,\theta)X(x,\theta)(\theta - t) + \frac{1}{2}\left.\frac{\partial C(x,x',\theta)}{\partial x'}\right|_{x'=x}\Big\} \\ &\approx \Big\{ X(x,t) + \frac{1}{2}\left.\frac{\partial C(x,x',t)}{\partial x'}\right|_{x'=x}\Big\}\tau + X'_x(x,t)X(x,t)\frac{\tau^2}{2}. \end{aligned}$$

Dividing this by τ and passing by (6.35) to the limit $\tau \to 0$, we obtain

$$A(x,t) = X(x,t) + \frac{1}{2}\left.\frac{\partial C(x,x',t)}{\partial x'}\right|_{x'=x}. \tag{6.38}$$

Note that (6.33) suggests symmetry of the function $C(x,x',t)$ under the transposition $x \rightleftarrows x'$. Using this, we can readily show that

$$\left.\frac{\partial C(x,x',t)}{\partial x'}\right|_{x'=x} = \frac{1}{2}\frac{\partial C(x,x,t)}{\partial x},$$

so that $A(x,t)$ can be rewritten as

$$A(x,t) = X(x,t) + \frac{1}{4}\frac{\partial C(x,x,t)}{\partial x}. \tag{6.39}$$

In problems where the random force $f(x,t)$ is independent of x, so that the factor at $\delta(t-t')$ in (6.33) is either a function of t alone (nonstationary force), or simply a constant (stationary force), the term with the derivative in (6.38) is absent, so that we get

$$A(x,t) = X(x,t),$$

i.e., $A(x,t)$ coincides with the deterministic term on the right-hand-side of the stochastic equation (6.32), as was the case in the earlier examples.

In calculating $B(x,t)$ we are in the case of $\overline{(\Delta x)^2}$ interested only in the zeroth- and first-order terms in τ. Hence, we can, in squaring Δx, confine ourselves to a first approximation in (6.36), by setting

$$\Delta x(t,t+\tau) \approx \int_t^{t+\tau} \{X(x,\theta) + f(x,\theta)\} d\theta, \quad x = x(t).$$

Then

$$\begin{aligned}\overline{(\Delta x)^2} &= \iint_t^{t+\tau} \{X(x,\theta)X(x,\theta') + \langle f(x,\theta)f(x,\theta')\rangle\} d\theta\, d\theta' \\ &\approx X^2(x,t)\tau^2 + C(x,x,t)\tau,\end{aligned}$$

hence

$$B(x,t) = C(x,x,t). \tag{6.40}$$

The function $B(x,t)$ in (6.34) defines the diffusion component of the probability flow, and thus coincides with the factor $C(x,x',t)$ in the covariance for the random action (6.33) at $x' = x$, i.e., with the measure of intensity of this action.

Results (6.38) and (6.40) are obtained by *assuming* that the delta-function $f(x,t)$ ensures the Markov character of the process $x(t)$. But the main idea, that the stochastic equations should for given statistics of all the initial random functions and variables completely determine the stochastic behavior of the solutions, is a more stringent requirement. Equations for distribution functions of the response $x(t)$ in the general multivariate case and in particular the Markov properties of $x(t)$, if any, should *follow* from the stochastic equations themselves. In recent years a number of important and general results in this domain have been obtained. For the case of a system of ordinary stochastic differential equations of first order with normal additive actions showing the delta-correlation in time, the mathematical theory was developed by *Ito* and *Mackin* [6.6] (the theory is also presented in [6.7]) and for sufficiently short correlations, by *Stratonovich* [6.8]. We will follow a more straightforward and physically clear theory due to *Klyatskin* and *Tatarskii* [6.9,10], giving here only its formulation and results[4].

Let an n-variate random process $\boldsymbol{x}(t) = \{x_i(t)\}$, $i = 1,2,\ldots,n$, satisfy the system of stochastic equations

$$\dot{x}_i(t) = X_i(\boldsymbol{x},t) + f_i(x,t), \tag{6.41}$$

[4] The proof (see [6.1]) requires more powerful and general mathematical techniques. It relies on the method developed by *Novikov* [6.11] for the more general case of stochastic equations, i.e., for random fields.

where X_i are deterministic functions, and f_i are random actions or forces which are *normally distributed* in the $(n+1)$-dimensional space of the parameters $(\boldsymbol{x}, t)$ (i.e., f_i are random *fields* in this space), with zero means $\langle f_i(\boldsymbol{x}, t)\rangle = 0$ and the covariance matrix

$$\psi_{ik}(\boldsymbol{x}, t; \boldsymbol{x}', t') = \langle f_i(\boldsymbol{x}, t) f_k(\boldsymbol{x}', t')\rangle. \tag{6.42}$$

Suppose that the ratios of the characteristic times ϑ_f/ϑ_x [ϑ_f is the longest of the correlation times for the functions $f_i(\boldsymbol{x}, t)$ and ϑ_x is the shortest of the times of variation of the response $\boldsymbol{x}(t)$] are sufficiently small and that the system enables the distribution functions of $\boldsymbol{x}(t)$ to be found using the small pertubation method. These functions are thus obtained as expansions in the powers of the small parameter ϑ_f/ϑ_x. In a zeroth approximation ($\vartheta_f = 0$) the covariance matrix of the forces $f_i(\boldsymbol{x}, t)$ is taken to be[5]

$$\psi_{ik}^{\text{eff}}(\boldsymbol{x}, t; \boldsymbol{x}', t') = C_{ik}(\boldsymbol{x}, \boldsymbol{x}', t)\delta(t - t') \tag{6.43}$$

where the functions C_{ik} are defined by demanding the equality of the integrals (with respect to t') of the true and effective covariance matrices (6.42,43):

$$C_{ik}(\boldsymbol{x}, \boldsymbol{x}', t) = \int_{-\infty}^{+\infty} \psi_{ik}(\boldsymbol{x}, t; \boldsymbol{x}', t') dt'. \tag{6.44}$$

Then in the zeroth approximation (6.43) the response $\boldsymbol{x}(t)$ *appears to be a continuous* (diffusion) *Markov process*. This means that the multivariate probability distributions for $\boldsymbol{x}(t)$ are expressed in terms of products of the univariate density of the state probability, $w_1(\boldsymbol{x}, t)$, by the chain of transition probabilities $v(t_k, \boldsymbol{x}_k | t_{k-1}, \boldsymbol{x}_{k-1})$. As a further result, we obtain here for $v(t, \boldsymbol{x} | t_0, \boldsymbol{x}_0)$ the Einstein-Fokker-Planck equation[6]

$$\frac{\partial v}{\partial t} = -\frac{\partial A_i(x, t) v}{\partial x_i} + \frac{1}{2}\frac{\partial^2 B_{ik}(\boldsymbol{x}, t) v}{\partial x_i \partial x_k}, \tag{6.45}$$

in which

$$A_i(\boldsymbol{x}, t) = X_i(\boldsymbol{x}, t) + \frac{1}{2}\frac{\partial C_{ik}(\boldsymbol{x}, \boldsymbol{x}', t)}{\partial x'_k}\bigg|_{\boldsymbol{x}'=\boldsymbol{x}},$$

[5] Strictly speaking, the delta-correlation (6.43) is incompatible with the normal distribution, which assumes a finite variance. But the resultant Einstein-Fokker-Planck equation (6.45) only includes the *integrals* (6.44) of covariances. Therefore, in the distribution functions of random forces f_i, the variances may be considered to be finite ($\vartheta_f \neq 0$), and ϑ_f to be small. The response $\boldsymbol{x}(t)$ is then almost the same as for $\vartheta_f = 0$.

[6] The summation from 1 to n is performed over each subscript that appears twice in any term.

$$B_{ik}(\boldsymbol{x},t) = C_{ik}(\boldsymbol{x},\boldsymbol{x},t). \tag{6.46}$$

For the transition probability density v the initial condition is

$$v(t_0,\boldsymbol{x}|t_0,\boldsymbol{x}_0) = \delta(\boldsymbol{x}-\boldsymbol{x}_0),$$

and for the state probability density w_1 that satisfies the same equation (6.45) the initial condition has the form

$$w_1(t_0,\boldsymbol{x}) = w_0(\boldsymbol{x}),$$

where w_0 is the specified initial distribution.

This general theorem thus leads to (6.45). It follows *directly from the stochastic equations* (6.41). Moreover, the treatment shown in [6.9] has the advantage that it allows us to obtain *corrections* to the zeroth (diffusion) approximation. The corrections are twofold. First, an additional term of the form $-\partial S'_k(\boldsymbol{x},t)/\partial x_k$ appears on the right-hand side of (6.45). This is a correction to the probability flow density (not smaller than ϑ_f by the order of magnitude). Second, the matrix C_{ik} in (6.46) is replaced by

$$\tilde{C}_{ik}(\boldsymbol{x},\boldsymbol{x}',t) = \int\limits_0^t \psi_{ik}(\boldsymbol{x},t;\boldsymbol{x}',t')dt'.$$

The distinction of $\tilde{C}_{ik}$ from C_{ik} is only notable for $t \lesssim \vartheta_f$. But suppression of the term with S'_k requires, generally speaking, *the intensity of the fluctuation forces* f_i to be limited. The general form of the expression for S'_k is fairly unwieldy [6.10], but an examination of its relevance in special cases may yield more precise applicability conditions for the diffusion approximation[7].

We may now clarify the statement, given in Sect. 5.10, that the propagation of rays in a random medium is only in the *small angle* approximation a Markov process.

If we take the ray length l as the independent variable, then the equations of geometric optics are written as

$$\frac{d\boldsymbol{r}(l)}{dl} = \boldsymbol{\tau}(l), \quad \frac{d\boldsymbol{\tau}}{dl} = \nabla\mu(\boldsymbol{r}) - \boldsymbol{\tau}(\boldsymbol{\tau},\nabla\mu) \equiv \boldsymbol{a}(\boldsymbol{r},\boldsymbol{\tau})$$

where $\boldsymbol{r}$ is the radius-vector of the ray point, $\boldsymbol{\tau}$ is the unit vector of the tangent to the ray at that point, $\mu(\boldsymbol{r}) = \ln n(\boldsymbol{r})$ and $n(\boldsymbol{r})$ is the refractive index of the medium. If we introduce the six-vectors $\boldsymbol{f} = \{0,\boldsymbol{a}\}$, $\boldsymbol{x} = \{\boldsymbol{r},\boldsymbol{\tau}\}$, $\boldsymbol{X} = \{\boldsymbol{\tau},0\}$, then the above equations take the form (6.41) with the independent variable l (instead of t). Let $\mu(\boldsymbol{r})$ be a Gaussian random field with the mean $\bar{\mu} = 0$. It seems that introducing for $\boldsymbol{f}$ the covariance of the form (6.43) enables us to

[7] The delta-correlation (6.43) is only a *necessary* condition for a process to be Markovian.

write the Einstein-Fokker-Planck equation (6.45) for the six-vector $\boldsymbol{x}(t)$, which combines the random coordinates of the ray points $\boldsymbol{r}(l)$ and the random direct cosines of the ray, $\boldsymbol{\tau}(l)$. But $\boldsymbol{a}$ depends on $\boldsymbol{x}$ and *not* on l, and so the covariance matrix of the forces $\boldsymbol{f}$ has the form $\psi_{ik}(\boldsymbol{x}, \boldsymbol{x}')$, i.e., they are independent of l and l'. In other words, the interval of correlation of $\boldsymbol{f}$ in the variable l is infinite, therefore it is impossible to write ψ_{ik} in the form (6.43) with the delta-function $\delta(l-l')$.

We can, however, describe the ray in another way. To this end, we regard as the independent variable not its length l, but the thickness z of the layer traversed by the ray in the inhomogeneous medium (the ray is incident normally on the layer). The ray point is then given by its *transverse displacement* $\boldsymbol{\varrho} = \{x, y\}$ from the point where the ray enters the layer, $x = y = 0$, and the direction by the transverse vector $\boldsymbol{\tau}_\perp = \{\tau_x, \tau_y\}$. These variables are thought of as functions of z. The respective equations of geometric-optics are then

$$\frac{d\boldsymbol{\varrho}(z)}{dz} = \frac{\boldsymbol{\tau}_\perp(z)}{\sqrt{1-\boldsymbol{\tau}_\perp^2}}, \quad \frac{d\boldsymbol{\tau}_\perp(z)}{dz} = \frac{\boldsymbol{a}_\perp(\boldsymbol{\varrho}, \boldsymbol{\tau}_\perp, z)}{\sqrt{1-\boldsymbol{\tau}_\perp^2}}, \tag{6.47}$$

where $\boldsymbol{a}_\perp = \{a_x, a_y\}$. This suggests that these equations can only be used up to the first point at which the ray is reflected by 90° about the z-axis (i.e., about its initial direction) because $\sqrt{1-\boldsymbol{\tau}_\perp^2}$ then becomes zero. This means that in the case of a random medium we should confine ourselves to an area where probabilities of negative τ_z are quite small, i.e., to small angles of deviation of the ray ($\tau_z \approx 1$). In this *small angle approximation* (6.47) becomes

$$\frac{d\boldsymbol{\varrho}(z)}{dz} = \boldsymbol{\tau}_\perp(z), \quad \frac{d\boldsymbol{\tau}_\perp(z)}{dz} = \nabla_\perp \boldsymbol{\mu}(\boldsymbol{\varrho}, z) \equiv \boldsymbol{a}_\perp(\boldsymbol{\varrho}, z).$$

The four-dimensional random process $\boldsymbol{x}(z) = \{\boldsymbol{\varrho}(z), \boldsymbol{\tau}_\perp(z)\}$ occurs with *transverse* random "forces" $\boldsymbol{f} = \{0, \boldsymbol{a}_\perp\} = \boldsymbol{f}(\boldsymbol{\varrho}, z)$, which now vary with the variable z and possess a covariance matrix $\psi_{ik}(\boldsymbol{\varrho}, z; \boldsymbol{\varrho}', z')$. Assuming a delta-correlation of inhomogeneities of the medium in z

$$\psi_{ik}(\boldsymbol{\varrho}, z; \boldsymbol{\varrho}', z') = C_{ik}(\boldsymbol{\varrho}, \boldsymbol{\varrho}', z)\delta(z-z'),$$

the process $\boldsymbol{x}(z)$ will be Markovian. This yields the Einstein-Fokker-Planck equation (5.91), which assumes in addition that $C_{ik} = B\delta_{ik} = \text{const}$.

For rotational Brownian motion of a spherical particle we have no such limitations, as the rotational moments on the particle are functions of the independent variable (time t) being at the same time absolutely independent of the state of the system (orientation of the particle axis).

6.5 Stochastic Equation for Random Actions with Arbitrary Distribution Laws

The results obtained in [6.12] (see also [6.13]) generalize the theorem on non-Gaussian random actions mentioned in the preceding section[8]. Further, the stochastic equations themselves are taken to be of a more general form than (6.41), namely, as a set of integro-differential equations

$$\dot{x}_i = X_i(\boldsymbol{x}, t) + \int D_{ik}(\boldsymbol{x}, \boldsymbol{y}, t) f_k(\boldsymbol{y}, t) d\boldsymbol{y}, \tag{6.48}$$

where X_i and the kernel D_{ik} are deterministic vector and tensor functions, respectively. Obviously, equations (6.48) reduce for $D_{ik}(\boldsymbol{x}, \boldsymbol{y}, t) = \delta_{ik}\delta(\boldsymbol{y} - \boldsymbol{x})$ to (6.41). Significantly, random actions $f_k(\boldsymbol{y}, t)$ enter *linearly*, as earlier, and the solution $\boldsymbol{x}(t)$ satisfies the *causality principle*: $\boldsymbol{x}(t)$ is functionally dependent on $f_k(\boldsymbol{x}, \tau)$ only for τ *before* time t ($t_0 \leq \tau \leq t$). At time $t = t_0$ the initial conditions are specified by $\boldsymbol{x}(0) = \boldsymbol{x}_0$.

It is found that given these conditions (6.48) has an *exact* corollary, which is an equation for the univariate probability density $w_1(\boldsymbol{x}, t)$, but this equation is not *closed* with respect to w_1, for it has a term functionally dependent on $f_k(\boldsymbol{x}, \tau)$ for all $t_0 \leq \tau \leq t$. Only when the *actions* $f_k(\boldsymbol{x}, t)$ are *delta-correlated* in t does the above-mentioned equation for $w_1(\boldsymbol{x}, t)$[9] become *closed* (it contains a univariate distribution function alone), and the process $\boldsymbol{x}(t)$ becomes *Markovian*: the n-variate probability density $w_n(\boldsymbol{x}_1, t_1 : \ldots ; \boldsymbol{x}_n, t_n)$ is expressed in terms of w_1 and the *transition probability* v by (4.3).

If the delta-correlated $f_k(\boldsymbol{x}, t)$ are Gaussian, then, according to the theorem given in the preceding section, the equation for w_1 or v will be the Einstein-Fokker-Planck equation (*diffusion* Markov process). If the random actions are pulse processes consisting of delta-impulses

$$f_k(\boldsymbol{x}, t) = \sum_{\nu} a_{k\nu} \delta(t - t_\nu)$$

(it should be noted that though such a process is delta-correlated, it is not bound to be Poissonian or stationary), then for w_1 or v we obtain the Kolmogorov-Feller equation (5.146) (*stepwise* Markov process).

We now turn to the afore mentioned advantage of stochastic differential equations, namely, the possibility of using them to tackle linear problems, in which the random force $f(t)$ is not delta-correlated, and consequently the system response $\xi(t)$ is no more a Markov process.

---

[8] The work [6.12] deals not only with random processes, but also with random fields. However, in this book we still limit ourselves to a discussion of the former.

[9] Or for the conditional probability density $v(t, \boldsymbol{x}|t_0, \boldsymbol{x}_0)$, if the initial condition at $t = t_0$ is

$$v(t_0, \boldsymbol{x}|t_0, \boldsymbol{x}_0) = \delta(\boldsymbol{x} - \boldsymbol{x}_0).$$

It follows from the above results that, if the random actions are not assumed to be delta-correlated, the general theory becomes exceedingly complex. We will only furnish such examples illustrating the significance of results that can be obtained under certain additional constraints. However, it should be stressed that now these constraints will not include as a necessary requirement the exact or approximate delta-correlation of random actions on the dynamic system.

1) Consider the first-order linear equation describing a dynamic system with a *parametric* action:

$$\dot{x} = [Q(t) + q(t)]x, \tag{6.49}$$

where $Q(t)$ is a deterministic function, and $q(t)$ is a random process with $\overline{q(t)} = 0$. This equation thus has the form of (6.32) with $X(x,t) = Q(t)x$ and $f(x,t) = q(t)x$. If we introduce the function $y = \ln x$, then we obtain the linear equation $\dot{y} = Q(t) + q(t)$ with an additive random force independent of the state y of the system. It is assumed that $q(t)$ is a *normal* process with an *arbitrary* covariance

$$\psi_q(t,t') = \langle q(t)q(t')\rangle. \tag{6.50}$$

According to a derivation due to V.I. Tatarskii, we can, without any additional requirements, obtain from (6.49) a closed equation for the conditional probability density $v(x,t|x_0,0)$ of the solution $x(t)$.

The solution to (6.49) with the initial condition $x(0) = x_0$ is

$$x(t) = x_0 \exp\left\{\int_0^t [Q(\tau) + q(\tau)]d\tau\right\}.$$

Therefore, the (conditional) characteristic function of the process $x(t)$ will be

$$\begin{aligned}\varphi(u,t) &= \langle \exp\{iux\}\rangle = \sum_{n=0}^{\infty} \frac{(iu)^n}{n!}\langle x^n(t)\rangle \\ &= \sum_{n=0}^{\infty} \frac{(iux_0)^n}{n!} \exp\left\{n\int_0^t Q(\tau)d\tau\right\}\left\langle \exp\left\{n\int_0^t q(\tau)d\tau\right\}\right\rangle.\end{aligned}$$

As the integral $\alpha = n\int_0^t q(\tau)d\tau$ is also a Gaussian quantity (and for *any* Gaussian quantity α at $\overline{\alpha} = 0$ we have $\langle e^{\alpha}\rangle = \exp\{\overline{\alpha^2}/2\}$), we obtain

$$\left\langle \exp\left\{n\int_0^t q(\tau)d\tau\right\}\right\rangle = e^{n^2F(t)}, \qquad \text{where}$$

$$F(t) = \frac{1}{2}\iint_0^t \psi_q(\tau_1,\tau_2)d\tau_1\,d\tau_2.$$

Thus

$$\varphi(u,t) = \sum_{n=0}^{\infty} \frac{(iux_0)^n}{n!} \exp\left\{n\int_0^t Q(\tau)d\tau + n^2 F(t)\right\} \tag{6.51}$$

and therefore

$$\frac{\partial\varphi(u,t)}{\partial t} = \sum_{n=0}^{\infty} \frac{(iux_0)^n}{n!}[nQ(t) + n^2\dot{F}(t)] \times \exp\left\{n\int_0^t Q(\tau)d\tau + n^2 F(t)\right\}. \tag{6.52}$$

Using the identities

$$nu^n = \left(u\frac{\partial}{\partial u}\right)u^n, \quad n^2 u^n = \left(u\frac{\partial}{\partial u}\right)^2 u^n$$

and expression (6.51) for φ, we can rewrite (6.52) to obtain

$$\frac{\partial\varphi(u,t)}{\partial t} = Q(t)\left(u\frac{\partial}{\partial u}\right)\varphi + \dot{F}(t)\left(u\frac{\partial}{\partial u}\right)^2\varphi. \tag{6.53}$$

Considering the symmetry of the covariance $\psi_q(\tau_1,\tau_2)$ under the transposition $\tau_1 \rightleftarrows \tau_2$, the coefficient $\dot{F}(t)$ can readily be presented as a *single* integral of ψ_q

$$\begin{aligned}\dot{F}(t) &= \frac{1}{2}\frac{\partial}{\partial t}\int_0^t d\tau_1 \int_0^t d\tau_2\, \psi_q(\tau_1,\tau_2) \\ &= \frac{1}{2}\int_0^t d\tau_2\, \psi_q(t,\tau_2) + \frac{1}{2}\int_0^t d\tau_1\, \psi_q(\tau_1,t) = \int_0^t \psi_q(t,\tau)d\tau.\end{aligned} \tag{6.54}$$

This solves essentially the problem, since we have arrived at the differential equation for the characteristic function. To deduce directly the equation for the probability density

$$v(x,t|x_0,0) = \frac{1}{2\pi}\int_{-\infty}^{+\infty} \varphi(u,t)e^{-iux}du,$$

we need only take the Fourier transformation of (6.54). We then multiply this equation by $\exp(-iux)/2\pi$, integrate with respect to u and, assuming that the integrated terms vanish as $|u| = \infty$, we obtain, by (6.54),

$$\frac{\partial v}{\partial t} = -Q(t)\frac{\partial xv}{\partial x} + \left[\int_0^t \psi_q(t,\tau)d\tau\right]\frac{\partial}{\partial x}\left(x\frac{\partial xv}{\partial x}\right). \tag{6.55}$$

It should be stressed once more that in deriving (6.55) we only assumed that the process $q(t)$ is normal. Equation (6.55) is a *generalized* Einstein-Fokker-Planck equation of the form (5.76), which is also valid for a non-Markov process. If, however, the random action $q(t)$ is delta-correlated,

$$\psi_q(t,\tau) = C(t)\delta(t-\tau), \quad \text{then} \tag{6.56}$$

$$\int_0^t \psi_q(t,\tau)d\tau = \frac{1}{2}C(t).$$

[The 1/2 arises because the delta-impulse lies on the boundary of the interval (0, t).] According to the general theorem given in Sect. 6.4, we thus come to the *Markov* process $x(t)$ described by the Einstein-Fokker-Planck equation

$$\frac{\partial v}{\partial t} = -Q(t)\frac{\partial xv}{\partial x} + \frac{C(t)}{2}\frac{\partial}{\partial x}\left(x\frac{\partial xv}{\partial x}\right). \tag{6.57}$$

Although (6.55) and (6.57) are similar in form, they differ in their meaning. It is only with the *delta-correlation* (6.56) that n-variate distribution functions of $x(t)$ may be expressed in terms of the product of identical conditional probabilities v, i.e., these latter are *transition* probabilities (satisfying the Smoluchowski equation).

If the parametric action $q(t)$ is *stationary*, so that

$$\psi_q(t,\tau) = \psi_q(t-\tau), \quad \text{then} \quad \int_0^t \psi_q(t-\tau)d\tau = \int_0^t \psi_q(\theta)d\theta.$$

At time $t \gg \vartheta_q$ [ϑ_q is the maximum time of correlation $q(t)$], the upper limit of integration can be made infinite. The factor in brackets in (6.55) will thus tend to a constant value with t, *as if* the covariance were $C\delta(t-\tau)$. In that case (6.55) will look like (6.57) with constant C and will really become the Einstein-Fokker-Planck equation. In other words, if the action $q(t)$ is stationary, the response $x(t)$ becomes with increasing t a Markov process. It is to be noted that in actual fact $\vartheta_q \neq 0$, and therefore the process will only approximately be Markovian – if its behavior is considered in time intervals $\Delta t \gg \vartheta_q$, not in $\Delta t \ll \vartheta_q$.

Unlike the simple example just considered, in a more general case[10] it is impossible (as indicated above) to derive a closed equation even for the univariate probability, unless we suppose that the random action is delta-correlated.

But with *delta-correlation* and initial equations, which are *linear* in x_i it is always possible to derive *closed* sets of equations for moments of any order, i.e., an individual set for the first moments $\langle x_i(t)\rangle$, for the second moments $\langle x_i(t)x_k(t)\rangle$, and so forth. Under such conditions these sets of equations are

[10] For example, for the set of linear parametric equations with parametric random actions $q_{ik}(t)$

$$\dot{x}_i = [Q_{ik}(t) + q_{ik}(t)]x_k, \quad i = 1, 2, \ldots, n.$$

not mutually linked. This was shown earlier [6.13] (see Exercise 6.6.7) and reestablished in a more general fashion in [6.15].

2) The potentialities of the theory of stochastic differential equations are also influenced, of course, by the degree of complexity of the dynamic system under consideration. We will now turn to the second example that is fairly important, namely, to *linear* systems subject to *random external forces* alone. The parameters of the system are taken to be deterministic.

We have thus a special case of (6.41)

$$\dot{x}_i(t) = a_{ik}(t)x_k(t) + f_i(t), \quad i,k = 1,2,\ldots,n, \tag{6.58}$$

where $a_{ik}(t)$ are deterministic functions, and $f_i(t)$ are random functions independent of the state of the system $\boldsymbol{x} = \{x_i\}$. No special assumptions as to the stochastic behavior of the forces $f_i(t)$ are made, but they are considered to be known.

By eliminating from (6.58) all $x_i(t)$, except for any one function [which is denoted by $x(t)$], we can also formulate the problem in the form of an equation of the nth order

$$x^{(n)} + a_1(t)x^{(n-1)} + \ldots + a_n(t) = f(t). \tag{6.59}$$

As was emphasized above, in the first stage we deal with stochastic equations as if they were equations for deterministic functions, i.e., we solve a "dynamic" problem. It is exactly this approach that is possible with linear equations, as their solution can always be written in the form of a certain, also linear, operator acting on the given force $f(t)$. We now make use of the source-type representation of the solution – the Duhamel integral,

$$x(t) = \int_0^t H(t,\theta)f(\theta)d\theta, \tag{6.60}$$

where $H(t,\theta)$ is the so-called *pulse response* of the dynamic system at hand, i.e., its response for $t \geq \theta$ to the force $f(t) = \delta(t-\theta)$, provided the system was at rest up to time θ. Accordingly, the partial solution (6.60) for an arbitrary force $f(t)$ corresponds to the initial conditions

$$x^{(k)}(0) = 0, \quad k = 0,1,\ldots,n-1.$$

If the coefficients in (6.59) are constants, i.e., if the dynamic system is *harmonic*, then $H(t,\theta) = H(t-\theta)$.

It follows from the linear nature of the relation (6.60) between $x(t)$ and $f(t)$ that the moments $x(t)$ of any integral order are expressed by quadrature in terms of mixed moments of the same order of $f(t)$. As the latter are assumed to be known, this gives us *all the moments* $x(t)$, which in all cases of practical interest also implies that we can in principle deduce n-variate distribution functions for $x(t)$. Accordingly,

$$\langle x(t)\rangle = \int_0^t H(t,\theta)\langle f(\theta)\rangle d\theta, \tag{6.61}$$

$$B(t,t') = \langle x(t)x(t')\rangle = \int_0^t H(t,\theta)d\theta \int_0^{t'} H(t',\theta')B_f(\theta,\theta')d\theta', \tag{6.62}$$

where

$$B_f(t,t') = \langle f(t)f(t')\rangle.$$

For a harmonic system the Duhamel integral can be written as

$$x(t) = \int_0^t H(t-\theta)f(\theta)d\theta = \int_0^t H(\theta)f(t-\theta)d\theta, \quad \text{hence} \tag{6.63}$$

$$\langle x(t)\rangle = \int_0^t H(\theta)\langle f(t-\theta)\rangle d\theta \quad \text{and} \tag{6.64}$$

$$\langle x(t)x(t+\tau)\rangle = \int_0^t H(\theta)d\theta \int_0^{t+\tau} H(\theta')\langle f(t-\theta)f(t+\tau-\theta')\rangle d\theta'. \tag{6.65}$$

If $f(t)$ is a normal random function, then, by (6.60), the response $x(t)$ will also be normally distributed, as it is in that case the sum (the limit of the sum) of normally distributed summands. Since an n-variate normal distribution contains as parameters only the first and second moments, it is sufficient here to know $\langle f(t)\rangle$ and $B_f(t,t')$. But in the general case, to find the univariate distribution of $x(t)$, the random function $f(t)$ must be specified exactly, i.e., all its n-variate distributions must be known. We will confine our discussion to the moments (6.64,65) and assume that $f(t)$ is a *stationary* random process with zero mean

$$\langle f(t)\rangle = 0, \quad \langle f(t)f(t')\rangle = \psi_f(t-t').$$

Then, by (6.64), we have $\langle x(t)\rangle = 0$ and, by (6.65),

$$\psi(t,\tau) = \int_0^t H(\theta)d\theta \int_0^{t+\tau} H(\theta')\psi_f(\tau+\theta-\theta')d\theta'. \tag{6.66}$$

The integrand is thus independent of t. As real systems have a "finite memory" (the response $H(\theta)$ tends to zero with increasing θ) it follows from (6.66) that the covariance ψ becomes with increasing t independent of t, i.e., in the system a (wide-sense) stationary process sets in (see Sect. 4.5) with the covariance

$$\psi(\tau) = \int_0^\infty H(\theta)d\theta \int_0^\infty H(\theta')\psi_f(\tau + \theta - \theta')d\theta'.$$

As the pulse response $H(\theta)$ is zero up to the time of the delta-impulse, i.e., $H(\theta) = 0$ for $\theta<0$, the lower limits of integration can be replaced by $-\infty$. Introducing the integration variable $\chi = \theta' - \theta$, instead of θ' we obtain that under steady-state conditions

$$\psi(\tau) = \int_{-\infty}^{+\infty} H(\theta)d\theta \int_{-\infty}^{+\infty} H(\theta + \chi)\psi_f(\tau - \chi)d\chi$$

which can be written as

$$\psi(\tau) = \int_{-\infty}^{+\infty} H_2(\chi)\psi_f(\tau - \chi)d\chi, \quad \text{where} \tag{6.67}$$

$$H_2(\chi) = \int_{-\infty}^{+\infty} H(\theta)H(\theta + \chi)d\theta. \tag{6.68}$$

The function H_2 determined by the pulse response H characterizes the dynamic system under consideration, whereas ψ_f characterizes the external force $f(t)$.

Note that the formula (6.67) is a composition of the functions H_2 and ψ_f by the definition of composition given in Sect. 3.2. However, (6.68) is not a composition as the integration variable θ enters into both factors in the integrand with the same sign.

To conclude our discussion, we give a simple illustration of the way in which a non-Markov process in a linear system can be reduced to a Markovian one with a higher dimensionality, as discussed in Sect. 5.1.

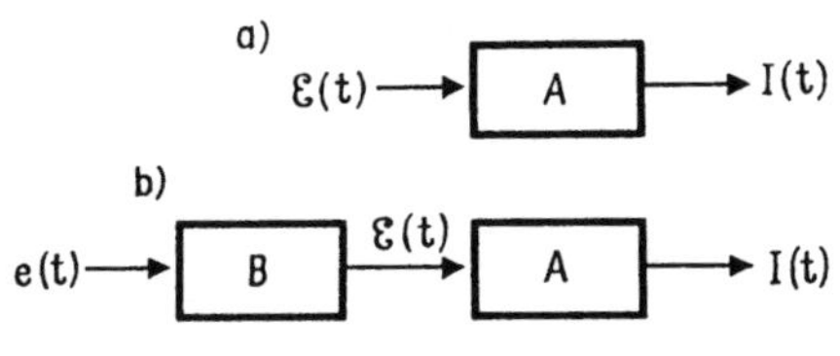

Fig. 6.1a,b. Reduction of a non-Markov process (a) to a Markov process with a higher dimensionality (b)

If a delta-correlated e.m.f. $\mathcal{E}(t)$ is fed into a linear electrical two-port A (Fig. 6.1a) then the output current $I(t)$ will be a component of the N_A-variate Markov process, where N_A is twice the number of degrees of freedom of the two-port (which can also be odd). Now let $\mathcal{E}(t)$ be non-delta-correlated, so that the current $I(t)$ is not a component of the N_A-variate Markov process. However, if a two-port B can be selected such that, when subjected to the delta-correlated noise $e(t)$ it gives at the output the e.m.f. $\mathcal{E}(t)$, then the *same* current $I(t)$ will be the response to the delta-correlated force in the system

$A + B$ (Fig. 6.1b). Thus, $I(t)$ will be a component of the $(N_B + N_A)$-variate Markov process.

Referring to the LR-cell considered earlier, the delta-correlated e.m.f. $e(t)$ is applied to the filter L_0, C_0, R_0 (Fig. 6.2). The state of the filter is described, say, by the charge q and current $\dot{q}$ (1 degree of freedom), and the state of the LR-cell by the current I (1/2 a degree of freedom). The current I will be a component of a three-dimensional Markov process with the transition probability $v(t, q, \dot{q}, I | t_0, q_0, \dot{q}_0, I_0)$. But if the delta-correlated e.m.f. acts directly on the LR-cell, then the current I will be a univariate Markov process with the transition probability $v(t, I | t_0, I_0)$.

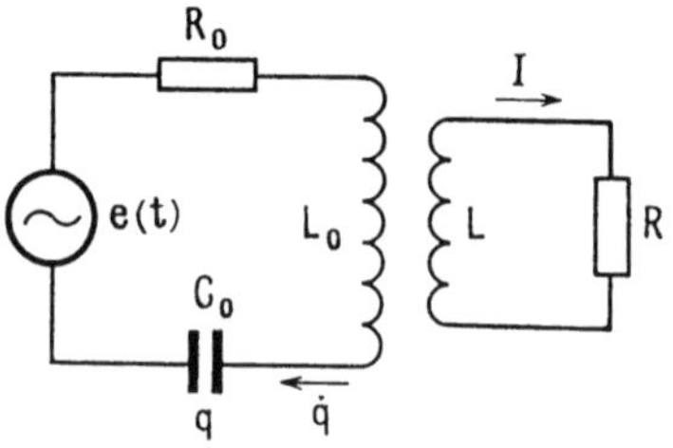

Fig. 6.2. Transition from a three- to a one-dimensional Markov process for the current in an RL-circuit

6.6 Exercises

6.6.1. Compute the covariance $\psi_I(t, \tau)$ for the current through the LR-circuit, subject to the stationary random e.m.f. $\mathcal{E}(t)$ with the covariance

$$\psi_{\mathcal{E}}(\tau) = \frac{\alpha C}{2} e^{-\alpha|\tau|}.$$

Solution. The equation for the current $I(t)$ is

$$\dot{I} + \beta I = \mathcal{E}(t)/L, \quad \beta = R/L.$$

The pulse response of the circuit is $H(t) = \exp(-\beta t)$ for $t \geq 0$, and zero for $t < 0$. By (6.66), we have

$$\begin{aligned}\psi_I(t, \tau) &= \int_0^t e^{-\beta\theta} d\theta \int_0^{t+\tau} e^{-\beta\theta'} \frac{\alpha C}{2L^2} e^{-\alpha|\tau+\theta-\theta'|} d\theta' \\ &= \frac{\alpha C}{2L^2(\alpha^2 - \beta^2)} \left\{ e^{-\beta\tau} \left[\frac{\alpha}{\beta}(1 - e^{-2\beta t}) - e^{-2\beta t} + e^{-(\alpha+\beta)t} \right] \right. \\ &\quad \left. + e^{-\alpha\tau} [1 - e^{-(\alpha+\beta)t}] \right\}. \end{aligned} \tag{6.69}$$

In particular, at $\tau = 0$, we get the current variance

$$D[I(t)] = \frac{\alpha C}{2L^2(\alpha^2 - \beta^2)} \left\{ \frac{\alpha}{\beta}(1 - e^{-2\beta t}) - e^{-2\beta t} + 2e^{-(\alpha+\beta)t} - 1 \right\}. \tag{6.70}$$

Both expressions hold for an *arbitrary* relationship between the circuit relaxation time $(1/\beta)$ and the correlation time of the random e.m.f. $(1/\alpha)$.

For $\beta t \gg 1$ we attain stationary values of ψ_I and $D[I]$

$$\psi_I(\tau) = \frac{\alpha C}{2L^2(\alpha^2 - \beta^2)}\left(\frac{\alpha}{\beta}e^{-\beta\tau} - e^{-\alpha\tau}\right), \quad D[I] = \frac{\alpha C}{2L^2\beta(\alpha+\beta)}.$$

As $\alpha \to \infty$ we come to the delta-correlated e.m.f. $\mathcal{E}(t)$ and the Markov process $I(t)$. Expressions (6.69,70) go over into

$$\psi_I(t,\tau) = \frac{Ce^{-\beta\tau}}{2L^2\beta}(1 - e^{-2\beta t}), \quad D[I(t)] = \frac{C}{2L^2\beta}(1 - e^{-2\beta t}).$$

The expression for $D[I]$, coincides of course, with (6.22).

If $\beta \to 0$, then, according to (6.70), we get

$$D[I(t)] = \frac{C}{L^2\alpha}(\alpha t + e^{-\alpha t} - 1),$$

which for $\alpha t \gg 1$ gives the diffusion law $D[I] = Ct/L^2$, i.e., the current

$$I(t) = \frac{1}{L}\int_0^t \mathcal{E}(t)dt$$

begins to behave as a function with independent increments.

6.6.2. Use the results of Sect. 5.8 to show that in a selective oscillatory circuit the envelope of the response to white noise satisfies a linear first-order stochastic equation.

Solution. The selective circuit is a Thomson oscillatory system described by the equation

$$\frac{d^2x}{dt'^2} + \mu h\frac{dx}{dt'} + x = \mu F(t'),$$

where $t' = \omega_0 t$, and μ is a small parameter. As stated,

$$\langle F(t_1')F(t_2')\rangle = C\delta(t_1' - t_2).$$

The equation for the circuit differs from (5.94) for the Thomson vacuum-tube oscillator only in that it has no nonlinear term and the term with dx/dt' has a different sign [μh is a decrement, in contrast to the increment μp in (5.94).] Accordingly, for the envelope $A(\theta)$ of the response $x = A(\theta)\cos[t' + \varphi(\theta)]$ we obtain the first of the Van der Pol equations (5.100) with $R(A) = -hA/2$, i.e.,

$$\frac{dA}{d\theta} + \frac{h}{2}A = -\frac{1}{2}F_\perp(\theta), \tag{6.71}$$

where $\theta = \mu t'$ is the slow dimensionless time.

If the component of the force $F_\perp(\theta)$ is delta-correlated, then according to the general theorem (Sect. 6.4), $A(\theta)$ will be a diffusion Markov process, whose covariance is, by (6.71), exponential. For a selective system we can, to a first approximation, regard $F_\perp(\theta)$ as a delta-correlated process [Ref. 6.4, Sect. 3.4].

The theorem relating to the Markov character of the envelope has been formulated in such a form (and proven in a fairly cumbersome way) by *Pierce* [6.15]. A proof using the Smoluchowski equation (or the Einstein-Fokker-Planck equation) has been suggested by *Helstrom* [6.16]. Among later works on the stochastic behavior of the envelope [6.17–20] are worth special mention.

6.6.3. Write the Einstein-Fokker-Planck equation for the oscillator with fluctuating natural frequency,

$$\ddot{x} + 2h\dot{x} + \omega_0^2[1 + f(t)]x = 0, \tag{6.72}$$

where $f(t)$ is a normal delta-correlated random process with

$$\langle f(t)\rangle = 0, \quad \langle f(t)f(t')\rangle = 2D\delta(t - t'), \quad D = \text{const.}$$

Solution. The conditions of the general theorem of Sect. 6.4 are met, and hence in the oscillator – a system with one degree of freedom – a diffusion Markov process occurs. Writing (6.72) as two first-order equations,

$$\begin{aligned} \dot{x} &= y \equiv X(x,y,t) + f_x(x,y,t), \\ \dot{y} &= -2hy - \omega_0^2[1 + f(t)]x \equiv Y(x,y,t) + f_y(x,y,t), \end{aligned} \tag{6.73}$$

we get

$$X = y, \quad Y = -2hy - \omega_0^2 x, \quad f_x = 0, \quad f_y = -\omega_0^2 x f(t).$$

Thus, in the correlation matrix of random impacts f_x and f_y, the only nonzero element is

$$\psi_{yy}(x,y,t;x',y',t') = \langle f_y(x,y,t)f_y(x',y',t')\rangle = 2D\omega_0^2 x x'\delta(t - t').$$

As a result, the Einstein-Fokker-Planck equation (6.45) for the transition probability density $v(x,y,t|x_0,y_0,0)$ has the form

$$\frac{\partial v}{\partial t} = -y\frac{\partial v}{\partial x} + \frac{\partial(2hy + \omega_0^2 x)v}{\partial y} + D\omega_0^4 x^2\frac{\partial^2 v}{\partial y^2}. \tag{6.74}$$

As the diffusion coefficient has a quadratic dependence on x, this equation can not be solved in the general form.

6.6.4. Use (6.74) in the above exercise to derive equations for the moments $M_{rs}(t) = \langle x^r(t) y^s(t) \rangle$. Show that these equations can be given as closed sets for the moments of each fixed order $r + s = n$. Examine the set of equations for the second moments

$$M_{20} = \overline{x^2}, \quad M_{11} = \overline{xy}, \quad M_{02} = \overline{y^2}$$

and find the condition of parametric excitation of the oscillator for the case in which the frequency fluctuations are the normal white noise.

Solution. We multiply (6.74) by $x^r y^s$ and integrate with respect to x and y from $-\infty$ to $+\infty$. Considering that

$$M_{rs}(t) = \iint\limits_{-\infty}^{+\infty} x^r y^s v(x, y, t | x_0, y_0, 0) dx\, dy,$$

and assuming that the integrated terms vanish for $|x| = \infty$ or $|y| = \infty$, we obtain for $M_{rs}(t)$ the following set of equations:

$$\begin{aligned} \dot{M}_{rs} &= rM_{r-1,s+1} - s(2hM_{rs} + \omega_0^2 M_{r+1,s-1}) \\ &\quad - \mathcal{D}\omega_0^4 s(s-1) M_{r+2,s-2}, \end{aligned} \tag{6.75}$$

having the initial conditions $M_{rs}(0) = x_0^r y_0^s$.

The sum of subscripts on all M in (6.75) is the same ($r + s = n$). It follows that for the moments of the nth order, equations (6.75) form a closed set of $n + 1$ equations. Putting in (6.75) $r = 1$, $s = 0$, and $r = 0$, $s = 1$ we derive equations for the first moments $M_{10} = \overline{x}$ and $M_{01} = \overline{y}$, i.e.,

$$\dot{\overline{x}} = \overline{y}, \quad \dot{\overline{y}} = -(2h\overline{y} + \omega_0^2 \overline{x}).$$

They also follow from the initial relations (6.73) in the absence of frequency fluctuations $[f(t) = 0]$. Thus, $\overline{x}$ and $\overline{y}$ vary according to a dynamic law.

For the second moments we deduce from (6.75) the system of three equations

$$\dot{M}_{20} = 2M_{11}, \quad \dot{M}_{11} = M_{02} - (2hM_{11} + \omega_0^2 M_{20}),$$

$$\dot{M}_{02} = -2(2hM_{02} + \omega_0^2 M_{11}) + 2\mathcal{D}\omega_0^4 M_{20}.$$

We seek a partial solution of these ordinary linear equations with constant coefficients in the form

$$M_{20} = C_1 e^{\lambda t}, \quad M_{11} = C_2 e^{\lambda t}, \quad M_{02} = C_3 e^{\lambda t}.$$

This yields for C_1, C_2, and C_3 the system of homogeneous algebraic equations

$$\lambda C_1 = 2C_2, \quad \lambda C_2 = C_3 - (2hC_2 + \omega_0^2 C_1),$$
$$\lambda C_3 = -2(2hC_3 + \omega_0^2 C_2) + 2D\omega_0^4 C_1.$$

For this system to have a nontrivial solution, its determinant must be equated to zero. Thus the characteristic equation for λ is obtained;

$$\lambda^3 + 6h\lambda^2 + 4(\omega_0^2 + 2h^2)\lambda + 4\omega_0^2(2h - \omega_0^2 D) = 0. \tag{6.76}$$

The necessary and sufficient conditions for stability, i.e., the absence of positive real parts of the roots λ_k of (6.76), are known to be the Routh-Hurwitz conditions. For the cubic equation

$$\lambda^3 + a_1\lambda^2 + a_2\lambda + a_3 = 0$$

they are

$$a_1 > 0, \quad a_2 > 0, \quad a_3 > 0, \quad a_1 a_2 - a_3 > 0.$$

For equation (6.76) these conditions will be

$$h > 0, \quad \omega_0^2 + 2h^2 > 0, \quad 2h - \omega_0^2 D > 0, \quad 4h(\omega_0^2 + 3h^2) + \omega_0^2 D > 0.$$

Only the third inequality can be violated, i.e., for

$$D > 2h/\omega_0^2. \tag{6.77}$$

and for the tiniest deviation of the initial conditions from zero the moments $\overline{x^2}$, $\overline{xy}$, and $\overline{y^2}$ and, in particular, the mean energy of the oscillator will grow in an exponential manner (parametric excitation).

The examination of equations for the moments of the order $n > 2$ indicates that the conditions of parametric excitation are different for different n. In particular, for the fourth moments this condition is weaker than (6.77), namely

$$D > \frac{2h}{3\omega_0^2} \frac{\omega_0^2 + 3h^2}{\omega_0^2 + 6h^2}.$$

If we assume that $f(t)$ is not normal, but suppose that the higher moments of $f(t)$ are expressed in terms of the delta-functions $\delta(t_2 - t_1)\delta(t_3 - t_1)\ldots\delta(t_n - t_1)$, then we can also obtain the same closed sets of equations (6.75) for the moments directly from the stochastic equations (6.73). Examination of the conditions of applicability of the diffusion approximation [6.10] indicates that, if $f(t)$ has a nonzero correlation time ϑ_f, the following conditions must be met:

$$\vartheta_f \ll T_0 = 2\pi/\omega_0, \quad \vartheta_f \ll 1/\omega_0^2 D, \quad t \gg \vartheta_f.$$

6.6.5. Write the Einstein-Fokker-Planck equation for the nonlinear oscillator

$$\ddot{x} + 2h\dot{x} + \frac{dU(x)}{dx} = f(t), \tag{6.78}$$

where the potential energy $U(x)$ contains, besides the term with x^2, terms with higher powers of x. The random force $f(t)$ is assumed to be normal and delta-correlated [6.10]. Examine the stationary solution to the Einstein-Fokker-Planck equation.

Solution. The set of equations for x and $y = \dot{x}$ has the form

$$\dot{x} = y = \frac{\partial H}{\partial y}, \quad \dot{y} = -2hy - \frac{\partial H}{\partial x} + f(t),$$

where $H(x, y) = y^2/2 + U(x)$ is a Hamiltonian for a closed ($f = 0$) and conservative ($h = 0$) system. This set of equations satisfies the conditions of the general theorem (Sect. 6.4). Equation (6.45) for the transition probability $v(x, y, t|x_0, y_0, 0)$ then becomes

$$\frac{\partial v}{\partial t} = -\left\{\frac{\partial v}{\partial x}\frac{\partial H}{\partial y} - \frac{\partial v}{\partial y}\frac{\partial H}{\partial x}\right\} + 2h\frac{\partial yv}{\partial y} + D\frac{\partial^2 v}{\partial y^2}, \quad \text{or}$$

$$\frac{\partial v}{\partial t} = -y\frac{\partial v}{\partial x} + \frac{dU}{dx}\frac{\partial v}{\partial y} + 2h\frac{\partial yv}{\partial y} + D\frac{\partial^2 v}{\partial y^2}.$$

[$2D$ is the constant in front of the delta-function in the covariance $f(t)$].

It is easily ascertained by substitution that the stationary solution $w_1(x, y)$ obtained as $t\to\infty$ has the form of the Gibbs distribution

$$w_1(x, y) = Ce^{-2hH(x,y)/D} = C \exp\left[-\frac{hy^2}{D} - \frac{2h}{D}U(x)\right] = w_x(x) \cdot w_y(y).$$

Thus, x and y are independent, and the distribution of the velocity $\dot{x} = y$ is Gaussian. Of course, the stationary solution only exists if the function $\exp[-2hU(x)/D]$ is integrable. The maxima and minima of the probability density of x,

$$w_x(x) = C_1 e^{-2hU(x)/D},$$

lie at points of minima and maxima, respectively of the potential energy $U(x)$, i.e., they coincide with stable and unstable equilibrium positions of the isolated dynamic system. For instance, if $U(x) = \omega_0^2 x^2/2 + \alpha x^4/4\,(\alpha{>}0)$, then there is only one stable equilibrium position $x = 0$. In that case the isolated conservative system is described by the Douffing equation

$$\ddot{x} + \omega_0^2 x + \alpha x^3 = 0.$$

But if $U(x) = -\omega_0^2 x^2/2 + \alpha x^4/4\ (\alpha>0)$, then $x=0$ is an unstable equilibrium, while the stable positions are $x = \pm\omega_0/\sqrt{\alpha}$. In this case $w_x(x)$ is a symmetric double-humped curve.

6.6.6. If in the preceding exercise we can ignore inertia (the term with the acceleration $\ddot{x}$) in equation (6.71), i.e., the equilibrium distribution of the velocity is reached very quickly, then the further behavior of the system is described by an equation of the first order

$$2h\dot{x} + \frac{dU(x)}{dx} = f(t).$$

Use the appropriate Einstein-Fokker-Planck equation to write the equation for the average time of reaching a boundary $a<x$ or $b>x$, and, in particular, to find the average transition time (Sect. 5.12) from one stable position to another for the case

$$U(x) = -\tfrac{1}{2}\omega_0^2 x^2 + \tfrac{1}{4}\alpha x^4.$$

Solution. In the case in question the Einstein-Fokker-Planck equation for the transition probability $v(x,t|x_0,0)$ will be

$$\frac{\partial v}{\partial t} = \frac{1}{2h}\frac{\partial}{\partial x}\left(\frac{dU}{dx}v\right) + \frac{D}{4h^2}\frac{\partial^2 v}{\partial x^2},$$

i.e., the coefficients $A(x,t)$ and $B(x,t)$ are independent of t :

$$A(x) = -\frac{1}{2h}\frac{dU}{dx}, \quad B(x) = \frac{D}{4h^2} = \text{const.}$$

Therefore, the equation (5.165) for the average time $\bar{t}(x)$ of reaching a boundary $a<x$ or $b>x$ from the initial position x will be

$$\frac{D}{8h^2}\frac{d^2\bar{t}}{dx^2} - \frac{1}{2h}\frac{dU}{dx}\frac{d\bar{t}}{dx} + 1 = 0.$$

We will make use of the solution to this equation in the case where we have the upper boundary $b>x$ alone, i.e., the solution (see Sect. 5.12)

$$\bar{t}_b(x) = 2\int_x^b e^{-q(z)}dz \int_{-\infty}^{z} d^{q(y)}\frac{dy}{B(y)}, \quad q(y) = 2\int_0^y \frac{A(x)}{B(x)}dx.$$

In the problem under consideration

$$q(y) = -\frac{4h}{D}\int_0^y \frac{dU}{dx}dx = -\frac{4h}{D}[U(y) - U(0)],$$

hence

$$\bar{t}_b(x) = 2\int\limits_x^b e^{4\bar{h}U(z)/D}dz \int\limits_{-\infty}^{z} e^{-4hU(y)/D}dy.$$

We can obtain the average transition time T_{12} from the state x_1 to the state x_2 ($x_1<x_2<b$) as a difference of average times of reaching the boundary b from x_1 and from x_2,

$$T_{12} = \bar{t}_b(x_1) - \bar{t}_b(x_2) = \frac{8h^2}{D}\int\limits_{x_1}^{x_2} e^{4hU(z)/D}dz \int\limits_{-\infty}^{z} e^{-4hU(y)/D}dy.$$

Of course, this quantity is independent of b.

In the special case where $U(y) = -\omega_0^2 y^2/2 + \alpha y^4/4$ and we are interested in finding the average transition time from one stable position $x_1 = -\omega_0/\sqrt{\alpha}$, to another stable position $x_2 = \omega_0/\sqrt{\alpha}$ we get

$$T_{12} = \frac{8h^2}{D}\int\limits_{-\omega_0/\sqrt{\alpha}}^{\omega_0/\sqrt{\alpha}} \exp\left\{-\frac{h}{D}(2\omega_0^2 z^2 - \alpha z^4)\right\}dz$$

$$\times \int\limits_{-\infty}^{z} \exp\left\{\frac{h}{D}(2\omega_0^2 y^2 - \alpha y^4)\right\}dy.$$

At sufficiently small D the exponential function in the integrand with respect to y has two symmetric sharp peaks at the points $\pm\omega_0/\sqrt{\alpha}$, so that the integral with respect to y is approximately equal to $C/2$ in the interval $(-\omega_0/\sqrt{\alpha}, \omega_0/\sqrt{\alpha})$, where

$$C = \int\limits_{-\infty}^{+\infty} \exp\left\{\frac{h}{D}(2\omega_0^2 y^2 - \alpha y^4)\right\}dy,$$

and the average transition time from one stable position to another will be

$$T_{12} \approx \frac{8h^2 C}{D}\int\limits_{0}^{\omega_0/\sqrt{\alpha}} \exp\left\{-\frac{h}{D}(2\omega_0^2 z^2 - \alpha z^4)\right\}dz.$$

6.6.7. Derive the first and second moments directly from the stochastic equation

$$\ddot{x} + p(t)\dot{x} + q(t)x = f(t), \tag{6.79}$$

where $f(t)$ is a stationary random process with $\langle f(t)\rangle = 0$, and $p(t)$, $q(t)$ are deterministic functions.

Solution. The operations of averaging and differentiating are commutative, and therefore the averaging of (6.79) gives for the moment $\overline{x}(t)$

$$\ddot{\overline{x}} + p(t)\dot{\overline{x}} + q(t)\overline{x} = 0.$$

We introduce the notation $\mu_{00} = x^2$, $\mu_{01} = \mu_{10} = x\dot{x}$, $\mu_{11} = \dot{x}^2$. Differentiating these μ_{kl} with respect to t and using (6.79) to eliminate $\ddot{x}$ gives

$$\dot{\mu}_{00} = 2x\dot{x} = 2\mu_{01},$$

$$\dot{\mu}_{01} = \dot{x}^2 + x\ddot{x} = \mu_{11} - p\mu_{01} - q\mu_{00} + xf,$$

$$\dot{\mu}_{11} = 2\dot{x}\ddot{x} = -2p\mu_{11} - 2q\mu_{01} + 2\dot{x}f.$$

Averaging these equations we obtain for the moments $m_{kl} = \overline{\mu}_{kl}$ the following relations:

$$\begin{aligned} \dot{m}_{00} &= 2m_{01}, \quad \dot{m}_{01} = m_{11} - pm_{01} - qm_{00} + \langle xf\rangle, \\ \dot{m}_{11} &= -2pm_{11} - 2qm_{01} + 2\langle\dot{x}f\rangle. \end{aligned} \tag{6.80}$$

We thus have to calculate the average values of $\langle x(t)f(t)\rangle$ and $\langle\dot{x}(t)f(t)\rangle$.

Now we express $x(t)$ in terms of the Duhamel integral (with zero initial conditions at $t = 0$) :

$$x(t) = \int_0^t H(t,\theta)f(\theta)d\theta.$$

If $x_1(t)$ and $x_2(t)$ are fundamental solutions to the homogeneous equation (6.79), then the pulse response will be written as

$$H(t,\theta) = \frac{1}{\Delta(\theta)}[x_1(\theta)x_2(t) - x_2(\theta)x_1(t)],$$

where $\Delta(\theta)$ is the Wronskian

$$\Delta(\theta) = \begin{vmatrix} x_1(\theta) & x_2(\theta) \\ \dot{x}_1(\theta) & \dot{x}_2(\theta) \end{vmatrix}.$$

Consequently,

$$H(t,t) = 0, \quad \left.\frac{dH(t,\theta)}{dt}\right|_{\theta=t} = 1. \tag{6.81}$$

We thus have

$$\langle x(t)f(t)\rangle = \int_0^t H(t,\theta)\langle f(\theta)f(t)\rangle d\theta = \int_0^t H(t,\theta)\psi_f(t-\theta)d\theta,$$

$$\langle \dot{x}(t)f(t)\rangle = \int_0^t \dot{H}(t,\theta)\langle f(\theta)f(t)\rangle d\theta = \int_0^t \dot{H}(t,\theta)\psi_f(t-\theta)d\theta.$$

These integrals can be expressed neither in terms of m_{kl}, nor in terms of the parameters ψ_f, save in the case where the force is delta-correlated

$$\psi_f(t-\theta) = \langle f(\theta)f(t)\rangle = 2D\delta(t-\theta).$$

Then

$$\langle x(t)f(t)\rangle = DH(t,t), \quad \langle \dot{x}(t)f(t)\rangle = D\frac{dH(t,\theta)}{dt}\bigg|_{\theta=t}$$

(the factor $1/2$ emerges, because the delta-function is nonzero at the boundary of the interval). By (6.81), we have

$$\langle xf\rangle = 0, \quad \langle \dot{x}f\rangle = D. \tag{6.82}$$

The significance of this result is that the response $x(t)$ at time t does not yet contain a component due to the action of the force $f(t)$ at the same moment. In contrast, the derivative "feels" the action of the force $f(t)$ instantaneously.

By (6.82) the equations (6.80) become a closed set

$$\dot{m}_{00} = 2m_{01}, \quad \dot{m}_{01} = m_{11} - pm_{01} - qm_{00},$$

$$\dot{m}_{11} = -2pm_{11} - 2qm_{01} + 2D,$$

that must be solved for the conditions $m_{kl}(0) = 0$.

This technique has been worked out in [6.14] for the general case of an equation of the nth order

$$p_n(t)x^{(n)} + p_{n-1}(t)x^{(n-1)} + \ldots + p_0(t)x = f(t). \tag{6.83}$$

In principle the technique remains the same, and leads in the case of a delta-correlated force $f(t)$ to a closed system of $n(n+1)/2$ equations for the moments $m_{kl}(t) = \langle x^{(k)}(t)x^{(l)}(t)\rangle$. In that case, it is necessary to calculate the values of n mixed moments $\langle x^{(k)}(t)f(t)\rangle$, $k = 0, 1, \ldots, n-1$.

If $f(t)$ is not delta-correlated, then the method can be applied to an equation of higher dimensionality, provided that we choose a linear differential operator $\hat{L}$, such that on the right-hand side of the new equation it will give the delta-correlated force $F(t) = \hat{L}\{f(t)\}$.

References

Preface

1. S.A. Akhmanov, A.S. Chirkin: *Statistical Phenomena in Nonlinear Optics* (Moscow University, Moscow 1971) (in Russian);
   S.A. Akhmanov, Y.Y. Dyakov, A.S. Chirkin: *Introduction to Statistical Radiophysics and Optics*, Vols. I, II (Springer, Berlin, Heidelberg, New York) in preparation
2. A.M. Yaglom: *An Introduction to the Theory of Stationary Random Functions* (Prentice-Hall, New York 1962)
3. A. Blanc-Lapierre, R. Fortet: *Théorie des fonctions aléatoires* (Paris 1953)
4. J.L. Doob: *Stochastic Processes* (Wiley, New York 1953)
5. M.S. Bartlett: *An Introduction to Stochastic Processes* (Cambridge University, Cambridge 1956)
6. Yu. A. Rozanov: *Stationary Random Processes* (Fizmatgiz, Moscow 1963) (in Russian)
7. A.A. Sveshnikov: *Applied Methods in Theory of Random Processes* (Nauka, Moscow 1968) (in Russian)
8. H. Cramér, M.R. Leadbetter: *Stationary and Related Stochastic Processes* (Wiley, New York 1967)
9. S. Karlin: *A First Course in Stochastic Processes* (Academic, New York 1966)
10. V.L. Lebedev: *Random Processes in Electric and Mechanic Systems* (Gostekhizdat, Moscow 1958) (in Russian)
11. A. van der Ziel: *Noise* (Prentice-Hall, New York 1954)
12. B.R. Levin: *Theory of Random Processes with Applications in Radio Engineering*, 2nd. ed. (Sovradio, Moscow 1960) (in Russian)
13. W.B. Davenport, Jr., W.L. Root: *An Introduction to the Theory of Random Signals and Noise* (McGraw, New York 1958)
14. A. van der Ziel: *Fluctuation Phenomena in Semi-Conductors* (Butterworths, London 1959)
15. R.L. Stratonovich: *Selected Topics in Theory of Fluctuations in Radio Engineering* (Sovradio, Moscow 1961) (in Russian)
16. D.A. Middleton: *An Introduction to Statistical Communication Theory* (McGraw, New York 1960)
17. V.I. Tikhonov: *Statistical Radio Engineering* (Sovradio, Moscow 1966) (in Russian)
18. A.N. Malakhov: *Fluctuations in Self-Oscillatory Systems* (Nauka, Moscow 1968) (in Russian)
19. B.R. Levin: *Theoretical Foundations of Statistical Radio Engineering*, Vol. 1 (Sovradio, Moscow 1974) (in Russian)

Chapter 1

1.1 S.M. Rytov, Yu.A. Kravtsov, V.I. Tatarskii: *Principles of Statistical Radiophysics 2, Correlation Theory of Random Processes* (Springer, Berlin, Heidelberg) in preparation

1.2 S.M. Rytov, Yu.A. Kravtsov, V.I. Tatarskii: *Principles of Statistical Radiophysics 3, Elements of Random Field Theory* (Springer, Berlin, Heidelberg) in preparation

1.3 S.M. Rytov, Yu.A. Kravtsov, V.I. Tatarskii: *Principles of Statistical Radiophysics 4, Wave Propagation Through Random Media* (Springer, Berlin, Heidelberg) in preparation

1.4 Ya.S. Shifrin: *Problems of Statistical Theory of Antennas* (Sovradio, Moscow 1971) (in Russian)

Chapter 2

2.1 M. Kendall, P. Moran: *Geometrical Probability* (Haffner, London 1963)
2.2 A.N. Kolmogorov: *Foundations of the Theory of Probability* (Chelsea, New York 1950)
2.3 H. Cramér, M.R. Leadbetter: *Stationary and Related Stochastic Processes* (Wiley, New York 1967)
2.4 L.I. Mandelshtam: *Lectures in Optics, Relativity and Quantum Mechanics* (Nauka, Moscow 1972) (in Russian)
2.5 R. Mizes: *Probability and Statistics* (Gostekhizdat, Moscow 1930) (in Russian)
2.6 M.A. Leontovich: *Introduction to Thermodynamics* (Nauka, Moscow 1983) (in Russian)
2.7 G.S. Gorelik: Usp. Fiz. Nauk **44,** 33 (1951)
2.8 R. Clausius: *Die mechanische Wärmetheorie*, 2nd ed., Vol. III (Braunschweig 1891) p. 204

Chapter 3

3.1 N. Campbell: Proc. Cambr. Phil. Soc. **15,** 117 (1909); **15,** 310 (1909)
3.2 H. Cramér: *Mathematical Methods of Statistics* (Princeton University Press, Princeton 1946)
3.3 E.N. Gilbert, H.O. Pollak: Bell Syst. Techn. Journ. **39,** 333 (1960)
3.4 W. Schottky: Phys. Rev. **28,** 74 (1926)
3.5 S.M. Rytov, Yu.A. Kravtsov, V.I. Tatarskii: *Principles of Statistical Radiophysics 2, Correlation Theory of Random Processes* (Springer, Berlin, Heidelberg) in preparation
3.6 A.N. Malakhov: Radiotekhn. Elektronika **4,** 54 (1959)
3.7 C. Domb: Phil. Mag. **41,** 969 (1950)
3.8 I. Bar-David: Proc. IEEE **56,** 2167 (1968)
3.9 H. Cramér: *Random Variables and Probability Distribution* (Cambridge University Press, Cambridge 1946)
3.10 R.L. Mitchell: J. Opt. Soc. Am. **58,** 1267 (1968)
3.11 D.K. McGraw, J.F. Wagner: IEEE Trans. IT-**14,** 110 (1968)
3.12 A.N. Malakhov: Izv. VUZ, Radiofiz. **16,** 1287 (1973)

Chapter 4

4.1 S.M. Rytov, Yu.A. Kravtsov, V.I. Tatarskii: *Principles of Statistical Radiophysics 2, Correlation Theory of Random Processes* (Springer, Berlin, Heidelberg) in preparation
4.2 A.Ya. Khintchine (Khinchin): Math. Ann. **109,** 604 (1934)
4.3 H. Cramér, M.R. Leadbetter: *Stationary and Related Stochastic Processes* (Wiley, New York 1967)
4.4 J.L. Doob: *Stochastic Processes* (Wiley, New York 1953)
4.5 S.M. Rytov: Radiotekhn. Elektronika **4,** 1415 (1959)
4.6 A.A. Bobrov, I.N. Verbitskaya: Nauchn. Ezhegodn. Odesskogo Univ. **2,** 79 (1961)
4.7 A.M. Yaglom: *An Introduction to the Theory of Stationary Random Functions* (Prentice-Hall, Englewood Cliffs, New York 1962)
4.8 S.M. Rytov, Yu.A. Kravtsov, V.I. Tatarskii: *Principles of Statistical Radiophysics 3, Elements of Random Fields* (Springer, Berlin, Heidelberg) in preparation
4.9 S.M. Rytov, Yu.A. Kravtsov, V.I. Tatarskii: *Principles of Statistical Radiophysics 4, Wave Propagation Through Random Media* (Springer, Berlin, Heidelberg) in preparation

Chapter 5

5.1 A. Einstein, M. Smoluchowski: *Brownian Motion*, Collected Papers (ONTI, Moscow 1936) (in Russian)
5.2 S. Karlin: *A First Course in Stochastic Processes* (Academic, New York 1966)
5.3 A.N. Kolmogorov: Math. Ann. **104,** 431 (1931)
5.4 M.S. Bartlett: *An Introduction to Stochastic Processes* (Cambridge University, Cambridge 1956)

5.5 M.A. Leontovich: Zh. Eksp. Teor. Fiz. **5,** 211 (1935)
5.6 J.W. Strutt (Lord Rayleigh): *The Theory of Sound*, Vol. 1 (1877)
5.7 M.A. Leontovich: *Statistical Physics* (Nauka, Moscow 1983) (in Russian)
5.8 S.O. Rice: Bell Syst. Techn. J. **23,** 282 (1944); **24,** 46 (1945)
5.9 B.R. Levin: *Theory of Random Processes with Applications in Radio Engineering*, 2nd ed. (Sovradio, Moscow 1960) (in Russian)
5.10 M. Nakagami: "The M-Distribution", in *Statistical Methods in Radio Wave Propagation* (Pergamon, Oxford 1960), p. 3
5.11 P. Beckmann: J. Res. NBS, sec. D, Radio Science **68,** 927 (1964)
5.12 W. Feller: Math. Ann. **109,** 619 (1934)
5.13 R.F. Pawula: IEEE Trans. IT-**13,** 33 (1967)
5.14 A.N. Kolmogorov: Mat. Ann. **113,** 766 (1937)
5.15 A.M. Yaglom: Matem. Sbornik **24,** 457 (1949)
5.16 R.L. Stratonovich: *Conditional Markov Processes and their Application to Optimal Control Theory* (Moscow University Press, Moscow 1066) (in Russian)
5.17 S.M. Rytov: *Correlation Theory of Random Processes*, Vol. II of S.M. Rytov, Yu.A. Kravtsov, V.I. Tatarskii: *Principles of Statistical Radiophysics* (Springer, Berlin, Heidelberg) in preparation
5.18 G.S. Gorelik: Zh. Eksp. Teor. Fiz. **20,** 351 (1950)
5.19 Yu.A. Krutkov: Dokl. Akad. Nauk SSSR **3,** 153 (1934); Dokl. Akad. Nauk SSSR **1,** 366 (1935)
5.20 S.M. Rytov, Yu.A. Kravtsov, V.I. Tatarskii: *Principles of Statistical Radiophysics 3, Elements of Random Fields* (Springer, Berlin, Heidelberg) in preparation
5.21 S.M. Rytov, Yu.A. Kravtsov, V.I. Tatarskii: *Principles of Statistical Radiophysics 4, Wave Propagation Through Random Media* (Springer, Berlin, Heidelberg) in preparation
5.22 V.I. Klyatskin, V.I. Tatarskii: Izv. VUZ, Radiofiz. **14,** 706 (1971)
5.23 W. Feller: Trans. Am. Math. Soc. **48,** 488 (1940)
5.24 D. Slepian: Ann. Math. Statist. **32,** 610 (1961)
5.25 A.A. Borovkov: Teoriya Veroyatnostei **10,** 331 (1965)
5.26 V.I. Tikhonov: *Surges of Random Processes* (Nauka, Moscow 1970) (in Russian)
5.27 L.S. Pontryaghin, A.A. Andronov, A.A. Vitt: Zh. Eksp. Teor. Fiz. **3,** 165 (1933)
5.28 D.A. Darling, A.J. Siegert: Ann. Math. Statist. **24,** 624 (1953)
5.29 J.L. Doob: *Stochastic Processes* (Wiley, New York 1953)
5.30 M.C. Wang, G.E. Uhlenbeck: Rev. Mod. Phys. **17,** 323 (1945)
5.31 F.T.S. Yu: J. Opt. Soc. Am. **59,** 343 (1969)
5.32 N.G. van Kampen, I. Oppenheim: J. Math. Phys. **13,** 842 (1972)

Chapter 6

6.1 S.M. Rytov, Yu.A. Kravtsov, V.I. Tatarskii: *Principles of Statistical Radiophysics 3, Elements of Random Fields* (Springer, Berlin, Heidelberg) in preparation
6.2 S.M. Rytov, Yu.A. Kravtsov, V.I. Tatarskii: *Principles of Statistical Radiophysics 4, Wave Propagation Through Random Media* (Springer, Berlin, Heidelberg) in preparation
6.3 P. Langevin: Compte Rendus **146,** 530 (1908)
6.4 S.M. Rytov, Yu.A. Kravtsov, V.I. Tatarskii: *Principles of Statistical Radiophysics 2, Correlation Theory of Random Processes* (Springer, Berlin, Heidelberg) in preparation
6.5 V.V. Vladimirskii, Ya.P. Terletskii: Zh. Eksp. Teor. Fiz. **15,** 258 (1945)
6.6 K. Ito, G. McKin: *Diffusion Processes and their Trajectories* (Springer, Berlin, Heidelberg, New York 1963)
6.7 Yu.A. Rozanov: *Random Processes* (Nauka, Moscow 1971) (in Russian)
6.8 R.L. Stratonovich: *Conditional Markov Processes and their Application to Optimal Control Theory* (Moscow University, Moscow 1966) (in Russian)
6.9 V.I. Klyatskin, V.I. Tatarskii: Izv. VUZ, Radiofiz. **14,** 706 (1971)
6.10 V.I. Klyatskin, V.I. Tatarskii: Usp. Fiz. Nauk **110,** 499 (1973)
6.11 E.A. Novikov: Zh. Eksp. Teor. Fiz. **47,** 1919 (1964)
6.12 V.I. Klyatskin: Izv. VUZ, Radiofiz. **18,** 1454 (1975)
6.13 V.I. Klyatskin, V.I. Tatarskii: Teor. Mat. Fiz. **17,** 273 (1973)
6.14 D.B. Duncan: J. Appl. Phys. **24,** 609 (1953)

6.15 J.N. Pierce: IRE Trans. IT-**4**, 163 (1958)
6.16 C.W. Helstrom: IRE Trans. IT-**5**, 139 (1959)
6.17 C.T. Isley: IRE Trans. IT-**5**, 186 (1959)
6.18 F.B. Bunkin: Radiotekhn. Elektronika **5**, 1555 (1960)
6.19 R. Esposito: Proc. IEEE **51**, 1792 (1963)
6.20 V.I. Tikhonov: Radiotekhn. Elektronika **6**, 1082 (1961)

Subject Index

Made in the USA
Las Vegas, NV
06 June 2026

48252185R00149